Problems and Solutions in Particle Mechanics

Online at: https://doi.org/10.1088/978-0-7503-6442-3

Problems and Solutions in Particle Mechanics

**Pradeep Kumar Sharma MSc, MBA (IIT), Mtech, MPhil, PhD, AMIE,
CEng(I) MIET, MInstP, SMIEEE**
Consultant physicist and researcher, (India)

IOP Publishing, Bristol, UK

ISBN   978-0-7503-6442-3 (ebook)
ISBN   978-0-7503-6438-6 (print)
ISBN   978-0-7503-6439-3 (myPrint)
ISBN   978-0-7503-6441-6 (mobi)

DOI   10.1088/978-0-7503-6442-3

Version: 20251001

IOP ebooks

British Library Cataloguing-in-Publication Data: A catalogue record for this book is available from the British Library.

Published by IOP Publishing, wholly owned by The Institute of Physics, London

IOP Publishing, No.2 The Distillery, Glassfields, Avon Street, Bristol, BS2 0GR, UK

US Office: IOP Publishing, Inc., 190 North Independence Mall West, Suite 601, Philadelphia, PA 19106, USA

This book is dedicated to my mother Mrs Mukta Manjari Sharma and my mother-inlaw Mrs Annapurna Dash for their constant support, motivation and blessings.

Contents

Preface

Overview of the series

From my studenthood to date I have been in search of a series of books on College Physics that carries every concept in depth along with the scholarly problems to cater for the needs of potential students preparing for top entrance examinations. I found that students and scholars were spending hundreds of hours gathering information from different sources, websites and books on each concept and chapter. This practice still continues today. This is due to the lack of a desired college physics book series. We have a lot of books in the market; each has its own merits and limitations. Practically, it is not possible to get everything in a single book. However, a potential student expects everything in a book such as varieties of adequate problems with detailed theories and examples and solved problems.

Students are confused by the scattering and variety of sources for the problems, as well as their provenance and validity of solutions. So, not just helpless students but also teachers need concise and precise books where maximum concepts are covered systematically. The proposed series of books is a sincere effort to minimize all the above limitations of existing books and resources and maximise the potential of the books in terms of content, quality, rigor, and depth in theories, questions, and problems.

There are no such books available in the market balanced with problems, theories, quality, and quantity of problems. In this series, we have attempted to balance the books with theories, examples, and problems using a systematic approach for concept building. By virtue of my experience and expertise, as well as suggestions and recommendations of my colleagues and hundreds of my gifted students, the proposed series of books is an outcome of my strong desire to transform physics-phobia into a physics-loving attitude for the students.

Readership

At present, sophomores prepare for entrance examinations to get into premier universities like Oxford, Cambridge, etc, in the United Kingdom; MIT, Princeton, etc, in the Unites States of America. Millions of students all over the world appear for Physics Olympiads (both national and international) and other international physics examinations such as Physics GRE, etc. In India, millions of students appear for national-level examinations for prestigious Indian Institutes of Technology (IITs) through a toughest Joint Entrance Examination called IIT-JEE with the lowest success rate in the world.

Moreover, this book is also useful for students preparing for the PhD qualifying examinations of top universities and Physics GRE examinations. In India, students preparing for UPSC examinations and semester examinations and physics majors in BSc and BEng (or any branch of engineering study) can use this book as a handbook of physics problems. This book will be ultimately useful for Indian students preparing for JEE (Mains and Advanced).

To educate potential students, teachers should be strong in concepts and problemsolving ability. I strongly hope that this text could be a valuable reference book for JEE-educators in India due to its content and quality, such as familiar problems, theories, creative problems, and problems with detailed solutions along with impressive coloured diagrams throughout. I have handpicked many problems from various sources. Each problem is observed being asked in different ways in various examinations all over the world. I spent adequate time in analysing each problem's most original version and then put it in my series. Furthermore, I prefer to put all possible questions linked with the original problems in one place so that the student does not have to waste their time thinking too much and wandering.

About this book

This book contains seven chapters: 1. Vectors; 2. Motion in one dimension; 3. Motion in two and three dimensions; 4. Newton's Laws of motion of a particle; 5. Friction; 6. Dynamics of circular motion; 7. Work, energy, power.

Vectors: This chapter deals with the *algebra* (addition, subtraction and multiplication) and *calculus* (differentiation and integration) of vector quantities. In physics, we use the vectors such as position, displacement, velocity and accelerations etc. The physical quantities such as displacement and relative velocity can be expressed by using the concept of vector-subtraction. Work is defined as the dot product of force and displacement; torque is defined as the cross (vector) product of position and force. Furthermore, in physics, we need to define one vector quantity as the rate of change of another vector quantity. For instance, *velocity* is expressed as the derivative or differentiation (rate of change) of position vector; acceleration is defined as the derivative of velocity vector. Force is defined as the rate of change of linear momentum. Torque is defined as the rate of change of angular momentum. So, vector mathematics is an ultimate tool to express the physical laws that govern all phenomena of Nature.

Motion in one dimension: Kinematics deals with the calculation of position, velocity and acceleration without worrying about the cause of motion. The starting point of the study of kinematics is to define the location of a particle by a vector called 'position vector'. Then we define the velocity of the particle as the rate of change of its position vector. Finally, we will measure the time rate of change of velocity to define acceleration of the particle. These three vectors such as position, velocity and acceleration are sufficient to describe the nature of motion of a particle in space. Then, we will use the general kinematics to discuss the motion in one dimension, that is, straight line motion. This chapter deals with uniformly accelerated motion, free fall, graphs (s–t, v–t etc.), graphical solution etc.

Motion in two and three dimensions: This chapter extends the concepts of one-dimensional kinematics to explain the motion of a particle in a plane (two-dimensions) and space (three-dimensions). The motion in two dimension can be described as the superposition of one dimensional motion along any two perpendicular axes (directions). For instance, a projectile motion can be expressed as the vector sum of a horizontal uniform motion and a vertical accelerated motion. A circular motion about the z-axis can be interpreted as the sum of two simple harmonic motions along the x- and the y-axis, say. So, circular and elliptical motion, projectile motion are two-dimensional motion; relative motion and motion of a bird in the sky can be a three-dimensional motion. By using the concept of position, velocity and acceleration we can study kinematics in two and three dimensions.

Newton's laws of motion of a particle: This chapter deals with the 'cause of motion' of particles (point mass). So, here you will learn *why* a particle moves and how to find the unknown forces acting on a particle and accelerations caused by the force acting on it. In this chapter of dynamics, we will study the concept of *force* and *mass*. This chapter will cover the three fundamental laws of motion formulated by Sir Isaac Newton based on experimental observations. This will form the basis of understanding the other chapters in mechanics. Here we deal with three mechanical systems such as a string–particle system (pulley connected with particles by strings), wedge–particle system and spring–particle system. In this chapter, we need to find the unknown forces such as tensions in the connected strings, reactions between the contacting surfaces and accelerations of the connected bodies. We have to write the force equations and constraint equations so that the number of equations and unknown quantities will be equal. At the end of the chapter, we explain the non-inertial frame and solve the problems by using the pseudo-force method.

Friction: This chapter talks about two types of friction, namely static and kinetic friction. Static friction prevents relative sliding, whereas kinetic friction opposes relative sliding between two surfaces. The laws of both types of friction are described. Using the laws of friction and all other concepts such as Newton's laws and kinematics, you will learn how to solve the dynamics of a mechanical system in the presence of friction.

Dynamics of circular motion: This chapter deals with centripetal force, the cause of circular motion. When a particle moves in a circle, it must accelerate towards the center of the circle. A force is required to do so which is called centripetal force. How can different forces such as gravity, normal reaction, tension, friction etc, be centripetal forces? We will describe all these in this chapter by taking the examples of banking in a curve, a motor cyclist in a hollow-drum, a vehicle negotiating a curve etc. The horizontal and also vertical circular motion are described. In non-uniform circular motion we have both centripetal and tangential acceleration. It is more interesting to learn the role of gravity on a swinging pendulum bob and the role of friction in producing a non-uniform circular motion of a coin placed on a rotating platform.

Work, energy, power: This chapter deals with three fundamental concepts: work, energy and power. Work involves a force and a displacement of an object. While work is done, energy is transferred from the *doer* (agent) to the mechanical system. So, *work* is the process of energy transfer.

In other words, *energy* is the ability of doing work. Energy can be *potential* and *kinetic*. The term potential energy defines the conservative force and distinguishes it from non-conservative forces. Furthermore, the concept of potential energy forms the basis of explaining the principle of a conservative mechanical system. In this chapter, the most powerful theorem is the work–energy theorem by which we can solve so many problems in mechanics, heat and thermodynamics an electromagnetism etc. Three types of equilibrium (stable, unstable and neutral) by using the concept of conservative force are carefully presented.

The *power* delivered by a force is defined as the rate at which the force is doing work. It helps us to understand why it is easier to climb by walking on a *zig-zag* road rather than a stiff one. Input and output power are explained to define the efficiency of a mechanical system.

How to use this book

Students should first complete the given theory with the examples and then try to solve the problems without referring the solution by their own way. Sometimes students find better methods for solving the problems and should not rely solely on the methods provided. However, if a student falters after repeated attempts, only then they can refer to the given solutions. Although I tried to create an error-free text, some unavoidable mistakes might appear in some subtle form each time. So, always there will be a scope of improvement of the content. I request my readers to critically analyse my work and give their valuable comments and suggestions for the overall improvement of this book and make this book error-less in due course of time.

Acknowledgments

It goes without saying that a teacher is worth millions of books. An ideal teacher is built by the association of potential students. I acquired my experience and expertise by the association of gifted students of premier institutes. So, first of all, I would like to express my gratitude to the directors of all the institutes where I worked. Most illustrious are Brilliant Tutorials Private Limited, Madras (Chennai), FIIT-JEE Ltd, New-Delhi, and Narayana Group of Educational Institutions, Hyderabad.

My high school science and mathematics teacher Mr Ramesh Chandra Behera imparted an ever-lasting style of solving problems and presented the theory in a handy and student-friendly way. I am also indebted to my professors Professor Sarat Mohapatra and Professor P K Dhir for imparting their deepest knowledge of electricity and magnetism and invaluable suggestions regarding my previous books on electromagnetism which helped me to improve the standard of this book. All the attributes of my gifted teachers (many of them have left their mortal bodies) and students (who are now global leaders in their own fields) are reflected in these books. So, I express my sincere thanks to all my revered teachers (gurus) and past students.

I am thankful to Professor T Surya Kumar, Professor Bhanumati and Professor G N Subramanian of BT who was well-versed with both physics and mathematics helped me to solve the standard problems in mechanics.

In FIIT-JEE Limited, New-Delhi, I thank my seniors Er Srikant Kumar and Er P K Mishra for imparting a standard subject knowledge enhancement programme that had a great impact on this book.

I thank the late Mr Prakash Chand Bathla for offering me to write five books under his nationally leading publication house (G R Bathla & Sons). Based upon my experience as a national level author and educator, I could attain a chance to write for an international publishing house (IOPP).

While I headed the department of Physics in Narayana Group of educational institutions, I would like to thank all my previous top students who edited my books. Furthermore, I express my deepest gratitude to the principals and deans of Nayayana Group, Hyderabad especially Mr Krishna Reddy and Mr Ramalinga Reddy, with whom I worked major portion of my professional career, for giving me operational freedom, status, stability and respect. There, I could complete some theories and examples of the present series in rudimentary form. My sincere thanks and admiration to some leading physics educators such as Er Aditya Sachan, Er L N Prusty, Er Sekhar Somnath, Mr Monoj Pandey and Mr S K Singh. I am also thankful to Professor Kundal Rao and Professor Raghunath of Narayana IIT and Professor Srinivasa Chary of Sri-Mega for their suggestions and inspirations for my publishing works.

I would like to express my profound gratitude to my wife Usha in supporting and bearing me in the pandemic in 2020 when I started conceptualizing this book-series. I remain obliged to the commissioning editor of IOP Publishing Mr John Navas for his insightful comments, suggestions, and expertise, which have enhanced the rigor and depth of this work. Furthermore, I thank Mr David McDade and Phoebe Hooper, who streamlined the publication work of this book.

I express my gratitude to my Ex-publisher Mr Monoj Bathla who suggested me to accept the offer of IOPP realising the suitability of the publisher with my work.

I sincerely thank Er. Bismay Parida (Readers institute, Balasore, Odisha State) for his continuous effort in typing the manuscript in time.

I thank Professor Peter Dobson (Oxford University) who taught me how to do the things with perfection and I am applying this idea in the present publication. I am deeply indebted to Dr Benjamin Hourahine, Professor Yu Chen of University of Strathclyde for imparting a standard knowledge of nanoscience so that I could include this fastest growing field in my problem book series.

One of my notable friends is Mr Rajinder Sehra, director of S&RJ Ltd and Foot Print Media Production near Glasgow. His constant motivation for writing this series is also praiseworthy.

Furthermore, I would like to thank a potential Physics educator Mr Mithilesh for finding time to review some of the problems in my book. At last, my sincere thanks to Er K K Khandelwal (a graduate from IIT, an ex-baeurocrat and a gifted senior Physics educator) for reviewing some of the controversial problems in my book.

This book would not have been possible without the collective contributions and support of all those mentioned above. Their guidance and encouragement have been instrumental in the completion of this significant milestone in my authorship. Taking this as a blessing of the Almighty, I pray for the attainment of knowledge that would be an ultimate solution to all problems of human being and other living entities.

Pradeep Kumar Sharma
20 April 2025

Author biography

Pradeep Kumar Sharma

Pradeep Kumar Sharma is a physics educator in India possessing more than three decades of experience in physics education and research in training the aspirants of the joint entrance examination conducted by prestigious Indian Institutes of Technology, popularly known as IIT-JEE. Many of his students also won gold and silver medals in national and international physics Olympiads. His vast experience as a potential teacher, team leader and head of the department in some premier institutes like Brilliant tutorials (Chennai), FIIT-JEE Ltd (New Delhi), Narayana Group (Andhra and Telangana) etc, made him extend his service as a consultant physicist to mentor both students and teachers of reputed groups in India. He has authored bestselling study materials and five books known as GRB Understanding Physics for the entrance examinations. He has been associating as a research scholar of physics education, nanoscience, metaphysics and management in some Indian and foreign universities such as Oxford University, Strathclyde University, Sofia University, Indian Institutes of Technology, Patna etc. Furthermore, he is continuing his research while affiliated with various national and international organizations such as IEEE (USA), IET (UK), IE(I), IOP(UK) etc. He has published dozens of papers in national and international journals like IEEE-Scopus journals and journals published by Institute of Physics (UK). He is currently busy in completing the problems and solutions of a series of six books which will be ready to publish very shortly. Also, he is planning to design a unique interactive study material in the mode of Active Teaching and Active Learning (ATAL) that will make the physics easier for the students to learn.

IOP Publishing

Problems and Solutions in Particle Mechanics

Pradeep Kumar Sharma

Chapter 1

Vectors

1.1 Introduction

Physics is the scientific study of fundamental constituents of matter and energy. We can analyze the motion of the particles (elements) of matter and energy to examine, understand and interpret the behavior of Nature (matter–energy) as a basis of observation and experiments. So, in physics there are the set of hypothesis, rules, laws and principles by using the physical quantities like mass, length, time, volume, pressure, density, energy, velocity, acceleration, force, electric and magnetic field intensities etc. Some of the physical quantities carry a sense of *direction* apart from their magnitudes which are called vectors; others are called scalars having only magnitudes such as mass, length, time etc.

Due to the presence of *direction*, the operation of vectors such as (addition, subtraction, multiplication, derivatives and integration etc) is totally different from that of the scalars. In other words, vector mathematics (both algebra and calculus) is different from scalar mathematics. In this chapter we will discuss the *algebra* (addition, subtraction and multiplication) and *calculus* (differentiation and integration) of vector quantities. In daily life we use the idea of vectors to locate any objects and places, measure the displacement, velocity and accelerations of the particles and bodies in physics. In kinematics, we add the displacements, velocities and forces etc, to find a resultant vector. The physical quantities such as displacement and relative velocity can be expressed by using the concept of vector subtraction. *Work* is defined as the dot product of force and displacement; *torque* is defined as the cross(vector) product of position and force.

Furthermore, in physics, we need to define one vector quantity as the rate of change of another vector quantity. For instance, *velocity* is expressed as the derivative or differentiation (rate of change) of a position vector; acceleration is defined as the derivative of a velocity vector. Force is defined as the rate of change of

doi:10.1088/978-0-7503-6442-3ch1 1-1

linear momentum. Torque is defined as the rate of change of angular momentum. So, vector mathematics (that deals with the addition, subtraction, multiplication, derivatives and integration of vectors) is an ultimate tool to express the physical laws that govern all phenomena of Nature.

1.2 Definition of vectors

The Greek word *vector* means that which carries. So, the physical quantities such as force etc, that carry a *direction* are known as vectors. When we say, 'a car is moving', we can raise two basic questions. First, which direction is the car moving? Second, how fast is the car moving? This means that, we need both magnitude (fastness = speed) and direction of motion of the car to express its motion by a vector quantity known as 'velocity'.

Hence, the physical quantities like position, velocity, acceleration, force etc, that need both magnitude and direction to express them completely are termed 'vectors'.

If you ask a shopkeeper 'Please give me 4 kg of rice towards the south, 5 litres of kerosene towards the north and 3 meters of rope towards the east' it seems ridiculous and does not make any sense! The addition of direction is unnecessary with these physical quantities (mass, volume, etc) which are known as scalars. A scalar is associated with a magnitude only.

Frame of reference: In particle kinematics we measure the displacement, velocity and acceleration of a particle. For this, an observer is taken as a reference point. Generally, we consider the observer at the origin of a coordinate system. The place where the observer (reference point) is placed (fixed) is known as 'reference frame'. Hence, reference frame is geometrically defined as a coordinate system.

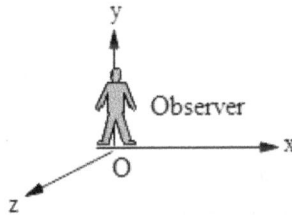

A reference frame is a coordinate system
with an observer is placed at the origin

Locating a particle, position vector: If we ask babies, where is the Moon? They will use their instinct to point a finger towards the Moon in the sky. This tells us that pointing of a finger indicates the location or position of an object. In other words, the position of a particle is associated with a direction. Furthermore, the distance of the object (Moon) and observer (baby) is also required to completely specify the location of the Moon.

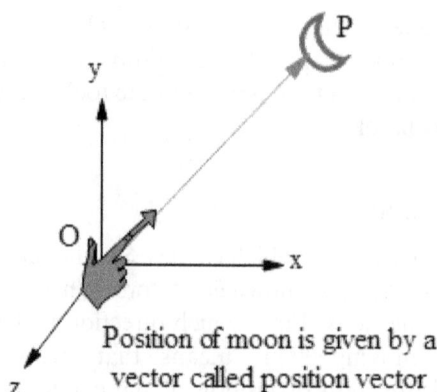

Position of moon is given by a
vector called position vector

From the above explanation we understand that since the position of a particle requires both magnitude (distance of separation between the object and observer) and direction for its complete specification, it is characterized by a vector quantity $\vec{r}$, termed as 'position vector' of the particle with respect to the origin (point of reference). It is the first fundamental vector quantity in physics.

In the Cartesian coordinate system, $\vec{r}$ can be expressed as

$$\vec{r} = x\hat{i} + y\hat{j} + z\hat{k},$$

where x, y and z are the components (projections) of $\vec{r}$ along $x-$, $y-$ and $z-$axes, respectively. In a polar coordinate system position, vector $\vec{r}$ can be given as

$$\vec{r} = r\hat{r},$$

where $r = |\vec{r}|$ and $\hat{r} = \frac{\vec{r}}{|\vec{r}|}$.

1.3 Representation of vectors

Generally, we represent vectors in two ways:

(i) **Symbolic representation:** If we put an arrow or cap over a letter A, it will look like $\vec{A}$ or $\hat{A}$. The arrow ($\rightarrow$) or cap ($\char`^$) represents the direction of the vector and the letter 'A' represents the magnitude of the vector.

(ii) **Graphical representation of displacement:** We can take the example of the vector *displacement* which is easiest to produce, define and understand. You can produce a displacement by just running a pencil on a piece of paper from P to Q. You can see that the directed arrow PQ is the displacement. It is a real vector that we can see. Please refer to chapter 2 of this book for a detailed discussion.

Pencil

Distance;actual path

$\vec{s}$ Displacement

P   Q

The directed arrow PQ is the actual displacement undergone by the tip of the pencil

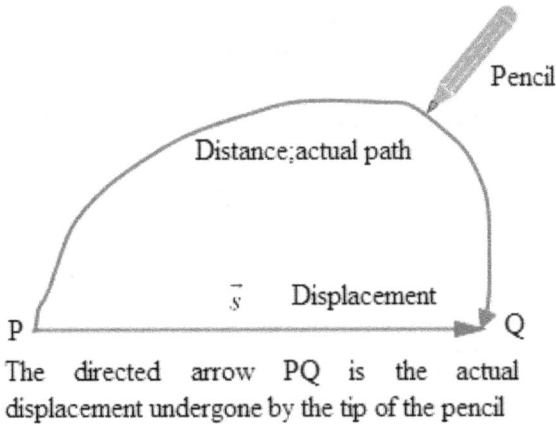

Let us represent the displacement PQ of magnitude 5 cm towards the east. For this, we take a graph paper and choose a scale of one small unit for 1 m of displacement. Then, we have five small divisions for 5 m of displacement. It means the magnitude of the displacement is equal to the length of five small units of the graph paper. If we change the scale, the length will be different. Since, the displacement is pointing due east, we put an arrow pointing towards the positive x-direction which represents east. Ultimately, we get a segmented arrow which represents the displacement of magnitude 5 m and directed towards the east. The length of the arrow is proportional to the magnitude of the displacement and the tip of the arrow shows the direction of the displacement.

Using the above representation, we can specify the displacement as, $\vec{d}$. The origin of the segmented arrow is the tail, and the tip of the arrow is known as the head of the vector.

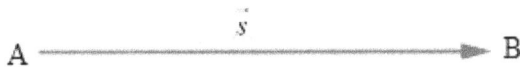

$\vec{s}$

A ———————————————————▶ B

The directed arrow AB is the displacement

N.B:

1. We can represent a vector by an arrow (a directed line segment) and label it by a letter capped by a symbolic arrow. The orientation of the arrow shows the direction of vector and the label $\vec{A}$ gives the name of the vector (like velocity $\vec{v}$, acceleration $\vec{a}$, force $\vec{F}$, position $\vec{r}$ etc); In general $\vec{A} = \vec{r}, \vec{v}, \vec{F}$ etc. In some books, vectors are represented by bold letters.

2. Remember that an arrow is not a vector; it simply represents a vector. If we choose a bent arrow, it may mean a sense of rotation of a point. Hence, a bent arrow is not a vector.
3. When we slide a vector in space keeping its orientation the same, the vector remains constant.

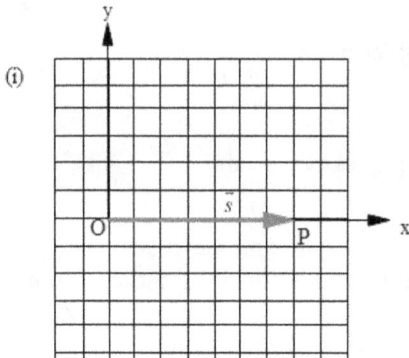

The arrow represents of the displacement in the graph paper

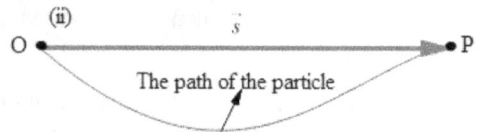

The actual displacement vector; this arrow is the displacement

The path of the particle

1.4 Types of vectors

After learning how to represent a vector, let us now categorize the vectors based upon their magnitude and direction (orientation in space) as follows:

(i) **Parallel vectors:** Two vectors $\vec{A}$ and $\vec{B}$ are parallel if they point in the same direction. Their magnitudes need not be equal.

(ii) **Equal vectors:** If two vectors $\vec{A}$ and $\vec{B}$ are parallel and equal, they must have the same magnitude and point in the same direction. That means if $\vec{A} = \vec{B}$, we have $|\vec{A}| = |\vec{B}|$ and $\vec{A}$ and $\vec{B}$ are parallel.

(iii) **Antiparallel vectors:** Two vectors $\vec{A}$ and $\vec{B}$ are antiparallel if they are directed opposite to each other. They need not have equal magnitudes.

(iv) **Negative vectors:** Two vectors $\vec{A}$ and $\vec{B}$ are negative when their magnitudes are equal and they point opposite to each other. This means that if $\vec{A} = \vec{B}$, then $|\vec{A}| = |\vec{B}|$ and the vectors are antiparallel.

(v) **Null (zero) vectors:** By definition, a null (zero) vector has magnitude 'zero' and has an arbitrary direction. This will be clear in the following example:

When two persons push an object with equal and opposite forces $\vec{F_1}$ and $\vec{F_2}$, the resultant force is $\vec{F} = \vec{F_1} + \vec{F_2} = 0$ because the object does not accelerate. Since the particle's acceleration is zero, logically we can say that it is impossible to define the direction of acceleration. In other words, we can say that the direction of a null vector is arbitrary (not specific).

The zero vector is the resultant
of two negative vectors

(vi) **Unequal vectors:** Two vectors $\vec{A}$ and $\vec{B}$ are unequal vectors, when they are different in either magnitude or direction or both magnitude and directions.

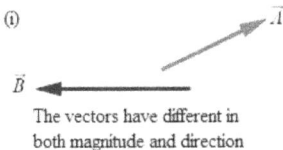

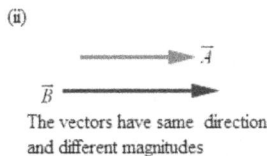

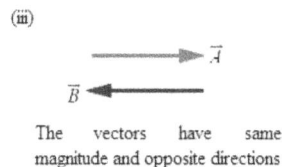

(i) The vectors have different in both magnitude and direction

(ii) The vectors have same direction and different magnitudes

(iii) The vectors have same magnitude and opposite directions

(vii) **Collinear vectors:** If two or more vectors $\vec{A}$, $\vec{B}$ and $\vec{C}$ etc, pass through the same straight line, they are said to be collinear vectors. For instance, when several persons pull a straight string at its different points, the tensions T_1, T'_2, T_2, T'_2,..., T'_3 etc in the string are collinear.

The vectors orient along same line

(viii) **Coplanar vectors:** If two or more vectors $\vec{A}$, $\vec{B}$ and $\vec{C}$, say, lie in the same plane, they are said to be coplanar. Here $\vec{A}$, $\vec{B}$ and $\vec{C}$ are coplanar as they act (lie) in the plane of the paper.

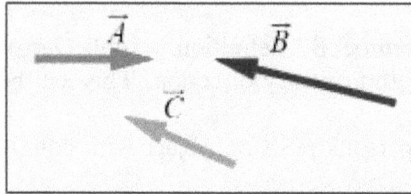

Theese three vectors lie in same plane

(ix) **Non-coplanar vectors:** When more than two vectors do not lie in the same plane they can be termed as non-coplanar vectors.
   In the figure below, since $\vec{A}$, $\vec{B}$ and $\vec{C}$ lie in the x–y, y–z and x–z planes, respectively, these vectors are non-coplanar.

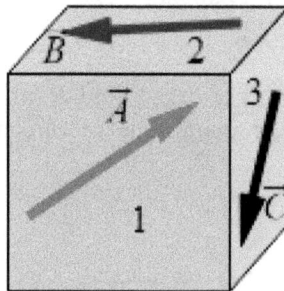

(x) **Orthogonal vectors:** If any two or three vectors are perpendicular to each other, we call them orthogonal vectors.
   Since, $\vec{A}$, $\vec{B}$ and $\vec{C}$ act along x-, y- and z-axes, respectively, they are orthogonal vectors. In other words, two orthogonal vectors when produced from a 90-degree angle at their point of intersection.

$$\vec{B}$$

$$\vec{A}$$

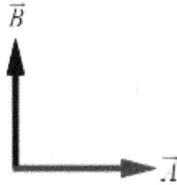

Two vectors are mutually
perpendicular to each other

(xi) **Polar vectors:** A vector such as force $\vec{F}$, velocity $\vec{v}$, electric field intensity $\vec{E}$ etc, having a fixed origin and different orientations in space is known as a polar vector. For instance, a position vector $\vec{r}$ is a polar vector velocity, momentum, force etc, are polar vectors. We will talk more about position vector in the next chapter.

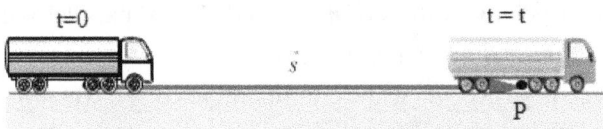

t=0   t = t

$\vec{s}$

P

The displacement has a well-defined initial and final point P

(xii) **Axial vectors:** Some vectors such as angular velocity $\vec{\omega}$, angular momentum $\vec{L}$ and torque $\vec{\tau}$, etc, are directed along the axis of rotation. They are known as axial vectors. However, we cannot locate the exact position (origin) of the axial vectors along the axis. Hence, the axial vectors are known as 'pseudo vectors'. Magnetic field density, angular velocity etc, are pseudo vectors.

You will learn that the cross product of two polar vectors must be a pseudo vector.

$$\vec{\omega}$$

Angular velocity is an axial vector

(xiii) **Unit vector:** When we divide the arrow which represents a vector of magnitude five units into five equal parts, each arrow represents a magnitude of one unit.

In general, when we divide any vector $\vec{A}$ by its magnitude $|\vec{A}|$, it gives us a unit vector of that vector denoted by the symbol $\hat{a} = \dfrac{\vec{A}}{|\vec{A}|}$.

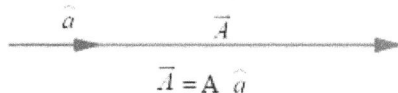

$\hat{a}$ $\vec{A}$

$$\vec{A} = A\,\hat{a}$$

So, we can write $\vec{A} = |\vec{A}|\, \hat{a}$; $\hat{a}$ shows the direction of $\vec{A}$.

When the unit vectors are directed along positive x, y and z directions we write them $\hat{i}$, $\hat{j}$ and $\hat{k}$ or ($\hat{a}_x$, $\hat{a}_y$ and $\hat{a}_z$), respectively. A unit vector does not have units and dimensions (M, L, T etc); it has a magnitude 'one' and it represents a 'direction'.

1.5 Multiplication of a vector by a scalar

A vector $\vec{A}$ multiplied with a scalar n gives a new vector $\vec{B}$, (say) which can be given as $\vec{B} = n\vec{A}$.

Case 1: If n is positive, $\vec{B}$ will be parallel to $\vec{A}$. n can be greater, equal and less than one. When $n = 1$, $\vec{A} = \vec{B}$, both are equal vectors as discussed earlier. For $n > 1$ and $n < 1$ we have illustrated the vectors in the figure below. For different positive values of n, $\vec{A}$ and $\vec{B}$ are parallel vectors

Case 2: If n is zero, $\vec{B} = \vec{0}$ (zero vector but not zero).

Case 3: If n is negative, $\vec{B}$ will be antiparallel to $\vec{A}$; when $n = -1$, we have $\vec{B} = -\vec{A}$. This means $\vec{A}$ and $\vec{B}$ are negative vectors. For the other cases ($|n| > 1$ and $|n| < 1$) the magnitudes of $\vec{A}$ and $\vec{B}$ are not equal as shown in the figure below.

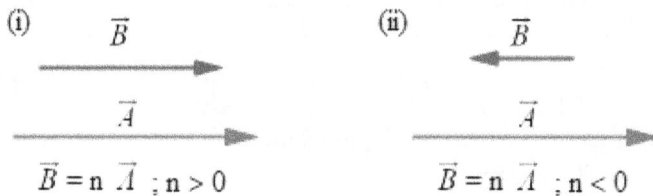

(i) $\vec{B}$ (ii) $\vec{B}$

$\vec{A}$ $\vec{A}$

$\vec{B} = n\,\vec{A}$; $n > 0$ $\vec{B} = n\,\vec{A}$; $n < 0$

Thus, we can conclude that:

For different negative values of n, $\vec{A}$ and $\vec{B}$ are antiparallel vectors

The multiplication of vector $\vec{A}$ by a scalar n yields a new vector $\vec{B}$ ($=n\vec{A}$) whose magnitude and direction are both controlled by the magnitude and sign of the scalar.

For $n > 0$, $\vec{B}$ is parallel to $\vec{A}$ and if $n < 0$, $\vec{B}$ is antiparallel to $\vec{A}$. Vector $\vec{A}$ is parallel to its unit vector, i.e., $\vec{A} \parallel \hat{a}$.

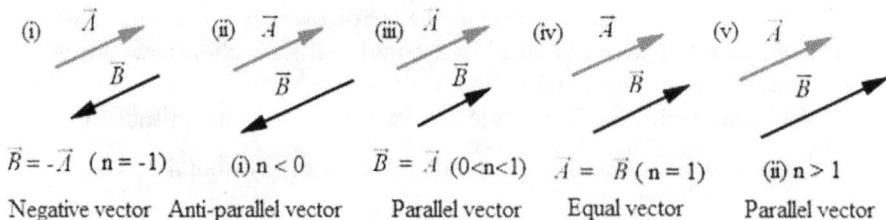

(i) $\vec{A}$ (ii) $\vec{A}$ (iii) $\vec{A}$ (iv) $\vec{A}$ (v) $\vec{A}$

$\vec{B}$ $\vec{B}$ $\vec{B}$ $\vec{B}$ $\vec{B}$

$\vec{B} = -\vec{A}$ $(n = -1)$ (i) $n < 0$ $\vec{B} = \vec{A}$ $(0 < n < 1)$ $\vec{A} = \vec{B}$ $(n = 1)$ (ii) $n > 1$

Negative vector   Anti-parallel vector   Parallel vector   Equal vector   Parallel vector

When $\vec{B} = n\,\vec{A}$, different magnitudes and directions of $\vec{B}$ for a given vector $\vec{A}$

1.6 Addition of vectors

If we pour 4 litres of oil into a pot containing 3 litres of oil, the net (resultant) amount of oil in the pot is (3 + 4 =) 7 litres. If we remove (subtract) 1 litre of oil from the pot we will have a net (3–1 =) 2 litres of oil in it.

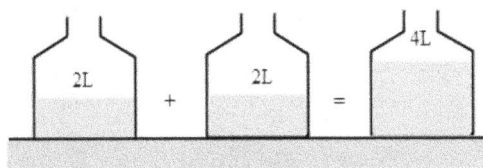

2L+2L=4L in Scalar algebra

What is the meaning of resultant in vector quantities? It is somewhat similar to that of the scalars. Let us understand this in a simple example of displacement. If you move from A to B, you undergo a displacement $\overrightarrow{AB} = \vec{a}$ (say); similarly, if you move from B to C, you are undergoing another displacement $\overrightarrow{BC} = \vec{b}$ (say), which means that you are putting or adding the second displacement $\overrightarrow{BC} = \vec{b}$ with the first displacement $\overrightarrow{AB} = \vec{a}$. As a result of, or addition of, these two successive displacements, you have arrived a point s which does not need any proof because we can see this geometrically. Then, your total or resultant or net displacement from A is $\overrightarrow{AC} = \vec{R}$, say. Then, we can write

$$\overrightarrow{AC}(=\vec{R}) = \overrightarrow{AB} + \overrightarrow{BC}$$

$$\Rightarrow \vec{R} = \vec{a} + \vec{b}$$

This tells us that the resultant displacement is the net displacement from the initial position after undergoing two or more successive displacements. This rule holds good for any vector.

(i)

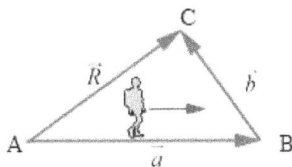

The man walks from A to B and B to C;Then the resultant dispacement is AC.

$$\vec{R} = \vec{a} + \vec{b}$$

(ii)

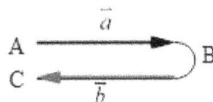

The man walks from A to B and B to C;Then the resultant dispacement is AC is equal to zero as he comes back to the same point;A and C coincide.

$$\vec{R} = \vec{a} + \vec{b}$$

We can see that the resultant displacement not only depends upon the magnitude of the displacements $\overrightarrow{AB} = \vec{a}$ and $\overrightarrow{BC} = \vec{b}$, but also the directions of these two displacements.

Let us now take another example by considering the vector as *force*. If we push a stationary object with a force of 3 N and simultaneously, another person pushes it with a force of the same magnitude in the opposite direction, the object does not move (more strictly, does not accelerate). That is because the net force acting on the object is zero, but according to the scalar addition it should be $(3 + 3 =)$ 6 N. Hence, vectors cannot be operated like scalars. The vectors obey different algebra which is known as 'vector algebra'. Let us talk about the 'algebra of vectors'.

$F_1 \longleftarrow \qquad \longrightarrow F_2$

Two equal and opposite forces are acting
on the boy to have a zero net force

The new concept of vector algebra redefines the vectors as the physical quantities having both magnitude and direction, so also they must obey the algebra of vectors (vector algebra). As a vector has one thing extra, that is, its direction, it makes all the difference between vector addition and scalar addition. When two or more vectors act at a point simultaneously, the net effect (or the resultant) of these vectors not only depends upon their magnitudes but also depends upon their directions.

Let us now discuss the resultant of any two vectors (of the same kind) acting at a point simultaneously.

(a) **Triangle law of vectors**

As you have seen, by moving from O to P, P to Q, you acquire a net displacement $\overrightarrow{OQ} = \overrightarrow{R}$ which is the addition or net or resultant of the displacements $\overrightarrow{OP} = \overrightarrow{A}$ and $\overrightarrow{PQ} = \overrightarrow{B}$. So, we can write that

$$\overrightarrow{R} = \overrightarrow{A} + \overrightarrow{B}$$

$\vec{R} = \vec{A} + \vec{B}$

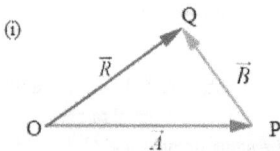

(i)

These three displacements are/ represent the sides of the triangle OPQ;so that $\vec{R} = \vec{A} + \vec{B}$

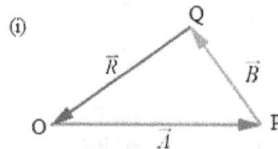

(i)

If you go from O to P, P to Q and Q to O along the triangle OPQ, the net displacement is zero;so, $\vec{A} + \vec{B} + \vec{C} = \vec{0}$

This relation tells us that, if any two sides of a triangle taken in either sense (anticlockwise sense, say) such as OP and PQ represent two displacements $\overrightarrow{OP} = \overrightarrow{A}$ and $\overrightarrow{PQ} = \overrightarrow{B}$, the closing side OQ taken in the reverse order (clockwise sense) must represent the resultant displacement $\overrightarrow{OQ} = \overrightarrow{R}$ as shown in the above figure (i).

Alternatively, if you move along the triangle OPQ in one sense, anticlockwise, say; you will move from O to P, P to Q and Q to O, undergoing three displacements $\overrightarrow{OP} = \overrightarrow{A}, \overrightarrow{PQ} = \overrightarrow{B}$ and $\overrightarrow{QO} = \overrightarrow{C}$, finally you have zero displacement because you come back to your starting (initial) point (position). So, we can write

$$\overrightarrow{OP} + \overrightarrow{PQ} + \overrightarrow{QO} = \overrightarrow{0}$$

Putting $\overrightarrow{QO} = \overrightarrow{C}$, we have

$$\overrightarrow{A} + \overrightarrow{B} + \overrightarrow{C} = \overrightarrow{0}$$

So, the sum of three successive displacements (vectors) is a zero displacement (vector). These two statements are known as the triangle law of vectors.

(b) **Polygon law of vectors**

We can extend the triangle law to a polygon law. For this purpose, just walk along a polygon of n sides in one sense (cyclic manner). For a round trip, you will get a zero net displacement. If the n sides are represented by the displacement vectors, $\overrightarrow{A_1}, \overrightarrow{A_2}, \ldots, \overrightarrow{A_n}$ respectively, the resultant of these vectors is a zero vector. So, we can write

$$\overrightarrow{A_1} + \overrightarrow{A_2} + \cdots + \overrightarrow{A_n} = \overrightarrow{0}$$

Alternatively, if you walk along a polygon of n sides undergoing the displacements $\overrightarrow{A_1}, \overrightarrow{A_2}, \ldots, \overrightarrow{A_{n-1}}$, respectively, the nth side (closing side) of the polygon, taken in reverse order will represent the resultant or net displacement.

$$\overrightarrow{R} = -\overrightarrow{A_n} = \overrightarrow{A_1} + \overrightarrow{A_2} + \cdots + \overrightarrow{A_{n-1}}$$

Proof: If n number of vectors $\overrightarrow{A_1}, \overrightarrow{A_2} \ldots \overrightarrow{A_{n-1}}$ are represented by a polygon of n sides, following the logic of triangle law of vectors, we have

$$\overrightarrow{A_1} + \overrightarrow{A_2} = \overrightarrow{R_1}$$
$$\overrightarrow{R_1} + \overrightarrow{A_3} = \overrightarrow{R_2}$$
$$\cdots$$
$$\cdots$$
$$\cdots$$
$$\overrightarrow{R_{n-3}} + \overrightarrow{A_{n-1}} = \overrightarrow{R}$$

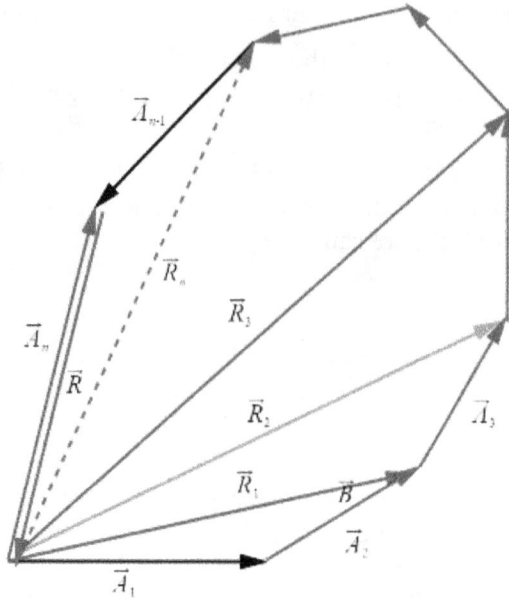

$$\vec{A}_1 + \vec{A}_2 + \vec{A}_3 + \ldots\ldots + \vec{A}_n = \vec{0}$$

Summing up both sides, we have

$$\vec{A}_1 + \vec{A}_2 + \vec{A}_3 + \cdots + \vec{A}_{n-1} = \vec{R}$$

This tells us that if $(n - 1)$ sides of a polygon of n sides represent $(n - 1)$ vectors, i.e., $\vec{A}_1, \vec{A}_2, \ldots, \vec{A}_{n-1}$, respectively, in the same or one cyclic order (sense), the closing side of that is, nth side of the polygon taken in the reverse order (sense), represents the resultant $\vec{R}$ of all the above vectors. This is known as the 'polygon law of vectors'. If we assume $-\vec{R} = \vec{A}_n$, substituting $-\vec{R} = \vec{A}_n$, in the above expression, we have $\vec{A}_1 + \vec{A}_2 + \cdots + \vec{A}_n = 0$. This means, if n vectors are represented by the sides of a polygon of n sides in a cyclic manner, the resultant of these vectors is a zero vector.

(c) **Parallelogram law of vectors**

The vector triangle can be transformed to a vector parallelogram. The adjacent sides of the parallelogram represent the displacements $\overrightarrow{OP} = \vec{A}$ and $\overrightarrow{PQ} = \vec{B}$. The diagonal of the parallelogram represents $\overrightarrow{OQ} = \vec{R}$. So, we can write that

$$\vec{R} = \vec{A} + \vec{B}$$

(i)

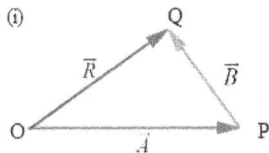

These three displacements are/ represent the sides of the triangle OPQ;so that $\vec{R}=\vec{A}+\vec{B}$

(ii)

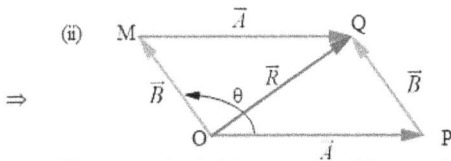

The vector triangle becomes a parallelogram; the adjacent sides of the parallelogram represent the displacements A and B; then, the diagonal represents the resultant $\vec{R}=\vec{A}+\vec{B}$

This law is valid for addition of any vector of the same physical quantity (for instance, 'force' can be added with 'force' but not with 'velocity'. This is called parallelogram law of vector addition which states that:

if the vectors $\vec{A}$, $\vec{B}$ are represented by the adjacent sides of a parallelogram, its diagonal passing through the point of application of the vectors $\vec{A}$ and $\vec{B}$ represents the resultant $\vec{R}$.

The magnitude of $\vec{R}$ is

$$|\vec{R}| = \sqrt{A^2 + B^2 + 2AB\cos\theta}.$$

The direction of $\vec{R}$ is given as

$$\phi = \tan^{-1}\frac{B\sin\theta}{A + B\cos\theta}$$

with vector $\vec{A}$. This is known as the 'parallelogram law of vector addition'.
Proof: Produce OP and drop a perpendicular SM from S onto OP.
In $\triangle OSM$,

$$R = OS = \sqrt{OM^2 + MS^2}, \quad \text{where} \quad OM = OP + PM$$

Equivalent parallelogram whose adjacent sides are represented by the vectors $\vec{A}$ and $\vec{B}$.
Then,

$$R = \sqrt{(OP + PM)^2 + MS^2}$$
$$= \sqrt{OP^2 + PM^2 + 2OP \cdot PM + MS^2}$$
$$= \sqrt{OP^2 + PS^2 + 2OP \cdot PM} \quad (\because MS^2 + PM^2 = PS^2)$$
$$= \sqrt{OP^2 + PS^2 + 2OP \cdot PS \cdot \cos\theta} \quad (\because \cos\theta = \frac{PM}{PS})$$
$$\Rightarrow R = \sqrt{A^2 + B^2 + 2AB\cos\theta}$$

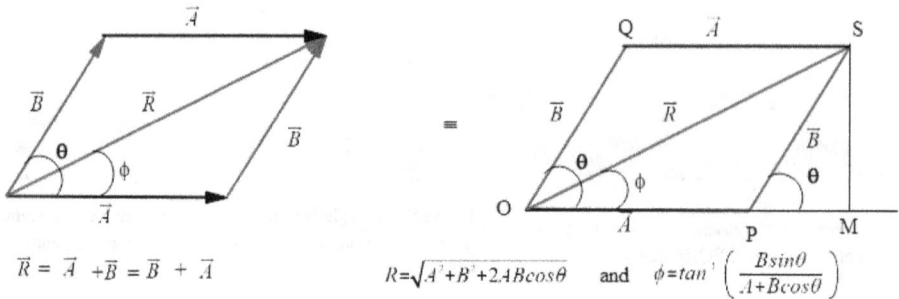

$$\vec{R} = \vec{A} + \vec{B} = \vec{B} + \vec{A}$$

$$R = \sqrt{A^2 + B^2 + 2AB\cos\theta} \quad \text{and} \quad \phi = \tan^{-1}\left(\frac{B\sin\theta}{A + B\cos\theta}\right)$$

If we take $-\vec{R} = \vec{C}$ (say), we have

$$\vec{A} + \vec{B} + \vec{C} = \vec{0}$$

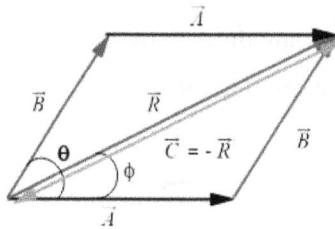

$$\vec{A} + \vec{B} + \vec{C} = \vec{0}$$

N.B: We have proved the laws of vector addition by taking the basic example of a displacement vector. In physics, we define the velocity as the rate of change in displacement. The acceleration is defined as the rate of change in velocity. Linear momentum is defined as the product of mass and velocity. Force is defined as mass times acceleration or rate of change of linear momentum. The strength or intensity of gravitational field, electric field and magnetic fields are defined in terms of the respective forces. Angular momentum is defined as the cross product of position vector and linear momentum. Torque is defined as the rate of change in angular momentum or cross product of position vector and force. Thus, all other vectors in physics such as velocity, acceleration, momentum, force, angular momentum, torque, gravitational field intensity, electric field intensity, magnetic field intensity etc, are derived from and defined in terms of displacement. Since the addition or subtraction laws hold good for displacements, all other vectors must obey the law of vector algebra (addition, subtraction and multiplication etc).

(d) **Lami's theorem**

If the vectors $\vec{A}$, $\vec{B}$ and $\vec{C}$ act at a point,

$$\frac{A}{\sin\alpha} = \frac{B}{\sin\beta} = \frac{C}{\sin\gamma},$$

where α, β and γ are the angles between, $\vec{B}$ and $\vec{C}$, $\vec{C}$ and $\vec{A}$, $\vec{A}$ and $\vec{B}$, respectively. This is known as Lami's theorem.

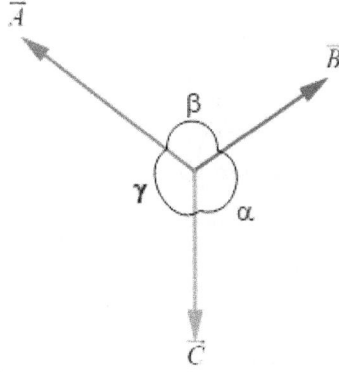

Proof: When three vectors $\vec{A}$, $\vec{B}$ and $\vec{C}$, say, acting at a point yield zero resultant, we can represent them by a triangle, as discussed in the previous section.

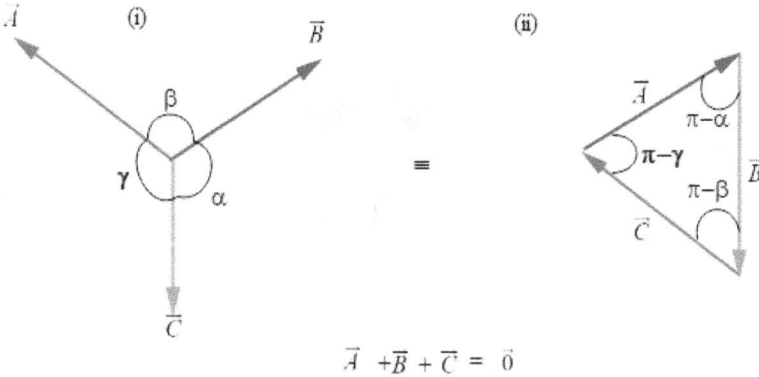

$$\vec{A} + \vec{B} + \vec{C} = \vec{0}$$

When $\vec{A}$, $\vec{B}$ and $\vec{C}$ represent the side of a triangle taken in the same sense, $\vec{A} + \vec{B} + \vec{C} = 0$.

Using the properties of a triangle, we can write

$$\frac{A}{\sin \alpha'} = \frac{B}{\sin \beta'} = \frac{C}{\sin \gamma'}$$

Using $\sin(180° - \theta) = \sin \theta$, the above equation can be expressed as

$$\frac{A}{\sin \alpha} = \frac{B}{\sin \beta} = \frac{C}{\sin \gamma}$$

Example 1 A bob is in equilibrium under the action of three forces. Find the relation between the forces. Put $\theta = 60°$.

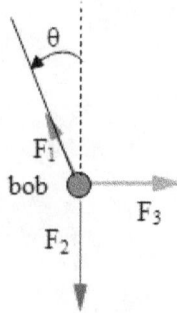

Solution

For equilibrium of the bob, the net resultant force is zero.

$$\overrightarrow{F_{net}} = \overrightarrow{F_1} + \overrightarrow{F_2} + \overrightarrow{F_3} = \overrightarrow{0}$$

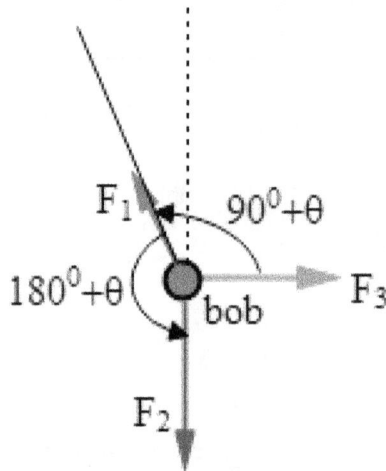

According to Lami's theorem,

$$\frac{F_1}{\sin 90°} = \frac{F_2}{\sin(90° + \theta)} = \frac{F_3}{\sin(180° - \theta)}$$

$$\Rightarrow \frac{F_1}{\sin 90°} = \frac{F_2}{\sin 150°} = \frac{F_3}{\sin 120°}$$

$$\Rightarrow F_1 = 2F_2 = \frac{2F_3}{\sqrt{3}}$$

$$\Rightarrow \sqrt{3}\,F_1 = 2\sqrt{3}\,F_2 = 2F_1 \quad \text{Ans.}$$

(e) **Rectangular components**
If R is the resultant of $\vec{A}$ and $\vec{B}$, that is, $\vec{R} = \vec{A} + \vec{B}$, $\vec{A}$ and $\vec{B}$ are called the components of R. When $\vec{A}$ and $\vec{B}$ are perpendicular to each other ($\theta = 90°$), the resulting parallelogram will be a rectangle.

Hence, A and B are called 'rectangular components' of $\vec{R}$ along the x- and y-axes, respectively, denoted by R_x and R_y, respectively. $R_x(=A) = R \cos \alpha$ and $R_y(=B) = R \sin \alpha$. Then

$$\vec{R} = \vec{A} + \vec{B}$$
$$= A\hat{i} + B\hat{j}$$
$$= R \cos \theta \hat{i} + R \sin \theta \hat{j}$$
$$= R(\cos \theta \hat{i} + \sin \theta \hat{j})$$

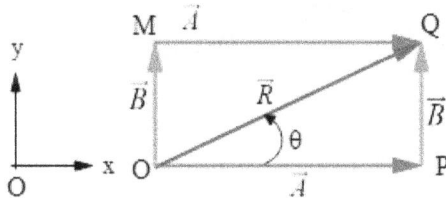

The rectangular components A and B of R can be given as $\vec{R} = \vec{A} + \vec{B} = A\hat{i} + B\hat{j} = R(\cos\theta\hat{i} + \sin\theta\hat{j})$

When the components of $\vec{R}$ lie in negative x- and y-directions, we must use the $-\hat{i}$ coordinate system in four different quadrants.

Let us consider the vector A_1. Resolving it into components, we have

$$\vec{A} = \vec{A_{i_x}} + \vec{A_{i_y}}$$
$$= \pm A_i \cos \theta_i \hat{i} \pm A_i \sin \theta_i \hat{j}$$

where θ_i is the acute angle between $\vec{A_i}$ and positive (or negative) x-axis. The positive sign is used for the components along the positive x-axis and vice versa. When the tails of many vectors lie at a point, the resultant of all vectors can be given as

$$\vec{A} = \sum \vec{A_i} = \sum \vec{A_{i_x}} + \vec{A_{i_y}}$$
$$= (\pm\sum A_i \cos \theta_i)\hat{i} + (\pm\sum A_i \sin \theta_i)\hat{j}$$
$$= \pm A_x \hat{i} \pm A_y \hat{j} \text{ (Let)}$$

or,

$$|\vec{A}| = \sqrt{A_x^2 + A_y^2} \text{ and } \phi = \tan^{-1}|\frac{A_y}{A_x}|,$$

relative to +ve or $-$ve x-axis.

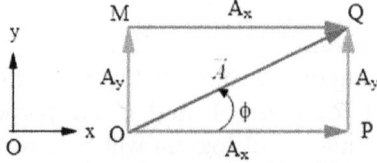

The rectangular components A_x and A_y of the vector $\vec{A}$

N.B.:

1. A_x and A_y are the components of the vector $\vec{A}$, known as vector components

2. $\vec{A}_{x_i} = |\vec{A}_{x_i}|(\pm \hat{i})$ and $\vec{A}_{y_i} = |\vec{A}_{y_i}|(\pm \hat{j})$; a +ve sign is used when the vectors, and $\vec{A}_y$ are directed along the +ve axis and a −ve sign is used when the vector points along the −ve axis.

Example 2 Three forces of magnitudes 20 N, 10 N and $20\sqrt{2}$ N are acting at the origin as shown in the figure below. Find the resultant force acting on the point O.

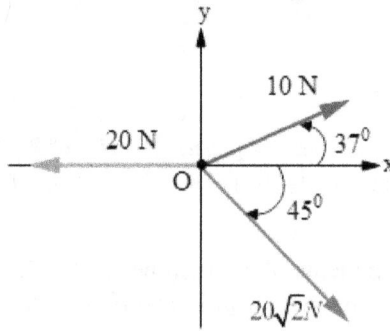

Solution
The forces acting at O are given as,

$$\vec{F_1} = 10(\cos 37°\hat{i} + \sin 37°\hat{j}) = 10\left(\frac{4}{5}\hat{i} + \frac{3}{5}\hat{j}\right) = 8\hat{i} + 6\hat{j} \cdot N$$

$$\vec{F_2} = -20\hat{i} \cdot N$$

$$\vec{F_3} = 20\sqrt{2}\cos 45°\hat{i} + \sqrt{2}\sin 45°(-\hat{j})$$
$$= 20\hat{i} - 20\hat{j}$$

The net force is $\vec{F} = \vec{F_1} + \vec{F_2} + \vec{F_3}$
Substituting the values of $\vec{F_1}$, $\vec{F_2}$ and $\vec{F_3}$, the net force is

$$\vec{F} = (8\hat{i} + 6\hat{j}) + (-20\hat{i}) + (20\hat{i} - 20\hat{j})$$
$$= 8\hat{i} - 20\hat{j} \cdot N. \text{ Ans.}$$

1.6.1 Rectangular components of a vector in three dimensions (optional)

Let us express a vector $\vec{A}$ oriented in space in terms of its components along the axes; for this we drop a perpendicular PM from the tip of the vector $\vec{A}$ onto the x–z-plane and onto the y-axis, as shown in the figure below. Then we drop perpendiculars MS and from M onto the x- and z-axes. In this way we now have three segments OS, OT and OP. These are termed as the projections or components of OP along x-, y- and z-axes, respectively. Since $OP = A$, we can write $OS = A_x$, $OT = A_y$ and $OR = A_z$ means $\vec{A}_x$, $\vec{A}_y$ and $\vec{A}_z$ are the vector components of vector $\vec{A}$ along the coordinates (x, y and z), respectively. Then,

$$\vec{A} = \vec{A}_x + \vec{A}_y + \vec{A}_z.$$

The vector $\vec{A}$ has three rectangular components A_x, A_y and A_z in the Cartesian coordinate system along the x-, y- and z-axes, respectively;

$$\vec{A} = A_x\hat{i} + A_y\hat{j} + A_z\hat{k}$$
$$= \vec{A}_x + \vec{A}_y + \vec{A}_z$$

Since $\vec{A}_x = A_x\hat{i}$, $\vec{A}_y = A_y\hat{j}$ and $\vec{A}_z = A_z\hat{k}$, we have

$$\vec{A} = A_x\hat{i} + A_y\hat{j} + A_z\hat{k}$$

The magnitude of $\vec{A}$ is $A = \sqrt{A_x^2 + A_y^2 + A_z^2}$.

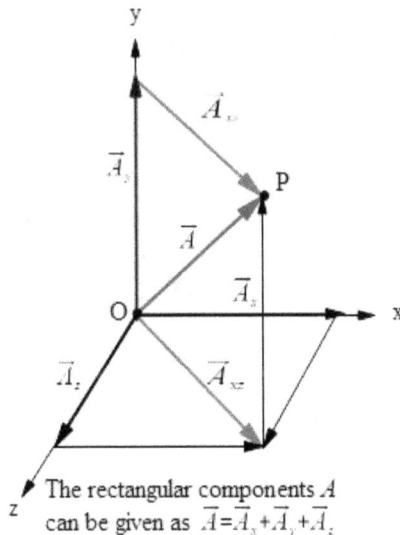

The rectangular components A can be given as $\vec{A} = \vec{A}_x + \vec{A}_y + \vec{A}_z$

Let us use the above ideas in the following example.

Example 3 A bird flies due east a distance of 100 m, then heads due north a distance of 50 m, it flies vertically up a distance of 20 m. Find the distance of the position of the bird relative to its initial position.

Solution

The final position $\vec{r}$ of the bird is followed by three successive displacements $\vec{s_1}$, $\vec{s_2}$ and $\vec{s_3}$, as shown in the figure below, where $\vec{s_1} = 100\hat{i}$, $\vec{s_2} = 50\hat{j}$ and $\vec{s_3} = 20\hat{k}$, as also shown in the figure.

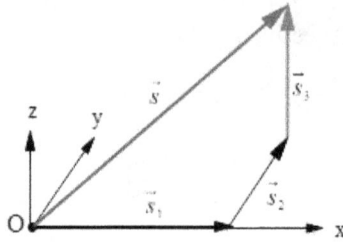

Hence, the position vector

$$\vec{r} = \vec{s_1} + \vec{s_2} + \vec{s_3} = 100\hat{i} + 50\hat{j} + 20\hat{k} \text{ m}$$

Then, $|\vec{r}| = \sqrt{100^2 + 50^2 + 20^2}$ m Ans.

1.7 Subtraction of vectors

The subtraction of a vector $\vec{B}$ from another vector $\vec{A}$ can be given as $\vec{A} - \vec{B}(=\vec{C}$, say). This means that if we take the negative of vector $\vec{B}$ and then add it with vector following the law of vector addition, we can find $\vec{A} + (-\vec{B}) = \vec{A} - \vec{B}$. The magnitude of $\vec{A} - \vec{B}$ is, $|\vec{A} - \vec{B}| = \sqrt{A^2 + B^2 - 2AB \cos \theta}$, where $\theta = $ angle between the vector $\vec{A}$ and $\vec{B}$.

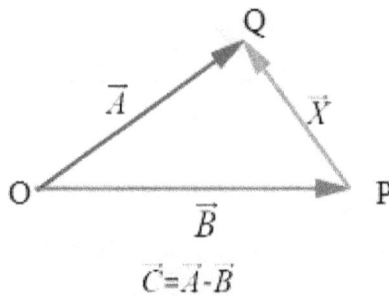

$\vec{C} = \vec{A} - \vec{B}$

The subtraction of two vectors is an addition of one vector with the negative of the other vector; $\vec{A} - \vec{B} = \vec{A} + (-\vec{B})$. Hence subtraction of two vectors is a vector quantity.

N.B.: If we combine the addition and subtraction of two vectors in parallelogram law of vectors, we can state that:

If the vectors $\vec{A}$, $\vec{B}$ are represented by the adjacent sides of a parallelogram, its diagonal passing through the tails of the vectors $\vec{A}$ and $\vec{B}$ represents the addition or resultant $\vec{R}_1$. Its diagonal passing through the tips of the vectors $\vec{A}$ and $\vec{B}$ represents the difference or subtraction of two vectors denoted as $\vec{R}_2$.

In one expression, the magnitudes of these diagonals can be given as

$$|\vec{R}| = \sqrt{A^2 + B^2 \pm 2AB \cos \theta}.$$

The direction of $\vec{R}$ is given as

$$\phi = \tan^{-1}\frac{B \sin \theta}{A \pm B \cos \theta}$$

with vector $\vec{A}$. This is a nut-shell formula of 'parallelogram law of vector addition and subtraction'.

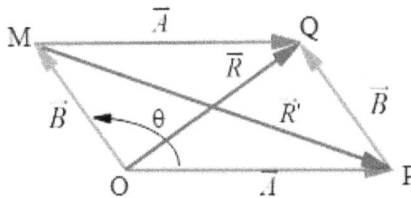

If the adjacent sides of the parallelogram represent the two vectors A and B, then, the diagonal passing through the point of application of the vectors represents the resultant of these two vectors and the other diagonal represents the difference or subtraction of the two vectors. $\vec{R}=\vec{A}+\vec{B}$; $\vec{R}'=\vec{A}-\vec{B}$

Example 4 If the magnitude of (a) addition of two vectors is equal to the magnitude of their subtraction, (b) subtraction of two vectors each of magnitude of A is equal to the magnitude of either vector, find the angle between the vectors.

Solution

(a) For two vectors $\vec{A}$ and $\vec{B}$,

$$|\vec{A} + \vec{B}| = |\vec{A} - \vec{B}|$$

$$\Rightarrow \sqrt{A^2 + B^2 + 2AB \cos \theta} = \sqrt{A^2 + B^2 - 2AB \cos \theta}$$

$$\Rightarrow \cos \theta = 0$$

$$\Rightarrow \theta = 90°$$

(b) For two vectors $\vec{A}$ and $\vec{B}$,

$$|\vec{A} - \vec{B}| = |\vec{A}|$$

$$\Rightarrow \sqrt{A^2 + B^2 - 2AB \cos \theta} = A$$

$$\Rightarrow B(1 - 2\cos\theta) = 0$$

$$\Rightarrow \theta = 60°.$$

1.8 Scalar product of two vectors

In physics we need to define some scalar quantities like work, power, flux, etc, as the product of two vectors. For this, we need to develop the idea of scalar (dot) product of two vectors. For example, work done W by a force $\vec{F}$ is defined as the product of the displacement of a point at which the force is acting and the component F_x of the force F along the displacement (direction of motion of the point). Symbolically,

$$W = |\vec{F_x}||\vec{s}|, \quad \text{where } |\vec{F_x}| = |\vec{F}|\cos\theta$$

This gives $W = |\vec{F}||\vec{s}|\cos\theta$. Since the scalar quantity W is defined as a product of two vectors $\vec{F}$ and $\vec{s}$, we can call this product the 'scalar product'. In short, substituting $\cos\theta$ by a dot($\cdot$), we write $W = \vec{F} \cdot \vec{s}$. So, the scalar product can also be termed as a 'dot product'.

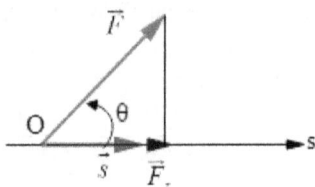

The work done by the force F
can be given as W=Fs cosθ

A scalar quantity C can be defined as the product of any two vectors $\vec{A}$ and $\vec{B}$ which can be given as $C = |\vec{A}||\vec{B}|\cos\theta = \vec{A} \cdot \vec{B}$, where θ is the smallest angle between the vectors $\vec{A}$ and $\vec{B}$.

We can note the following conditions:

1. When θ is a acute, $\vec{A} \cdot \vec{B}$ is +ve.
2. When θ is 90°, $\vec{A} \cdot \vec{B}$ is zero.
3. When θ is obtuse, $\vec{A} \cdot \vec{B}$ is −ve.
4. If $\hat{i}, \hat{j}$ and $\hat{k}$ are mutually perpendicular vectors, then,

$$\hat{i} \cdot \hat{i} = (1)(1)\cos 0° = 1$$

similarly, $\hat{i} \cdot \hat{i} = \hat{j} \cdot \hat{j} = \hat{k} \cdot \hat{k}$

and

$$\hat{i}.\hat{j} = (1)(1)\cos 90° = 0$$

$$\therefore \hat{i} \cdot \hat{j} = \hat{j} \cdot \hat{k} = \hat{k} \cdot \hat{i} = 0$$

Application of scalar product:

(i) Work done: $W = \vec{F}.\Delta\vec{r} = |\vec{F}||\Delta\vec{r}|\cos\theta$

(ii) Angle between the vectors is given by

$$\vec{A}.\vec{B} = |\vec{A}||\vec{B}|\cos\theta$$

$$\Rightarrow\theta = \cos^{-1}\frac{\vec{A}.\vec{B}}{|\vec{A}||\vec{B}|}$$

(iii) If $\vec{A} = A_1\hat{i} + A_2\hat{j} + A_3\hat{k}$ and $\vec{B} = B_1\hat{i} + B_2\hat{j} + B_3\hat{k}$

$$\vec{A}\cdot\vec{B} = A_1B_1 + A_2B_2 + A_3B_3$$

and

$$|\vec{A}| = \sqrt{A_1^2 + A_2^2 + A_3^2}$$

$$|\vec{B}| = \sqrt{B_1^2 + B_2^2 + B_3^2}$$

$$\Rightarrow\theta = \cos^{-1}\frac{\vec{A}.\vec{B}}{|\vec{A}||\vec{B}|} = \cos^{-1}\frac{A_1B_1 + A_2B_2 + A_3B_3}{(\sqrt{A_1^2 + A_2^2 + A_3^2})(\sqrt{B_1^2 + B_2^2 + B_3^2})}$$

(iv) Component or projection of one vector along another vector: Component of vector $\vec{B}$ along vector $\vec{A}$.

$$|\vec{B}|\cos\theta\hat{A} = \left(\frac{\vec{A}.\vec{B}}{|\vec{A}|}\right)\hat{A}$$

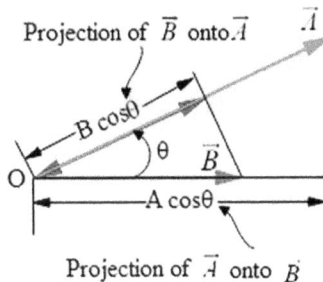

Projection of $\vec{B}$ onto $\vec{A}$

Projection of $\vec{A}$ onto $\vec{B}$

Example 5 Find the component of $\vec{A} = 3\hat{i} + 4\hat{j}$ along the direction of vector $\vec{B} = 5\hat{i} + 12\hat{j}$.

Solution

The component of $\vec{A}$ along $\vec{B}$ is

$$(|\vec{A}| \cos \theta)\hat{B} = |\frac{\vec{A}.\vec{B}}{|\vec{B}|}| \hat{B}$$

$$= \left(\frac{(3\hat{i} + 4\hat{j}).(5\hat{i} + 12\hat{j})}{\sqrt{25 + 144}} \right) \frac{(5\hat{i} + 12\hat{j})}{\sqrt{25 + 144}}$$

$$= \frac{(15 + 48)}{13 \times 13}(5\hat{i} + 12\hat{j}) = \frac{315\hat{i} + 756\hat{j}}{169} \text{ Ans.}$$

1.9 Vector product of two vectors

In the last section, we developed the idea of scalar product where we define some scalar quantities as the scalar (dot) product of two vectors. On the other hand, from a physical point of view, we need to define some vector quantities like torque, angular momentum as the product of two vectors. For this, we need to develop the idea of 'vector product'. Let us take an example from our daily experience.

If we want to define a vector as the product of two suitable vectors, we need to develop a general formula (expression) for the vector product.

If a vector $\vec{C}$ is to be defined (or expressed) as the product of two vectors $\vec{A}$ and $\vec{B}$, it is given as $\vec{C} = \vec{A} \times \vec{B} = |\vec{A}||\vec{B}| \sin \theta \hat{n}$ so as to satisfy the right-hand thumb or right-hand screw rule. The magnitude of $\vec{C}$ is given as

$$|\vec{C}| = \left(\vec{A} \times \vec{B} \right) = |\vec{A}||\vec{B}| \sin \theta,$$

where $0° \leqslant \theta \leqslant 180°$.

The direction of $\vec{C}$ is $\hat{n}$ which is perpendicular to both the vectors $\vec{A}$ and $\vec{B}$ or, in other words, it is perpendicular to the plane formed by $\vec{A}$ and $\vec{B}$.

Substituting $|\vec{A}| \sin \theta = p_1$ or $|\vec{B}| \sin \theta = p_2$ we can write $|\vec{C}| = p_1 |\vec{B}| = p_2 |\vec{A}|$, where p_1 and p_2 are the lengths of the perpendicular dropped from the tips of the vectors $\vec{A}$ and $\vec{B}$ onto the vectors $\vec{B}$ and $\vec{A}$, respectively, as shown in the following figure.

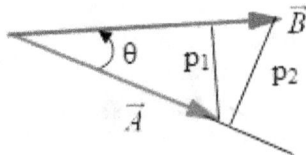

$p_1 = A\sin\theta$; $p_2 = B\sin\theta$

The direction of $\vec{C}$ is given by the 'right-hand screw rule' or 'right-hand thumb rule' as described below.

Right-hand thumb rule: If you want to find the direction of $\vec{A} \times \vec{B}$, first of all, curl or wrap the fingers of your right-hand by swinging (turning) the fingers from $\vec{A}$ to $\vec{B}$ following the shorter (smaller) angle θ between these vectors. Then the extended thumb will show the direction of $\vec{A} \times \vec{B}$. This is what we call right-hand thumb rule. I present hereby another way to use the above rule.

Following the above rule, if we reverse the order of the cross product, its direction of vector (or cross) product reverses. Hence $\vec{A} \times \vec{B} = -\vec{B} \times \vec{A}$. Hence the vector product is anti-commutative.

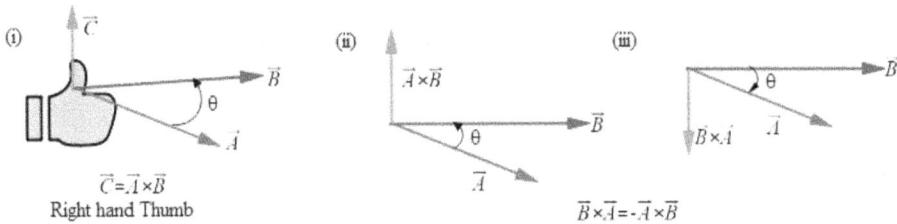

(i) $\vec{C} = \vec{A} \times \vec{B}$
Right hand Thumb

(ii) $\vec{A}$

(iii) $\vec{B} \times \vec{A} = -\vec{A} \times \vec{B}$

Right-hand screw rule: When we rotate the screw in the sense of measuring the smaller angle between $\vec{A}$ and $\vec{B}$, the direction of motion of the screw inside the wood gives the direction of $\vec{A} \times \vec{B}$. When we reverse the order of the vector product, the sense of measuring the angle is reversed.

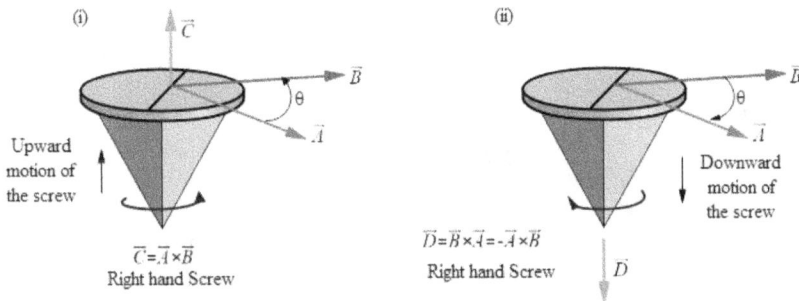

(i) $\vec{C} = \vec{A} \times \vec{B}$
Right hand Screw
Upward motion of the screw

(ii) $\vec{D} = \vec{B} \times \vec{A} = -\vec{A} \times \vec{B}$
Right hand Screw
Downward motion of the screw

Example 6 Find the cross product of two vectors $\vec{A}$ and $\vec{B}$ which are (i) parallel and equal (ii) antiparallel and negative (iii) orthogonal.

Solution

(i) If $\vec{A}$ is parallel (or equal) to $\vec{B}$, $\theta = 0°$. Hence $|\vec{A} \times \vec{B}| = AB \sin 0° = 0$

(ii) If $\vec{A}$ is antiparallel (or negative) to $\vec{B}$, $\theta = 180°$
   Hence, $|\vec{A} \times \vec{B}| = AB \sin 180° = 0$

(iii) If $\vec{A}$ is perpendicular to $\vec{B}$, $\theta = 90°$.

Then $|\vec{C}| = |\vec{A} \times \vec{B}| = AB \sin 90° = AB$ and $|\vec{C}| = (\vec{A} \times \vec{B})$ is directed outward obeying right-hand thumb rule. $\vec{A} \times \vec{B}$ must be perpendicular to the plane containing $\vec{A}$ and $\vec{B}$. Ans.

N.B.: Let us note the following points:

Since the cross product of two parallel vectors is a zero vector (but not zero), we have

$$\pm\hat{i} \times \hat{i} = \pm\hat{j} \times \hat{j} = \pm\hat{k} \times \hat{k} = \hat{0}.$$

By the right-hand thumb rule, the cross product of the orthogonal unit vectors can be given as

$$\hat{i} \times \hat{j} = \hat{k}, \hat{j} \times \hat{k} = \hat{i} \text{ and } \hat{k} \times \hat{i} = \hat{j}$$

Reversing the orders, we can write

$$\hat{j} \times \hat{i} = -\hat{k}, \hat{k} \times \hat{j} = -\hat{i} \text{ and } \hat{i} \times \hat{k} = -\hat{j}.$$

Example 7 Find $\vec{A} \times \vec{B}$, if $\vec{A} = A_x\hat{i} + A_y\hat{j} + A_z\hat{k}$ and $\vec{B} = B_x\hat{i} + B_y\hat{j} + B_z\hat{k}$.
   Solution
   Substituting $\vec{A} = A_x\hat{i} + A_y\hat{j} + A_z\hat{k}$ and $\vec{B} = B_x\hat{i} + B_y\hat{j} + B_z\hat{k}$ in the formula $\vec{C} = \vec{A} \times \vec{B}$, we have

$$\vec{C} = \left(A_x\hat{i} + A_y\hat{j} + A_z\hat{k}\right) \times \left(B_x\hat{i} + B_y\hat{j} + B_z\hat{k}\right)$$
$$= A_xB_x\hat{i} \times \hat{i} + A_xB_y\hat{i} \times \hat{j} + A_xB_z\hat{i} \times \hat{k} + A_yB_x\hat{j} \times \hat{i} + A_yB_y\hat{j} \times \hat{j}$$
$$+ A_yB_z\hat{j} \times \hat{k} + A_zB_x\hat{k} \times \hat{i} + A_zB_y\hat{k} \times \hat{j} + A_zB_z\hat{k} \times \hat{k}$$

Since

$$\hat{i} \times \hat{j} = \hat{k}, \hat{j} \times \hat{k} = \hat{i}, \hat{k} \times \hat{i} = \hat{j} \text{ and } \hat{i} \times \hat{i} = \hat{j} \times \hat{j} = \hat{k} \times \hat{k} = 0,$$

we have

$$\vec{C} = A_xB_y\hat{k} - A_xB_z\hat{j} - A_yB_x\hat{k} + A_yB_z\hat{i} + A_zB_x\hat{j} - A_zB_y\hat{i}$$

Rearranging the terms, we have

$$\vec{C} = \left(A_yB_z = A_zB_y\right)\hat{i} - (A_xB_z - A_zB_x)\hat{j} + \left(A_xB_y - A_yB_x\right)\hat{k}$$

Now we can put in a compact form which is easier to remember as the following determinant.

$$\vec{A} \times \vec{B} = \begin{vmatrix} \hat{i} & \hat{j} & \hat{k} \\ A_x & A_y & A_z \\ B_x & B_y & B_z \end{vmatrix}$$

1.9.1 Applications of vector product

1. Torque (moment of force):

   Here $\vec{F}$ is a force acting on a body free to rotate about O, and let $\vec{r}$ is the position vector of any point P on the line of action of the force.

   $$\text{Torque}(\vec{\tau}) = \vec{r} \times \vec{F};$$

   where $\vec{r} = x\hat{i} + y\hat{j} + z\hat{k}$ and $\vec{F} = F_x\hat{i} + F_y\hat{j} + F_z\hat{k}$

   $$(\vec{\tau}) = \begin{vmatrix} \hat{i} & \hat{j} & \hat{k} \\ x & y & z \\ F_x & F_y & F_z \end{vmatrix}$$

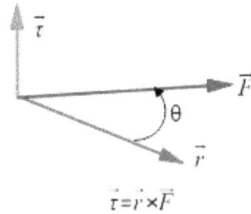

$\vec{\tau} = \vec{r} \times \vec{F}$

2. Angular momentum:

   $$\vec{L} = \vec{r} \times \vec{P}$$

   where $\vec{P} = m\vec{v}$ is the linear momentum of the particle and $\vec{r}$ is the position vector of the particle.

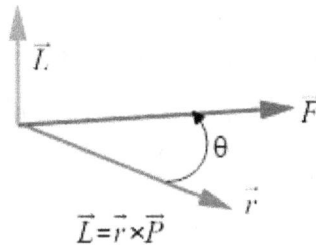

$\vec{L} = \vec{r} \times \vec{P}$

3. Unit vector along the direction perpendicular to plane of $\vec{A}$ and $\vec{B}$.

   $$\hat{n} = \frac{\vec{A} \times \vec{B}}{|\vec{A} \times \vec{B}|}$$

4. Area of parallelogram: $|\vec{A} \times \vec{B}|$ represents the area of a parallelogram whose sides are $|\vec{A}|$ and $|\vec{B}|$.

5. Area of a triangle: $\frac{1}{2}|\vec{A} \times \vec{B}|$ represents the area of a triangle when $|\vec{A}|$ and $|\vec{B}|$ are two sides of the triangle.

1.10 Some fundamental laws of vector algebra

Commutative law:

   (i) Addition: Since $\vec{A} + \vec{B} = \vec{B} + \vec{A}$, addition of two vectors is commutative.

   (ii) Subtraction: Since $\vec{A} - \vec{B} = -(\vec{B} - \vec{A})$ subtraction of two vectors is anti-commutative.

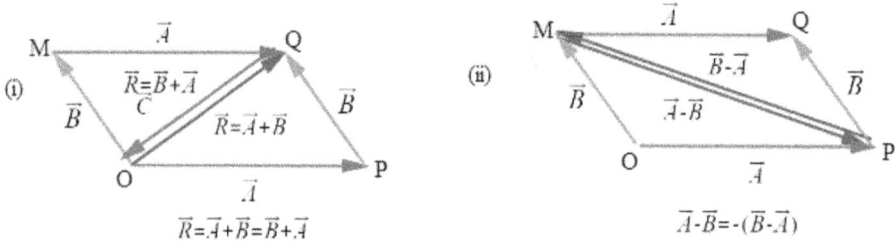

$$R = \vec{A} + \vec{B} = \vec{B} + \vec{A}$$

$$\vec{A} - \vec{B} = -(\vec{B} - \vec{A})$$

   (iii) Scalar product: Since $\vec{A} \cdot \vec{B} = AB \cos \theta = AB \cos \theta = \vec{B} \cdot \vec{A}$, a scalar product is commutative. It means, $\vec{A} \cdot \vec{B} = \vec{B} \cdot \vec{A}$

   (iv) Vector product: Since $\vec{A} \times \vec{B} = -\vec{B} \times \vec{A}$, a vector product is anti-commutative.

1.10.1 Associative law

$$(\vec{A} + \vec{B}) + \vec{C} = \vec{A} + (\vec{B} + \vec{C}), \text{ when}$$

we add $\vec{A}$ with $\vec{B}$ to get $(\vec{A} + \vec{B})$, then add $(\vec{A} + \vec{B})$ with $\vec{C}$ to finally obtain $(\vec{A} + \vec{B}) + \vec{C}$.

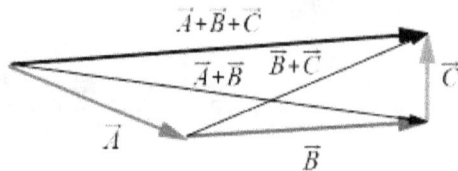

$$(\vec{A} + \vec{B}) + \vec{C} = \vec{A} + (\vec{B} + \vec{C})$$

Again we add $\vec{B}$ and $\vec{C}$ to get $(\vec{B}+\vec{C})$, then add $\vec{A}$ with $(\vec{B}+\vec{C})$ to obtain $\vec{A}+(\vec{B}+\vec{C})$. By following the above two orders of vector addition, we have $(\vec{A}+\vec{B})+\vec{C} = \vec{A}+(\vec{B}+\vec{C})$. This is known as associative law.

1.10.2 Distributive law

The following identities can also proved.

$$\vec{A}\cdot\left(\vec{B}+\vec{C}\right) = \vec{A}\cdot\vec{B}+\vec{A}\cdot\vec{C}$$

$$\vec{A}\times\left(\vec{B}+\vec{C}\right) = \vec{A}\times\vec{B}+\vec{A}\times\vec{C}$$

$$a\left(\vec{A}+\vec{B}\right) = a\vec{A}+a\vec{B}$$

If the component vectors $\vec{A}$ and $\vec{B}$ increase by a, their resultant will increase by the same factor and vice versa where 'a' is a scalar.

1.10.3 Condition of collinearity of vector $\vec{A}$ and $\vec{B}$

$$\vec{A} = A_1\hat{i} + A_2\hat{j} + A_3\hat{k}$$

$$\vec{B} = B_1\hat{i} + B_2\hat{j} + B_3\hat{k}$$

If $\vec{A}$ and $\vec{B}$ are collinear then the angle between them (θ) is 0 or π.

$$\therefore \sin\theta = 0$$

$$\therefore \vec{A}\times\vec{B} = 0 = \begin{vmatrix} \hat{i} & \hat{j} & \hat{k} \\ A_1 & A_2 & A_3 \\ B_1 & B_2 & B_3 \end{vmatrix}$$

$$\Rightarrow \hat{i}(A_2 B_3 - B_2 A_3) + \hat{j}(B_1 A_3 - A_1 B_3) + \hat{k}(A_1 B_2 - B_1 A_2) = 0$$

$$\Rightarrow \frac{A_1}{B_1} = \frac{A_2}{B_2} = \frac{A_3}{B_3}.$$

1.11 Condition for coplanar vectors

(i) If the vectors $\vec{A}$, $\vec{B}$ and $\vec{C}$ are of the same nature (like vectors) any one vector can be expressed as the linear combination of the other two vectors. If we choose $\vec{C}$, it can be expressed as the linear combination of $\vec{A}$ and $\vec{B}$. Mathematically,

$$\vec{C} = p\vec{A} + q\vec{B}$$

where p and q are scalars.

Example 8 If $\vec{A} = \hat{i} + 2\hat{j} - 3\hat{k}$, $\vec{B} = 3\hat{i} - \hat{j} + \hat{k}$ and $\vec{C} = \hat{i} + \hat{j} - n\hat{k}$, find '$n$' the vectors to be coplanar.

Solution

Using $\vec{C} = p\vec{A} + q\vec{B}$, substituting $\vec{A}$, $\vec{B}$ and $\vec{C}$, we have

$$p(\hat{i} + 2\hat{j} - 3\hat{k}) + q(3\hat{i} - \hat{j} + \hat{k}) = \hat{i} + \hat{j} - n\hat{k}.$$

Comparing the coefficients of $\hat{i}$, $\hat{j}$ and $\hat{k}$, we have

$$p + 3q = 1, \ 2p - q = 1 \ \text{and} \ -3p + q = -n$$

Solving first two equations, we have

$$p = \frac{4}{7} \ \text{and} \ q = \frac{1}{7}.$$

Substituting p and q in third equation, we have $n = \frac{11}{7}$ Ans.

N.B.: Since $(\vec{A} \times \vec{B})$ is perpendicular to both $\vec{A}$ and $\vec{B}$, if $\vec{A}$, $\vec{B}$ and $\vec{C}$ are coplanar, $(\vec{A} \times \vec{B})$ must be perpendicular to $\vec{C}$. Hence, $(\vec{A} \times \vec{B}) \cdot \vec{C} = 0$.

Basically $(\vec{A} \times \vec{B}) \cdot \vec{C}$ represents the volume of a parallelepiped represented by the vectors $\vec{A}$, $\vec{B}$ and $\vec{C}$. If the volume of the parallelpiped is zero, $\vec{A}$, $\vec{B}$ and $\vec{C}$ must lie in the same plane.

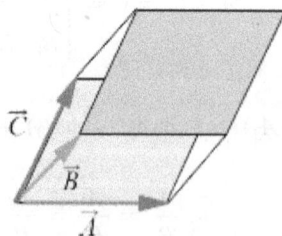

Volume of the Parallelopiped is given as

$$V = \vec{A} \times \vec{B} . \vec{C}$$

1.12 Derivative (differentiation) of a vector

The direction of the vector $\vec{A}$ is given by the unit vector $\hat{r}$. If this vector changes with time, it can be given as

$$\vec{A}(t) = A\hat{r}(t)$$

The rate of change of this vector is given as

$$\frac{d\vec{A}(t)}{dt} = \frac{dA\hat{r}(t)}{dt} = A\frac{d\hat{r}(t)}{dt} + \hat{r}(t)\frac{dA(t)}{dt}$$

The first term is the rate of change of the magnitude A of the vector, that is, $\frac{dA(t)}{dt}$. The second term is $\frac{d\hat{r}(t)}{dt}$; it is termed as rate of change of the unit vector which is given as

$$\frac{d\hat{r}(t)}{dt} = Lim_{\Delta t \to 0}\frac{\hat{r}(t + \Delta t) - \hat{r}(t)}{\Delta t}$$

We can see that the differential change or elementary change (change in unit vector during a very small-time interval dt) is perpendicular to the unit vector $\hat{r}(t)$ that points in the θ-axis that lies in the plane of revolution of the tip of the unit vector $\hat{r}(t)$. We can see that the rate of change in the unit vector is

$$\frac{d\hat{r}(t)}{dt} = \frac{d\theta}{dt}\hat{\theta},$$

where $\frac{d\theta}{dt}$ = angular velocity ω of rotation of the unit vector or vector $\vec{A}(t)$.

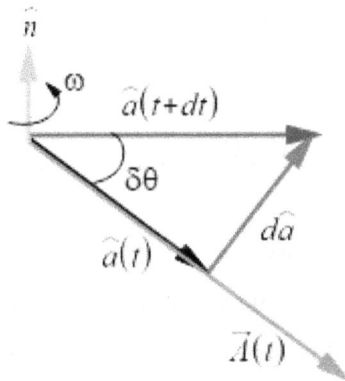

Example 9 A vector changes with time as $\vec{A}(t) = A(\cos \omega t \hat{i} + \sin \omega t \hat{j})$. Find the derivative of the vector.

Solution

Method 1:

The derivative of $\vec{A}(t) = A(\cos \omega t \hat{i} + \sin \omega t \hat{j})$ is given as

$$\frac{d\vec{A}(t)}{dt} = \frac{dA\hat{a}(t)}{dt} = A\frac{d\hat{a}(t)}{dt} + \hat{a}(t)\frac{dA(t)}{dt}$$

Since A is a constant quantity, $dA/dt = 0$. So, we can write

$$\frac{d\vec{A}(t)}{dt} = A\frac{d\hat{a}(t)}{dt}$$

Since $\frac{d\hat{a}(t)}{dt} = \frac{d\theta}{dt}\hat{\theta}$, we have $\frac{d\vec{A}(t)}{dt} = A\frac{d\theta}{dt}\hat{\theta}$

$$\Rightarrow \frac{d^2\vec{A}(t)}{dt^2} = A\frac{d\theta}{dt}\frac{d\hat{\theta}}{dt}$$

Putting $\frac{d\hat{\theta}}{dt} = -\frac{d\theta}{dt}\hat{a}$, we have

$$\frac{d^2\vec{A}(t)}{dt^2} = -A\frac{d\theta}{dt}\frac{d\theta}{dt}\hat{a} = -A\left(\frac{d\theta}{dt}\right)^2\hat{a} = -\left(\frac{d\theta}{dt}\right)^2\vec{A} = -\omega^2\vec{A} \text{ Ans.}$$

Method 2:

The derivative of $\vec{A}(t) = A(\cos \omega t \hat{i} + \sin \omega t \hat{j})$ is given as

$$\frac{d\vec{A}(t)}{dt} = \frac{d\{A(\cos \omega t \hat{i} + \sin \omega t \hat{j})\}}{dt} = \omega A(-\sin \omega t \hat{i} + \cos \omega t \hat{j})$$

$$\Rightarrow \frac{d^2\vec{A}(t)}{dt^2} = \frac{d\{\omega A(-\sin \omega t \hat{i} + \cos \omega t \hat{j})\}}{dt} = -\omega^2 A(\cos \omega t \hat{i} + \sin \omega t \hat{j})$$

Putting $A(\cos \omega t \hat{i} + \sin \omega t \hat{j}) = \vec{A}$, we have

$$\Rightarrow \frac{d^2\vec{A}(t)}{dt^2} = -\omega^2\vec{A} \text{ Ans.}$$

1.13 Integration of a vector

Space integral of a vector: Let us take the example of work done. We know the work done by a force $\vec{F}$ acting on a particle when it undergoes a displacement $\vec{dr}$ is

$$dW = \vec{F} \cdot d\vec{r}$$

The total work done W can be given by integrating the elementary work dV. Then,

$$W = \int dW = \int \vec{F} \cdot \vec{dr}$$

This is known as 'line integral of force'.

The line integral of any vector $\vec{A}$ can be given as

$$\int \vec{A} \cdot \vec{dl},$$

where $\vec{A} \cdot \vec{dl}$ is the dot product of $\vec{A}$ and $\vec{dl}$.

To evaluate the integral, we need to express $\vec{A}$ and $\vec{dl}$ in the Cartesian coordinate system.

Substituting

$$\vec{A} = A_x\hat{i} + A_y\hat{j} + A_z\hat{k} \text{ and } \vec{dl} = dx\hat{i} + dy\hat{j} + dz\hat{k},$$

we have

$$\int \vec{A} \cdot \vec{dl} = \int \left(A_x\hat{i} + A_y\hat{j} + A_z\hat{k}\right) \cdot (dx\hat{i} + dy\hat{j} + dz\hat{k})$$

By using the concept of dot product, we have

$$\int \vec{A} \cdot \vec{dl} = \int \left(A_x dx + A_y dy + A_z dz\right)$$

This gives $\int \vec{A} \cdot \vec{dl} = \int A_x dx + \int A_y dy + \int A_z\, dz$ which is the sum of three simple integrals where A_x, A_y and A_z are the components of $\vec{A}$ along x-, y- and z-axes, respectively.

Surface integral of a vector: Apart from the space integral, we have another integral called the surface integral of any vector. For instance, flux of liquid passing through an area can be stated as $\phi = \int \vec{v} \cdot \vec{dA}$, where $\vec{v}$ is velocity of the liquid. Similarly, electric current passing through any surface can be given as $i = \int \vec{j} \cdot \vec{dA}$, where $\vec{J}$ is current density. Electric flux ϕ_{el} and magnetic flux ϕ_m passing through an area A can be defined in a similar way; $\phi_{el} = \int \vec{D} \cdot \vec{dA}$ and $\phi_m = \int \vec{B} \cdot \vec{dA}$, where $\vec{D}$ and $\vec{B}$ are densities of electric and magnetic field, respectively.

Generalizing the above surface integrals of the vectors, $\vec{v}$, $\vec{J}$, $\vec{D}$ and $\vec{B}$, we can define the surface integral of any vector $\vec{B}$ as $\int \vec{B} \cdot \vec{dA}$, where $\vec{B} \cdot \vec{dA}$ is the dot product of $\vec{B}$ and $\vec{dA}$. Now, you can evaluate the integral by using Cartesian, cylindrical and spherical coordinate system. In Cartesian coordinate system, we have

$$\vec{B} = B_x\hat{i} + B_y\hat{j} + B_z\hat{k} \text{ and } \vec{dA} = dA_x\hat{i} + dA_y\hat{j} + dA_z\hat{k}.$$

Then, we have

$$\int \vec{B} \cdot d\vec{A} = \int B_x dA_x + \int B_y dA_y + \int B_z dA_z$$

We will explain how to evaluate $\int \vec{B} \cdot d\vec{A}$ by using other coordinate systems in later chapters.

Example 10 Find the line integral of the force $\vec{F} = (x\hat{i} + y\hat{j} + z\hat{k})N$ along the straight line joining the points $P(1m, 2m, 3m)$ to $Q(-1m, 1m, 2m)$. This can be called the work done by the force in shifting the particle from point P to Q along the straight line PQ.
  Solution

$$\int \vec{F} \cdot d\vec{l} = \int F_x \, dx + \int F_y \, dy + \int F_z \, dz,$$

where $F_x = x$, $F_y = y$ and $F_z = z$.
  Then, we have

$$\int \vec{F} \cdot d\vec{l} = \int_1^{-1} x \, dx + \int_2^1 y \, dy + \int_3^2 z \, dz$$
$$= \frac{x^2}{2}\Big|_1^{-1} + \frac{y^2}{2}\Big|_2^1 + \frac{z^2}{2}\Big|_3^2 = -4 \, J.$$

Problem 1 Two vectors each of magnitude A act at an angle θ between them. Find the magnitude and direction of their resultant.
  Solution
  The magnitude of resultant is

$$R = \sqrt{A^2 + B^2 + 2AB \cos \theta}$$
$$= \sqrt{A^2 + A^2 + 2AA \cos \theta}$$
$$= \sqrt{2A^2(1 + \cos \theta)}$$
$$= 2A \cos \frac{\theta}{2} \text{ Ans.}$$

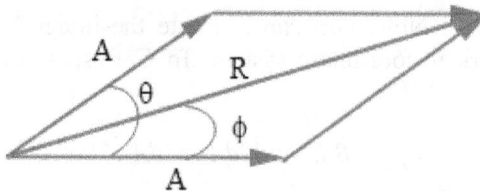

The direction of the resultant is given by the angle made by the resultant is given as

$$\phi = \tan^{-1}\frac{B\sin\theta}{A + B\cos\theta} = \tan^{-1}\frac{A\sin\theta}{A + A\cos\theta}$$

$$= \tan^{-1}\frac{\sin\theta}{1 + \cos\theta} = \tan^{-1}\frac{\sin(\theta/2)}{\cos(\theta/2)}$$

$$= \tan^{-1}\tan(\theta/2) = \theta/2$$

So, the resultant makes equal angles of $\frac{\theta}{2}$ to both vectors Ans.

Problem 2 The resultant of $\vec{P}$ and $\vec{Q}$ is doubled, $\vec{R}$ is doubled, when $\vec{Q}$ is reversed, $\vec{R}$ is again doubled, find $P: Q: R$.

Solution

Let θ be the angle between $\vec{P}$ and $\vec{Q}$. Then

$$R^2 = |\vec{P} + \vec{Q}|^2 = P^2 + Q^2 + 2PQ\cos\theta \tag{1.1}$$

If $\vec{Q}$ is doubled, $\vec{R}$ is doubled. That means, the magnitude of $2\vec{Q}$ and $\vec{P}$ is $2R$

$$(2R)^2 = P^2 + (2Q)^2 + 2P(2Q)\cos\theta$$

This yields, $4R^2 = P^2 + 4Q^2 + 4PQ\cos\theta \tag{1.2}$

When $\vec{Q}$ is reversed, $\vec{R}$ is doubled. Hence, the magnitude of resultant of $\vec{P}$ and $(\vec{P} - \vec{Q})$ is $2R$.

Then, $(2R)^2 = P^2 + Q^2 + 2PQ\cos(180° - \theta)$

This yields $4R^2 = P^2 + Q^2 - 2PQ\cos\theta \tag{1.3}$

Equations $(1.1) - (1.3)$ yields,

$$PQ\cos\theta = \frac{-3R^2}{4} \tag{1.4}$$

Equation $(1.1) + (1.3)$ yields,

$$P^2 + Q^2 = \frac{5R^2}{2} \tag{1.5}$$

Equation $(1.2) + (1.4)$ yields,

$$P^2 + 4Q^2 = 7R^2 \tag{1.6}$$

Solving equations (1.5) and (1.6) we obtain $Q = \sqrt{\frac{3}{2}}R$ and $P = R$.

Hence, $P: Q: R = \sqrt{2}: \sqrt{3}: \sqrt{2}$ Ans.

Problem 3 Four forces act along the sides of a smooth square from $ABCD$ is the order $A \rightarrow B$, $B \rightarrow C$, $C \rightarrow D$ and $D \rightarrow A$. If the magnitudes of the forces are F_1, F_2, F_3 and F_4, respectively, find the resultant force acting on the frame. Assume $F_1 = 10$ N, $F_2 = 20$ N, $F_3 = 30$ N and $F_4 = 40$ N.

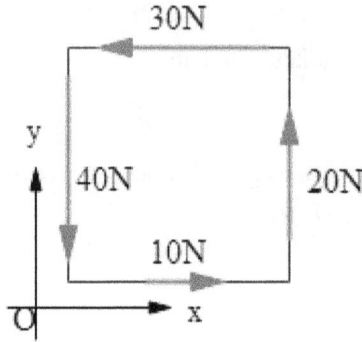

Solution
Let us consider the x–y coordinate system. The resultant of all the forces is

$$\vec{F} = \vec{F_1} + \vec{F_2} + \vec{F_3} + \vec{F_4}.$$

After bringing the tails of all the vectors to a point O and substituting $\vec{F_1} = 10\hat{i}N$, $\vec{F_2} = 20\hat{j}\,N$, $\vec{F_3} = -30\hat{i}N$ and $\vec{F_4} = -40\hat{j}\,N$, we have

$$\vec{F} = -20(\hat{i} + \hat{j})N$$

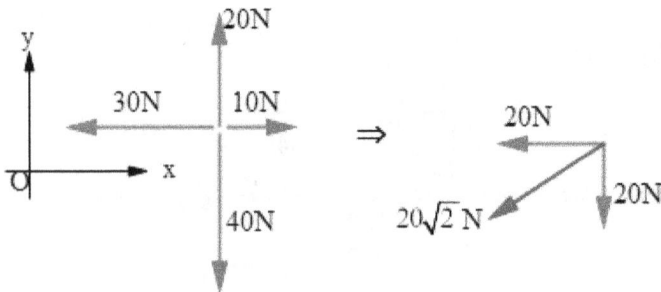

Hence the magnitude of $\vec{F}$, that is, $|\vec{F}| = 20\sqrt{2}$ is directed in third quadrant, as shown in the above figure. Ans.

N.B: You should not be tempted to write the net force as zero because the forces are passing through the sides of the square but they are not represented by these sides. Had the forces been represented by the sides of the square, their magnitudes would have been equal and hence the resultant of these four forces would have been zero.

Problem 4 *(Oblique components)*

Find the components of a vector $\vec{R}$ along two straight lines situated at both sides of the vector $\vec{R}$ making angles α and β with it.

Solution

Since the components $\vec{A}$ and $\vec{B}$ of the given vector $\vec{R}$ are chosen in the given directions, $\vec{R}$ is the resultant of $\vec{A}$ and $\vec{B}$; then, $\vec{R} = \vec{A} + \vec{B}$. Hence R is represented by the diagonal OQ of $OPQS$ where the component vectors $\vec{A}$ and $\vec{B}$ are represented by the adjacent sides OP and OS, respectively according to the parallelogram law of vectors. Then we convert the parallelogram $OSQP$ to the vector triangle OPQ which represents the vectors $\vec{R}$, $\vec{A}$ and $\vec{B}$ as shown in the figure below. By converting to a scalar triangle OPQ as shown in the figure, we can write

$$OQ = OP \cos \beta + PQ \cos \alpha = B \cos \beta + A \cos \alpha \qquad (1.7)$$

Furthermore, $PM = OP \sin \beta = PQ \sin \alpha$

$$\Rightarrow B \sin \beta = A \sin \alpha \qquad (1.8)$$

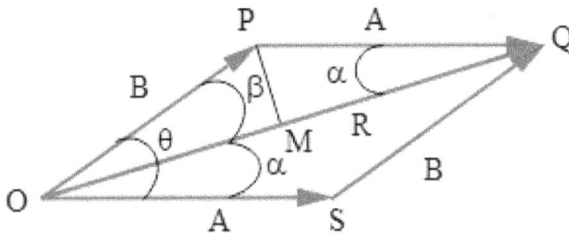

Solving equations (1.7) and (1.8), we have

$$A = \frac{R \sin \beta}{\sin(\alpha + \beta)} \quad \text{and} \quad B = \frac{R \sin \alpha}{\sin(\alpha + \beta)} \quad \text{Ans.}$$

Alternative solution

Using Lami's theorem, we have

$$\frac{A}{\sin \beta} = \frac{B}{\sin \alpha} = \frac{R}{\sin\{180° - (\alpha + \beta)\}}.$$

Problem 5 An insect moves in a circular path of radius R through an angle θ. Find its displacement.

Solution

When the insect moves from position 1 to position 2, the displacement $\vec{s}$ = change in position vector $\Delta \vec{r}$. Since $\Delta \vec{r} = \vec{r_2} - \vec{r_1}$, the magnitude of displacement is $|\Delta \vec{r}| = \sqrt{r_1^2 + r_2^2 - 2 r_1 r_2 \cos \theta}$.

Since the insect moves in a circular path, $r_1 = r_2 = R$.

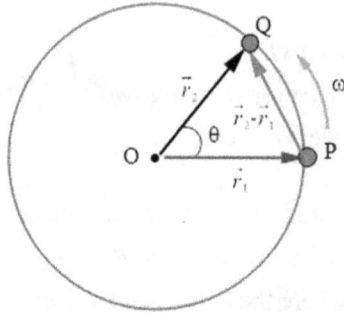

So, the magnitude of change in position vector is

$$|\Delta \vec{r}| = |\vec{r_2} - \vec{r_1}| = \sqrt{R^2 + R^2 - 2RR \cos \theta}$$
$$= \sqrt{2R^2(1 - \cos \theta)}$$
$$= 2R \sin \frac{\theta}{2} \text{ Ans.}$$

Problem 6 Two forces of magnitudes A and B, say, make an angle of 60° with each other. When they act on a point mass the net force is X, say. If one of the forces is reversed, the new resultant is Y, say. If $X/Y = \sqrt{7/3}$, find A/B.

Solution

When the forces act on a point mass the net force is X. So, we can write

$$X = \sqrt{A^2 + B^2 + 2A \cdot B \cos \theta} \qquad (1.9)$$

If one of the forces is reversed $\theta' = 180° - \theta$, the new resultant will be Y.

$$Y = \sqrt{A^2 + B^2 - 2A \cdot B \cos \theta} \qquad (1.10)$$

Using the last two equations,

$$\frac{\sqrt{A^2 + B^2 + 2A \cdot B \cos \theta}}{\sqrt{A^2 + B^2 - 2A \cdot B \cos \theta}} = X/Y = \sqrt{7/3}$$

$$\Rightarrow \frac{A^2 + B^2 + 2A \cdot B \cos \theta}{A^2 + B^2 - 2A \cdot B \cos \theta} = 7/3$$

$$\Rightarrow \frac{A^2 + B^2 + A \cdot B}{A^2 + B^2 - A \cdot B} = 7/3$$

$$\Rightarrow \frac{k^2 + 1 + k}{k^2 + 1 - k} = 7/3. \text{ Where } k = A/B$$

$$\Rightarrow 2k^2 - 5k + 2 = 0 \Rightarrow k = 2 \text{ or } \frac{1}{2} \text{ Ans.}$$

Problem 7 If two vectors $\vec{A}$ and $\vec{B}$ represent the adjacent sides of a parallelogram, find the angle between the diagonals of the parallelogram. Put $A/B = \eta$.

Solution

We know that

$$(\vec{A}+\vec{B}) \cdot (\vec{A}+\vec{B}) = A^2 - B^2 \qquad (1.11)$$

The magnitude of resultant $\vec{A}+\vec{B}$ is given by,

$$|\vec{A}+\vec{B}| = \sqrt{A^2 + B^2 + 2A \cdot B \cos \theta} = \sqrt{A^2 + B^2 + 2A \cdot B \cos 60°}$$

$$\Rightarrow |\vec{A}+\vec{B}| = \sqrt{A^2 + B^2 + A \cdot B} \qquad (1.12)$$

The magnitude of resultant $\vec{A}-\vec{B}$ is given by,

$$|\vec{A}-\vec{B}| = \sqrt{A^2 + B^2 - 2A \cdot B \cos \theta} = \sqrt{A^2 + B^2 - 2A \cdot B \cos 60°}$$

$$\Rightarrow |\vec{A}-\vec{B}| = \sqrt{A^2 + B^2 - A \cdot B} \qquad (1.13)$$

The angle β between $\vec{A}+\vec{B}$ and $\vec{A}+\vec{B}$ is given as

$$(\vec{A}+\vec{B}) \cdot (\vec{A}+\vec{B}) = |\vec{A}+\vec{B}||\vec{A}-\vec{B}| \cos \beta \qquad (1.14)$$

Using the last four equations, we have

$$\left(\sqrt{A^2 + B^2 - A \cdot B}\right)\left(\sqrt{A^2 + B^2 - A \cdot B}\right)\cos \beta = (A^2 - B^2)$$

$$\Rightarrow \beta = \cos^{-1}\left\{\frac{(A^2 - B^2)}{\left(\sqrt{A^2 + B^2 + A \cdot B}\right)\left(\sqrt{A^2 + B^2 - A \cdot B}\right)}\right\}$$

$$\Rightarrow \beta = \cos^{-1}\left\{\frac{(A^2 - B^2)}{\sqrt{A^4 + B^4 + A^2 \cdot B^2}}\right\}$$

$$\Rightarrow \beta = \cos^{-1}\left\{\frac{(A/B)^2 - 1)}{\sqrt{(A/B)^4 + (A/B)^2 + 1}}\right\}$$

$$\Rightarrow \beta = \cos^{-1}\left\{\frac{\eta^2 - 1)}{\sqrt{\eta^4 + \eta^2 + 1}}\right\} \text{ Ans.}$$

Problem 8 If $\hat{i}$ and $\hat{n}$ represent the incident and ray and the normal to a plane mirror, respectively, find the reflected ray.

Solution

The direction of the incident ray is given by the unit vector $\hat{i}$ and the direction of the incident ray is given by the unit vector $\hat{r}$.

$$\overrightarrow{OP} = \hat{r} - \hat{i} = 2\sin(\pi/2 - \theta)\hat{n} = 2\cos\theta\hat{n}$$

$$\Rightarrow \hat{r} = \hat{i} + 2\cos\theta\hat{n} \tag{1.15}$$

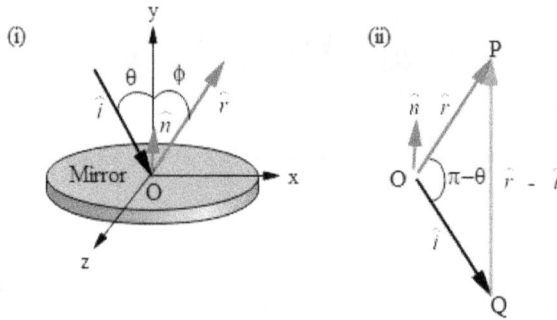

Furthermore, the dot product of $\hat{i}$ and $\hat{n}$ is given as

$$\hat{i}.\hat{n} = \cos(\pi - \theta) = -\cos\theta \tag{1.16}$$

Eliminating $\cos\theta$ from equation (1.15) by (1.16), we have

$$\hat{r} = \hat{i} + 2(-\hat{i}\cdot\hat{n})\hat{n}$$

$$\Rightarrow \hat{r} = \hat{i} - 2(\hat{i}\cdot\hat{n})\hat{n} \text{ Ans.}$$

Problem 9 Two vectors $\overrightarrow{A}$ and $\overrightarrow{B}$ act at a point. If the resultant is perpendicular to the vector, find the (a) angle between $\overrightarrow{A}$ and $\overrightarrow{B}$, (b) magnitude of the resultant, (c) angle between the resultant and the vector $\overrightarrow{B}$, and (d) angle between the diagonals of the parallelogram. Put $\overrightarrow{B} = \eta\overrightarrow{A}$ for (a).

Solution

(a) The direction of the resultant is given by the angle made by the resultant

$$\phi = \tan^{-1}\frac{B\sin\theta}{A + B\cos\theta} = \tan^{-1}\frac{A\sin\theta}{A + \eta A\cos\theta}$$

Since the resultant vector is perpendicular to either vector, putting $\phi=90°$, we have

$$A + \eta A \cos \theta = 0$$

$$\Rightarrow \cos \theta = -1/\eta$$

$$\Rightarrow \theta = \cos^{-1}(-1/\eta) \text{ Ans.}$$

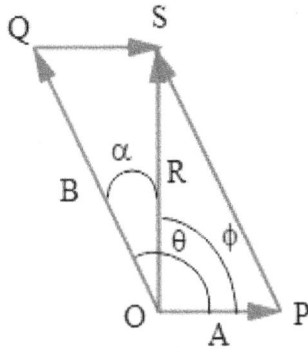

(b) Applying the Pythagoras theorem, the resultant vector is given as

$$R = \sqrt{B^2 - A^2}$$

(c) The angle between the resultant and the vector $\vec{B}$ is

$$\theta = \sin^{-1}(A/B)$$

(d) The angle β between $\vec{A} + \vec{B}$ and $\vec{A} + \vec{B}$ is given as

$$(\vec{A} + \vec{B}) \cdot (\vec{A} + \vec{B}) = |\vec{A} + \vec{B}||\vec{A} - \vec{B}| \cos \beta$$

$$\left(\sqrt{A^2 + B^2 + 2A \cdot B \cos \theta}\right)\left(\sqrt{A^2 + B^2 - 2A \cdot B \cos \theta}\right)\cos \beta = (A^2 - B^2)$$

$$\Rightarrow \beta = \cos^{-1}\left\{\frac{(A^2 - B^2)}{\sqrt{(A^2 + B^2)^2 - (2A \cdot B \cos \theta)^2}}\right\}$$

Putting $B \cos \theta = -A$ and simplifying the factors, we have

$$\Rightarrow \beta = \cos^{-1}\left\{\frac{(A^2 - B^2)}{\sqrt{B^4 - 3A^4 + 2A^2 \cdot B^2}}\right\} \text{ Ans.}$$

Problem 10 If the position of the vertices of a triangle ABC are $\vec{a}$, $\vec{b}$ and $\vec{c}$, respectively, find the area of the triangle.

Solution

The area of the triangle is given as

$$A = \frac{1}{2}\,|\overrightarrow{AB} \times \overrightarrow{AC}| \tag{1.17}$$

From vector subtraction, we have

$$\overrightarrow{AB} = \vec{b} - \vec{a} \tag{1.18}$$

Similarly,

$$\overrightarrow{AC} = \vec{c} - \vec{a} \tag{1.19}$$

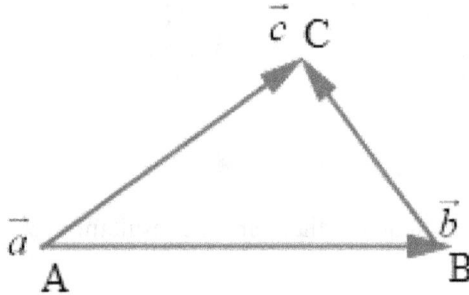

Using the last three equations, we have

$$A = \frac{1}{2}\,|(\vec{b} - \vec{a}) \times (\vec{c} - \vec{a})| = \frac{1}{2}\,|(\vec{b} \times \vec{c} - \vec{b} \times \vec{a} - \vec{a} \times \vec{c} + \vec{a} \times \vec{a})|$$

$$\Rightarrow A = \frac{1}{2}\,|(\vec{b} \times \vec{c} - \vec{b} \times \vec{a} - \vec{a} \times \vec{c}|$$

$$\Rightarrow A = \frac{1}{2}\,|\vec{b} \times \vec{c} + \vec{a} \times \vec{b} + \vec{c} \times \vec{a}|$$

$$\Rightarrow A = \frac{1}{2}\,|\vec{a} \times \vec{b} + \vec{b} \times \vec{c} + \vec{c} \times \vec{a}| \text{ Ans.}$$

Problem 11 The position of the vertices of a triangle ABC are $\vec{a}$, $\vec{b}$ and $\vec{c}$, respectively. If the diagonals of the parallelogram $ABCD$ are perpendicular to each other, find the (a) area of the parallelogram position of D and (b) position of D.

Solution

(a) The area of the parallelogram is given as

$$A = |\overrightarrow{AB} \times \overrightarrow{AC}| \tag{1.20}$$

From vector subtraction, we have

$$\overrightarrow{AB} = \overrightarrow{b} - \overrightarrow{a}$$ (1.21)

Similarly,

$$\overrightarrow{AC} = \overrightarrow{c} - \overrightarrow{a}$$ (1.22)

Using the last three equations, we have

$$A = |(\overrightarrow{b} - \overrightarrow{a}) \times (\overrightarrow{c} - \overrightarrow{a})| = \frac{1}{2} |(\overrightarrow{b} \times \overrightarrow{c} - \overrightarrow{b} \times \overrightarrow{a} - \overrightarrow{a} \times \overrightarrow{c} + \overrightarrow{a} \times \overrightarrow{a})|$$

$$\Rightarrow A = |(\overrightarrow{b} \times \overrightarrow{c} - \overrightarrow{b} \times \overrightarrow{a} - \overrightarrow{a} \times \overrightarrow{c}|$$

$$\Rightarrow A = |\overrightarrow{b} \times \overrightarrow{c} + \overrightarrow{a} \times \overrightarrow{b} + \overrightarrow{c} \times \overrightarrow{a}|$$

$$\Rightarrow A = |\overrightarrow{a} \times \overrightarrow{b} + \overrightarrow{b} \times \overrightarrow{c} + \overrightarrow{c} \times \overrightarrow{a}| \text{ Ans.}$$

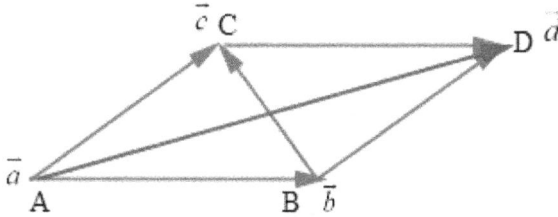

(b) By vector addition, we have

$$\overrightarrow{AD} = \overrightarrow{AB} + \overrightarrow{AC}$$

$$\Rightarrow \overrightarrow{d} - \overrightarrow{a} = (\overrightarrow{b} - \overrightarrow{a}) + (\overrightarrow{c} - \overrightarrow{a})$$

$$\Rightarrow \overrightarrow{d} = \overrightarrow{b} - \overrightarrow{a} + \overrightarrow{c} \text{ Ans.}$$

Problem 12 If the diagonals of a parallelogram are represented by the vectors $\overrightarrow{a}$, $\overrightarrow{b}$ respectively, find the area of the parallelogram.
Solution

(a) The area of the parallelogram is given as

$$A = |\overrightarrow{AB} \times \overrightarrow{AC}| = |\overrightarrow{a} \times \overrightarrow{b}|$$ (1.23)

By vector subtraction, we have

$$\vec{c} = \vec{b} - \vec{a} \qquad (1.24)$$

Using vector addition, we have

$$\vec{d} = \vec{a} + \vec{b} \qquad (1.25)$$

Using the last two equations, we have

$$\vec{a} = \frac{1}{2}(\vec{d} - \vec{c}) \qquad (1.26)$$

$$\vec{b} = \frac{1}{2}(\vec{c} + \vec{d}) \qquad (1.27)$$

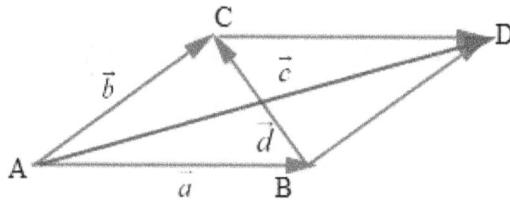

Using equations(1.23), (1.26) and (1.27), we have

$$A = |\vec{a} \times \vec{b}| = \left| \frac{1}{2}(\vec{d} - \vec{c}) \times \frac{1}{2}(\vec{d} - \vec{c}) \right|$$

$$\Rightarrow A = \left| \frac{1}{4}(\vec{d} \times \vec{d} + \vec{d} \times \vec{c} - \vec{c} \times \vec{d} - \vec{c} \times \vec{c}) \right|$$

Since $\vec{c} \times \vec{c} = 0$ and $\vec{d} \times \vec{d} = 0$, we have

$$A = \left| \frac{1}{4}(\vec{d} \times \vec{c} - \vec{c} \times \vec{d}) \right|$$

Since $\vec{d} \times \vec{c} = -\vec{c} \times \vec{d}$, we have

$$A = \frac{1}{4}2 \,|(\vec{d} \times \vec{c})| = \frac{1}{4} \,|-2(\vec{c} \times \vec{d})| = \frac{1}{2} \,|\vec{c} \times \vec{d}| \text{ Ans.}$$

Problem 13 In the following vector diagram, if AD, AE, AC and DC are represented by the vectors $\vec{a}$, $\vec{b}$, $\vec{c}$ and $\vec{d}$ respectively and $CE = \lambda CD$, find the relation between $\vec{a}$, $\vec{b}$, $\vec{c}$.

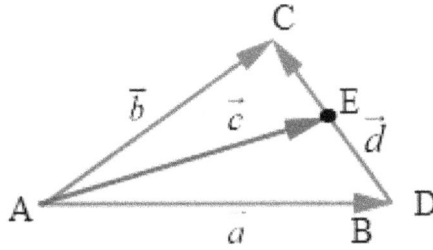

Solution

By vector subtraction, we have

$$\vec{d} = \vec{b} - \vec{a} \tag{1.28}$$

Furthermore,

$$\vec{EC} = \lambda\vec{d} = (\vec{b} - \vec{c}) \tag{1.29}$$

Using the last two equations,

$$\lambda(\vec{b} - \vec{a}) = (\vec{b} - \vec{c})$$

$$\Rightarrow \lambda\vec{b} - \lambda\vec{a} = \vec{b} - \vec{c}$$

$$\Rightarrow \vec{c} = \vec{b}(1 - \lambda) + \lambda\vec{a} \text{ Ans.}$$

Problem 14 A uniform magnetic field $\vec{B} = (2\hat{i} + 3\hat{j} - \hat{k})T$ is present in space. Using the formula of flux $\phi = \vec{B} \cdot \vec{A}$, find the magnetic flux passing through the loop of area vector $\vec{A} = (\hat{i} - \hat{j} + 2\hat{k})m^2$.

Solution

The magnetic flux passing through the loop is

$$\phi = \vec{B} \cdot \vec{A} = (2\hat{i} + 3\hat{j} - \hat{k}) \cdot (\hat{i} - \hat{j} + 2\hat{k})$$

$$\Rightarrow \phi = (2\hat{i} + 3\hat{j} - \hat{k}) \cdot (\hat{i} - \hat{j} + 2\hat{k})$$
$$= 2 \times 1 + 3 \times (-1) + (-1)(+2) = 2 - 3 - 2 = -3 \text{ Weber Ans.}$$

Problem 15 A uniform magnetic field $\vec{B} = (2\hat{i} + 3\hat{j} - \hat{k})T$ is present in a space. Using the formula of energy $U = -\vec{\mu} \cdot \vec{B}$, find the magnetic energy possessed by the current loop of magnetic moment vector $\vec{\mu} = (\hat{i} - \hat{j} + 2\hat{k})$ Amp m^2.

Solution

The magnetic energy possessed by the current loop is

$$U = -\vec{\mu}.\vec{B}$$

$$\Rightarrow U = -(\hat{i} - \hat{j} + 2\hat{k}) \cdot (2\hat{i} + 3\hat{j} - \hat{k})$$
$$= -\{(1 \times 2 + (-1) \times 3 + (-1)2 = -(2 - 3 - 2) = 3 \text{ Joule Ans.}$$

Problem 16 A force $\vec{F} = (\hat{i} + 3\hat{j} - 5\hat{k})N$ acts on a point that undergoes a displacement from a point P having the position $\vec{a} = (\hat{i} - \hat{j} + 2\hat{k})$ to another point Q having position vector $\vec{b} = (-2\hat{i} + 3\hat{j} - 4\hat{k})$. Using the formula of work $W = -\vec{F} \cdot \vec{s}$, find the work done by the force.

Solution

The work done by the force is

$$W = \vec{F} \cdot \vec{s} = \hat{F}(\vec{b} - \vec{a})$$

$$\Rightarrow W = (\hat{i} + 3\hat{j} - 5\hat{k}) \cdot \{(-2\hat{i} + 3\hat{j} - 4\hat{k}) - (\hat{i} - \hat{j} + 2\hat{k})\}$$

$$\Rightarrow W = (\hat{i} + 3\hat{j} - 5\hat{k}) \cdot (-3\hat{i} + 4\hat{j} - 6\hat{k})$$
$$= (1 \times (-3) + 3 \times 4 + (-5)(-6) = (-3 + 12 + 30) = 39J \text{ Ans.}$$

Problem 17 A force $\vec{F} = (2\hat{i} + 3\hat{j} - 4\hat{k})N$ acts on a point P having the position $\vec{a} = (\hat{i} - \hat{j} + \hat{k})$. Find the torque of the force about a point Q having position vector $\vec{b} = (-2\hat{i} - \hat{j} + 3\hat{k})$. Use the formula of torque $\vec{\tau} = \vec{r} \times \vec{F}$.

Solution

The torque of the force is

$$\vec{\tau} = \vec{r} \times \vec{F} = (\vec{b} - \vec{a}) \times \vec{F}$$

$$\Rightarrow \vec{\tau} = \{(-2\hat{i} - \hat{j} + 3\hat{k}) - (\hat{i} - \hat{j} + \hat{k})\} \times (2\hat{i} + 3\hat{j} - 4\hat{k}).$$

$$\Rightarrow \vec{\tau} = (-3\hat{i} + 2\hat{k}) \times (2\hat{i} + 3\hat{j} - 4\hat{k}).$$

$$\Rightarrow \vec{\tau} = -6\hat{i} - 8\hat{j} - 9\hat{k}N \cdot m \text{ Ans.}$$

Problem 18 Two vectors of magnitude A and B are acting at a point. If the magnitude of the vector A increases by a factor μ and the vector B increases by x, the new resultant's direction remains the same as the initial resultant's direction. Find the value of B.

Solution

The initial and final magnitudes of the vectors are $(A, A' = \mu A)$, $(B, B' = B + x)$, respectively. The resultants $\overrightarrow{R}$ and $\overrightarrow{R}'$ point in the same direction. By the parallelogram law of vector addition, we have

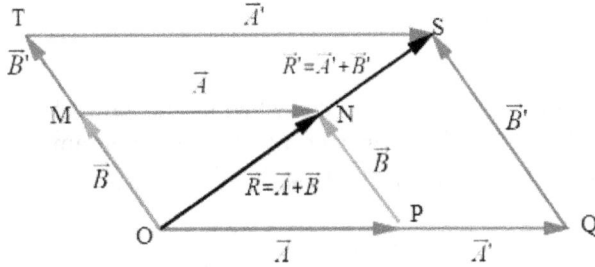

$\overrightarrow{R} = \overrightarrow{A} + \overrightarrow{B}$ and $\overrightarrow{R}' = \overrightarrow{A}' + \overrightarrow{B}'$ as described by the parallelograms $OPNM$ and $OQST$, respectively. Using the properties of the triangles OPN and OQS, we have

$$OQ/OP = QS/PN$$

$$\Rightarrow A'/A = B'/B$$

$$\Rightarrow \mu A/A = (B + x)/B$$

$$\Rightarrow \mu = (B + x)/B$$

$$\Rightarrow (\mu - 1)B = x$$

$$\Rightarrow B = x/(\mu - 1) \text{ Ans.}$$

N.B: The value of A can be any, whereas the value of B is definite.

Problem 19 Two vectors of magnitude A and B are acting at a point making an angle θ with each other. If the magnitude of the vector A increases by a factor η and the vector B remains unchanged. Find the angle between the two resultants. Find the value of B.

Solution

The initial and final magnitudes of the vectors are $(A, A' = \eta A)$, $(B, B' = B)$, respectively. In this case, the resultants $\overrightarrow{R}$ and $\overrightarrow{R}'$ point in different directions. Let the angles made by the resultant vectors with the vector $\overrightarrow{A}$ be ϕ and β, respectively. By parallelogram law of vector addition, we have

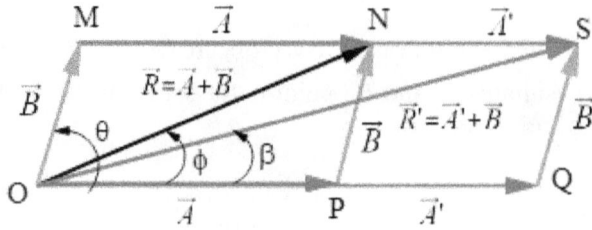

$\vec{R} = \vec{A} + \vec{B}$ and $\vec{R}' = \vec{A}' + \vec{B}'$ as described by the parallelograms $OPNM$ and $OQST$, respectively. Using the properties of the parallelogram law of vectors, the angle between the first resultant $\vec{R}$ and vector $\vec{A}$ we have

$$\phi = \tan^{-1}\frac{B \sin \theta}{A + B \cos \theta}$$

Similarly, the angle between the first resultant $\vec{R}'$ and vector $\vec{A}'$ we have

$$\beta = \tan^{-1}\frac{B \sin \theta}{A' + B \cos \theta}$$

$$\Rightarrow \beta = \tan^{-1}\frac{B \sin \theta}{\eta A + B \cos \theta}$$

Then, the angle between the two resultants $\vec{R}$ and $\vec{R}'$ is

$$\phi - \beta = \tan^{-1}\frac{B \sin \theta}{A + B \cos \theta} - \tan^{-1}\frac{B \sin \theta}{\eta A + B \cos \theta} \quad \text{Ans.}$$

IOP Publishing

Problems and Solutions in Particle Mechanics

Pradeep Kumar Sharma

Chapter 2

Motion in one dimension

2.1 Introduction

Kinematics is a subfield of physics and mathematics. It describes the motion of a body (object) ignoring agents such as the *force (push or pull)* or *torque (twist)* that can cause the motion. Any object or a body (rigid or non-rigid) is a system of particles. A particle is ideally a point mass. A point is a geometrical sysmbol without any dimensions; it has only a position. Generally, a body can have two types of motion, namely translation and rotation. Translational motion is defined for a particle or point mass or any point of an object. The rotation of a body is the description of its angular motion which is explained in the book *Problems and Solutions on Rotational Mechanics* of this series.

More generally, we have three types of motion; translational, rotational and vibrational. For instance, sliding of an object is translational motion, spinning of a cricket ball is rotational and to and fro (periodic) motion of a simple pendulum is oscillatory (or vibrational). All objects ranging from electrons to galaxies are in (relative) motion.

The *particle* or point mass is a physical concept and the idea of a *point* is a mathematical concept. The kinematics of a particle describes the motion of very small objects. For instance, motion of an electron relative to a nucleus, motion of a planet relative to the Sun etc. An element of an object can also be called a particle which is mathematically expressed as a point. When an object undergoes any type of motion, all its particles (points) move in arbitrary paths. Hence, to analyze the motion of an object, we need to understand the motion of any point (particle) of the object. The study of motion of a particle moving in an arbitrary path is called curvilinear kinematics, or three-dimensional motion, of motion in space or kinematics of a particle in a curve.

Although there are different types of paths or trajectory or curves made by a moving particle, there are several familiar types of motion based upon the trajectory. Some of them are straight-line motion, parabolic motion, elliptical motion, circular motion, spiral motion, helical motion etc, as shown in the above figures.

doi:10.1088/978-0-7503-6442-3ch2

In this chapter, first of all, we will talk about the motion of a particle in a curve. The starting point of the study of kinematics is to define the location of a particle by a vector $\vec{r}$ called 'position vector'. Then we will define the velocity $\vec{v}$ of the particle as the rate of change of its position vector. Finally, we will measure the time rate of change in velocity to define acceleration $\vec{a}$ of the particle. These three vectors $\vec{r}$, $\vec{v}$ and $\vec{a}$ are sufficient to describe the nature of motion of a particle in space. Then, we will use the general (curvilinear) kinematics to discuss different types of motion of particles, i.e., straight-line motion, circular motion, projectile (parabolic) motion etc.

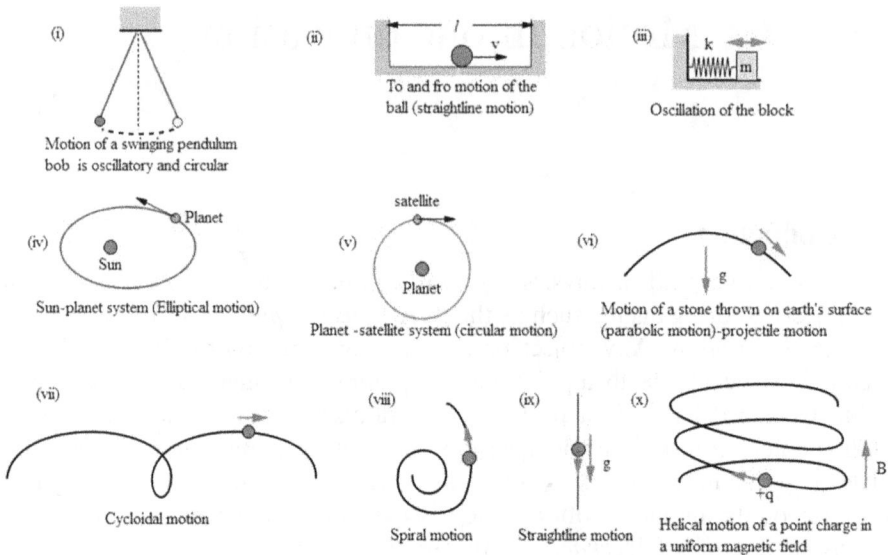

(i)

Motion of a swinging pendulum
bob is oscillatory and circular

(ii)

To and fro motion of the
ball (straightline motion)

(iii)

Oscillation of the block

(iv)

Sun-planet system (Elliptical motion)

(v)

Planet -satellite system (circular motion)

(vi)

Motion of a stone thrown on earth's surface
(parabolic motion)-projectile motion

(vii)

Cycloidal motion

(viii)

Spiral motion

(ix)

Straightline motion

(x)

Helical motion of a point charge in
a uniform magnetic field

To solve the problems in kinematics of a particle, generally, we begin with declaring the initial (at time $t = 0$) parameters (data) such as position, velocity etc, of the particle. Then, using the concept of geometry, the final (at time t) parameters such as (position, velocity and acceleration) of the particle can be determined. In particle kinematics, we do not discuss the effect of the forces. However, the study of 'action of the forces on the bodies' falls within kinetics or dynamics, but not kinematics. The combination of kinematics and dynamics of a particle is called mechanics of a particle or 'particle-mechanics'. In mechanics we use the geometry, calculus and algebra of vectors and scalars.

2.2 Position, displacement and distance

Particle: Kinematics deals with the geometrical property of motion of a particle. A particle, ideally means a point mass. However, in practice, a particle need not be a tiny (point) object. For instance, an electron can be regarded as a point object (particle) compared to an atom. An atom can be considered as a particle when compared with a grain of matter. The celestial bodies like stars and planets etc, can be practically assumed as particles when viewed from Earth as they occupy negligible volumes compared to entire space of the universe. The 'particle' is a physical concept represented by a mathematical concept 'point'.

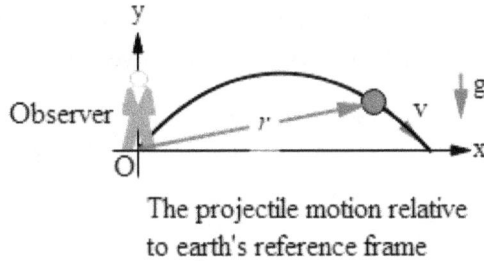

The projectile motion relative
to earth's reference frame

Frame of reference: In particle kinematics we measure the displacement, velocity and acceleration of a particle. For this, an observer is taken as a reference point. Generally, we consider the observer placed at the origin of a coordinate system. The place where the observer (reference point) is placed (fixed) is known as 'reference frame'. Hence, reference frame is geometrically defined as a coordinate system.

Locating a particle, position vector: While a bird is flying, if you ask a baby, where the bird is, he will promptly point his finger towards the bird. This tells us that pointing of a finger indicates the location or position of a point mass. In other words, the position of a particle is associated with a direction. Furthermore, the distance of the object (bird) and observer (baby) is also required to completely specify the location of the bird.

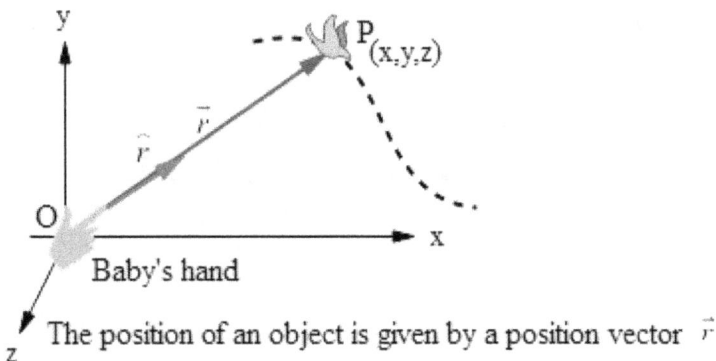

The position of an object is given by a position vector $\vec{r}$

From the above explanation we understand that the position of a particle requires both magnitude (distance of separation between the object and observer) and direction for its complete specification; so, it is characterized by a vector quantity $\vec{r}$, termed as 'position vector' of the particle with respect to the origin (point of reference). It is the first fundamental vector quantity in physics.

In the Cartesian coordinate system, $\vec{r}$ is given as

$$\vec{r} = x\hat{i} + y\hat{j} + z\hat{k},$$

where x, y and z are the components (projections) of $\vec{r}$ along the x-, y- and z-axes, respectively. In a polar coordinate system position vector $\vec{r}$ can be given as

$$\vec{r} = r\hat{r},$$

where $r = |\vec{r}|$ and $\hat{r} = \frac{\vec{r}}{|\vec{r}|}$

Example 1 The top of a tree of height 20 m is located at a distance of 60 m to the east and 30 m to the north from the origin. Find the position vector of the top of the tree. Assume yourself located at the origin and horizontal plane as the x–z plane.

Solution

The position of the top P of the tree is given as

$$\vec{r} = x\hat{i} + y\hat{j} + z\hat{k}$$
$$= (OM)\hat{i} + (PQ)\hat{j} - (ON)\hat{k}$$
$$= 60\hat{i} + 20\hat{j} - 30\hat{k}$$

The distance $r = OP$ is given as

$$r = \sqrt{60^2 + 20^2 + 30^2} = 70 \text{ m}$$

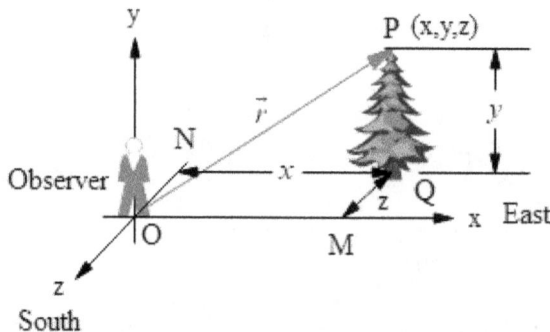

Rest and motion: When a particle is at rest, its position does not change. In other words, when neither the magnitude nor the direction of position vector changes, the particle is at rest. However, when the particle moves, its position vector changes either its magnitude, or direction or both magnitude and direction.

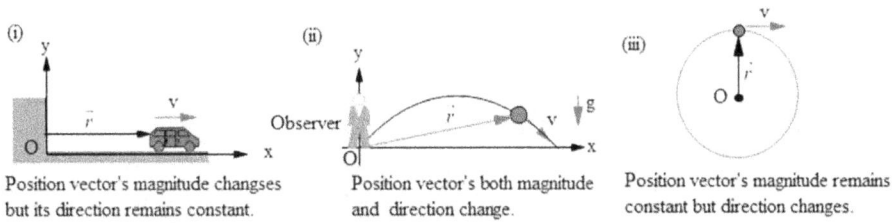

(i)	(ii)	(iii)

Position vector's magnitude changses but its direction remains constant.

Position vector's both magnitude and direction change.

Position vector's magnitude remains constant but direction changes.

Displacement: When a particle moves from position 1 to position 2 in an arbitrary path during a time interval Δt, let its position vector of the particle change from $\vec{r_1}$ at point 1 to $\vec{r_2}$ at point 2. In other words, during the motion of the particle, the position vector changes from $\vec{r_1}$ to $\vec{r_2}$ during time Δt. Hence, the motion of the particle is associated with a change in position vector which is given as

$$\vec{s} = \Delta\vec{r} = \vec{r_2} - \vec{r_1}$$

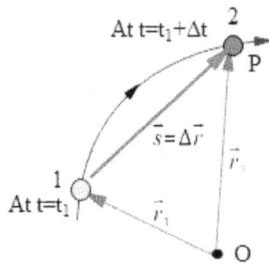

The particle undergoes a displacement from position 1 to position 2

It is worth noting that, during the motion of a particle, either magnitude or direction or both magnitude and direction of its position vector change. This means that the motion of a particle is defined as the change in its position vector.

Example 2 A mosquito flies from a point A $(1, -2, 3)$ m to a point B $(-2, 1, 2)$ m. Find the displacement of the mosquito.

Solution

The displacement of the mosquito is

$$\vec{s} = \Delta\vec{r} = \vec{r_B} - \vec{r_A},$$

where $\vec{r_B} = \hat{i} - 2\hat{j} - 3\hat{k}$ and $\vec{r_A} = -2\hat{i} + \hat{j} + 2\hat{k}$.

$$\Rightarrow s = (-2\hat{i} + \hat{j} + 2\hat{k}) - (\hat{i} - 2\hat{j} + 3\hat{k})$$
$$= (-3\hat{i} + 3\hat{j} - \hat{k}) \, \text{m}.$$

2.2.1 Distance

The 'distance D' is the length of the actual path traversed by the particle. It is a scalar quantity. It always increases with time, whereas magnitude of displacement may increase, decrease or remain constant with time.

In general, $D \geqslant s$; this means that the distance is greater than or equal to the magnitude of displacement. In the motion in one direction, $D = s$ but in other cases, distance is greater than displacement.

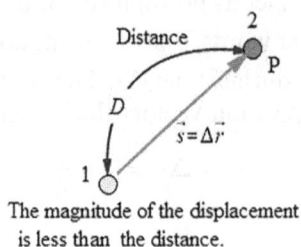

The magnitude of the displacement
is less than the distance.

Example 3 If you walk 12 m to the east and then 8 m due north in a straight line, find the magnitude of total displacement and distance covered.

Solution

Since you walk a distance of 12 m due east, $\vec{s_1} = 12\hat{i}$

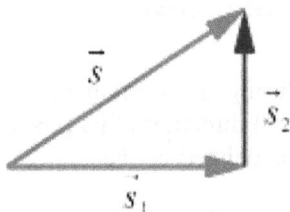

Then, in walking due north through a distance of 8 m, we have $\vec{s_2} = 8\hat{j}$.
The resultant displacement is

$$\vec{s} = \vec{s_1} + \vec{s_2} = 12\hat{i} + 8\hat{j}$$

The magnitude of $\vec{s}$ is

$$s = \sqrt{12^2 + 8^2} = 13 \text{ m}$$

The total distance covered is

$$D = s_1 + s_2 = 12 + 8 = 20 \text{ m}.$$

2.3 Velocity and speed

When a particle moves, its position vector $\vec{r}$ changes in either magnitude or direction or both. However, the distance covered by the particle always increases with time. We need to find the rate which the position vector $\vec{r}$ and distance D changes with time. The time rate of change in position vector $\vec{r}$ and distance D are called velocity and speed, respectively.

Average velocity: Let the particle P pass through two points or positions 1 and 2 at times $t = t_1$ and $t = t_2$, respectively, while moving in an arbitrary path. During the time interval $\Delta t (= t_2 - t_1)$ the position vector of the particle changes from $\vec{r_1}$ to $\vec{r_2}$ by $\Delta \vec{r} (= \vec{r_2} - \vec{r_1})$.

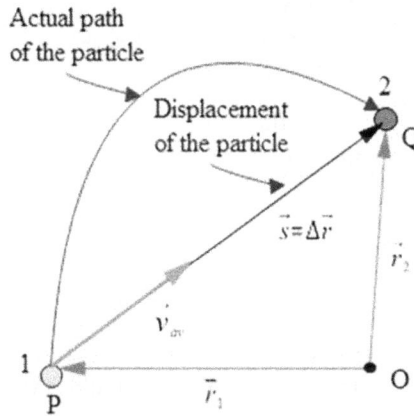

Actual path
of the particle

Displacement
of the particle

$\vec{s} = \Delta \vec{r}$

$\vec{r_2}$

$\vec{v}_{av}$

$\vec{r_1}$

The average velocity points in the direction of displacement.

Hence, on average, the position vector changes with time at a rate $\frac{\Delta \vec{r}}{\Delta t}$. This is known as the 'average time rate of change in position vector' or 'average velocity' of the particle denoted as $\vec{v}_{av}$.

$$\vec{v}_{av} = \frac{\Delta \vec{r}}{\Delta t} = \frac{\vec{r_2} - \vec{r_1}}{t_2 - t_1} = \frac{\vec{S}}{\Delta t}$$

The average velocity of a particle for any time interval is defined as the change in its position vector (displacement) divided by the time interval. So, the average velocity $\vec{v}_{av}$ is parallel to the displacement $\Delta \vec{r} (= \vec{s})$. In other words, average velocity points in the direction of displacement.

Average speed: In a round trip the as particle comes back to its initial position, its displacement will be zero. So, its average velocity will be zero. However, the distance always increases with time for a moving particle. So, average speed is a non-zero scalar.

We can find the average speed by dividing the total distance D by the time taken Δt to cover that distance. It is given as

$$u_{av} = \frac{\text{Distance covered}}{\text{Time lapsed}} = \frac{D}{\Delta t}$$

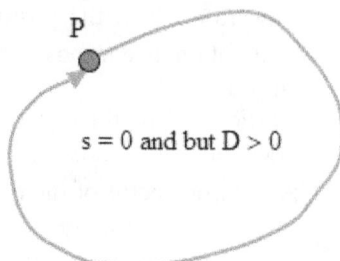

In a round trip, the average velocity is zero
but always the average speed is non-zero

The average speed of a particle is the ratio of its total distance covered (but not displacement) and the total time taken by it. Only for one-directional motion, average speed will be equal to the magnitude of average velocity.

For any time interval, $\frac{u_{av}}{v_{av}} = \frac{D}{s}$

Since, $D \geqslant s$, we have $u_{av} \geqslant v_{av}$.

Example 4 To reach your school at 10.30 AM, let your start be at 10 AM. First you move 800 m to the east, then 600 m to the north and finally 200 m to the west. Find the
  (a) average velocity,
  (b) average speed.

Solution
Let us take $+x$- and $+y$-directions as east and north, respectively.
  (a) As you move 800 m to the east, then 600 m to the north and finally 200 m due west,
  put $\vec{s_1} = 800\hat{i}$, $s_2 = 600\hat{j}$ and $\vec{s_3} = -200\hat{i}$.
  Then, the net displacement is
$$\vec{s} = \vec{s_1} + \vec{s_2} + \vec{s_3}$$
$$= 800\hat{i} + 600\hat{j} - 200\hat{i} = 600\hat{i} + 600\hat{j}$$

The time taken to reach the school is equal to 30 minutes $= 30 \times 60 = 1800$ s. Hence, the average velocity is

$$\vec{v_{av}} = \frac{\vec{s}}{\Delta t} = \frac{(600\hat{i} + 600\hat{j})\text{m}}{(30 \times 60)\text{s}} = \frac{1}{3}(\hat{i} + \hat{j})\text{ms}^{-1} \text{ Ans.}$$

(b) The total distance covered is

$$D = s_1 + s_2 + s_3 = (800 + 600 + 200) \text{ m} = 1600 \text{ m}$$

Hence, the average speed is $v_{av} = \frac{D}{\Delta t} = \frac{1600}{30 \times 60} = \frac{8}{9}ms^{-1}$ Ans.

2.3.1 Instantaneous velocity

As mentioned earlier, the average velocity is zero in a round trip. Does it mean that the particle is not moving? Of course not. So, the average velocity cannot always describe the motion of a particle. Then we need to define the velocity of a particle at any point of time (or instant) or instantaneous velocity to describe the motion of the particle.

The word *instant* literally means a very small-time interval. An *instant* means Δt tends to zero; $(\Delta t \rightarrow 0)$. In other words, *instant* may mean an infinitesimal time dt which replaces the term '$\Delta t \rightarrow 0$'. This idea of *instant* was introduced by Newton to find the instantaneous velocity (or simply 'velocity') of a particle. Let us proceed with this idea.

To find the velocity at any time t, first of all we allow the particle to move for an additional time Δt and measure the displacement $\overrightarrow{\Delta r}$ of the particle from a given point P. Then we will go on reducing the time interval Δt to zero. While doing so, we observe that the displacement vector $\overrightarrow{s} = (\overrightarrow{\Delta r})$ gradually leans towards the tangent drawn at the point P. Eventually, the average velocity $\overrightarrow{v}_{av}$ will tend to be tangential because it always points in the direction of displacement $\overrightarrow{\Delta r}$. So, $\overrightarrow{v}_{av}$ tends to be tangential when Δt tends to zero. Furthermore, the magnitude of $\overrightarrow{v}_{av}$ will also be finite. This limiting value (of magnitude and direction) which is known as 'instantaneous velocity' at the point P is shown in the figure below.

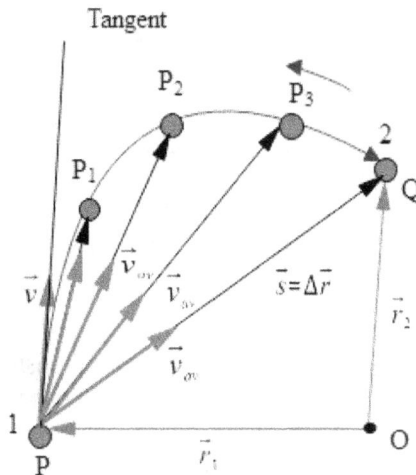

The average velocity tends to be an instantaneous velocity which is finite and tangential at the point P

So, we can express instantaneous velocity $\vec{v}$ as

$$\vec{v} = \lim_{\Delta t \to 0} \vec{v}_{av} = \lim_{\Delta t \to 0} \frac{\Delta \vec{r}}{\Delta t} = \lim_{t_2 \to t_1} \frac{\vec{r_2} - \vec{r_1}}{t_2 - t_1}$$

If we know all positions of the particle as the function of time, we can write $\vec{r_1} = \vec{r}(t)$ and $\vec{r_2} = \vec{r}(t + \Delta t)$. Now the above expression can be given as:

$$\vec{v} = \lim_{\Delta t \to 0} \frac{\vec{r}(t + \Delta t) - \vec{r}(t)}{\Delta t} = \frac{d\vec{r}}{dt}$$

$$\Rightarrow \vec{v} = \frac{d\vec{r}}{dt}$$

Velocity as the slope of an r–t graph: We can see in the velocity–time graph as shown in the figure that, for any time interval Δt, the average velocity $\vec{v}_{av} = \frac{\Delta \vec{r}}{\Delta t}$ is represented by $\tan \theta$. So, the slope of the line AB gives us the average velocity in the r–t graph as shown in figure.

When we reduce the time interval gradually to zero, the straight line AB tends to be a tangent AC at the time t. This means, $\lim_{\Delta t \to 0} \frac{\Delta \vec{r}}{\Delta t} = \vec{v} = \frac{d\vec{r}}{dt}$ can be represented by $\tan \theta$, that is, the slope of the tangent drawn at the time 't' under consideration.

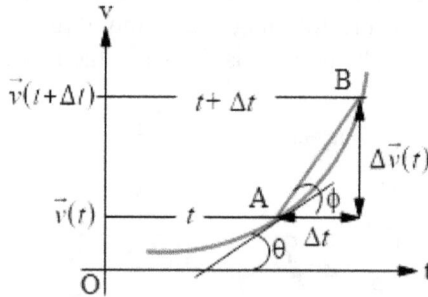

Average acceleration a_{av} = slope of AB $\equiv \tan\phi$ and slope of the tangent
of v-t graph at the point A = instantaneous acceleration a $\equiv \tan \theta$

From the above discussions, we can note the following points.
1. Slope of the tangent drawn in the r–t graph at any instant gives (defines) the velocity at that instant. In other words, instantaneous velocity is equal to the slope of a position–time graph.
2. The instantaneous velocity is tangential to the path.
3. The magnitude of velocity, that is, speed is a finite quantity. The instantaneous velocity is a measure of motion of a particle which tells us 'how fast'

and 'in which direction' the particle moves. The instantaneous velocity is always tangential to the path followed by the particle.

As explained earlier, the distance covered by a particle increases gradually with time. Following this fact, during a small time interval, a particle covers a small distance. If the time interval is infinitesimal, the distance covered by the particle will be also infinitesimal. The ratio of infinitesimal distance and infinitesimal time interval is a finite quantity. We call it 'instantaneous speed' denoted by the letter u.

Instantaneous speed is given as

$$u = \lim_{\Delta t \to 0} \frac{\Delta D}{\Delta t}$$

$$\Rightarrow u = \frac{dD}{dt}$$

For an elementary time, $\Delta t \to 0$, the elementary distance dD is numerically equal to the elementary displacement ds.

So, the instantaneous speed and instantaneous velocity is given as

$$\frac{dD}{dt} = \left| \frac{d\vec{s}}{dt} \right|$$

$$\Rightarrow u = v$$

Instantaneous speed is equal to the magnitude of instantaneous velocity. It gives us an idea how fast the particle is moving. The speedometer of a vehicle measures the instantaneous speed.

Example 5 Galileo proved that an object in gravity falls obeying the relation $y = 4.9t^2$. Newton determined the instantaneous velocity of the particle at any instant by his differential calculus as described above. Let us the find the velocity at the time $t = 2$ s.

Solution

Using the of calculus formula derived by Newton, the instantaneous velocity is

$$v = \frac{dy}{dt}$$

$$\Rightarrow v = \frac{d}{dt}(4.9t^2)$$

$$\Rightarrow v = 4.9\frac{d}{dt}(t^2)$$

$$\Rightarrow v = 4.9 \times (2t)$$

$$\Rightarrow v = 9.8t$$

Putting $t = 2$, $v = 9.8(2) = 19.6 \text{ m s}^{-1}$

2.3.2 Acceleration

When a particle changes its velocity, it is said to be accelerating. The magnitude or direction or both magnitude and direction of velocity $\vec{v}$ may change.

2.3.2.1 Average acceleration

Let a particle move with velocities $\vec{v_1}$ and $\vec{v_2}$ at points 1 and 2 at times $t_1(=t)$ and $t_2(=t + \Delta t)$, respectively. So, the particle changes its velocity by $\Delta\vec{v}(=\vec{v_2}-\vec{v_1})$ during a time interval Δt. Dividing the change in velocity by time interval, we have $\frac{\Delta\vec{v}}{\Delta t}$. This is what call average rate of change in velocity or average acceleration denoted as $\vec{a}_{av}$. So, we can write

$$\vec{a}_{av} = \frac{\Delta\vec{v}}{\Delta t} = \frac{\vec{v_2}-\vec{v_1}}{t_2 - t_1}$$

The average acceleration of a particle between any two points is equal to the ratio of change in its velocity and the time elapsed between these points. Since, $\vec{a}_{av}$ is parallel to $\Delta\vec{v}$, average acceleration points in the direction of change in velocity.

2.3.2.2 Instantaneous acceleration

If the particle has the same velocity at any two points A and B, say, in its path, the change in its velocity is zero. Then the average acceleration of the particle over the time interval Δt is zero. Does it $(\vec{a}_{av})$ really confirm that the particle is not changing its velocity (accelerating) between A and B? In fact, the average acceleration deals with the net (total) change in velocity and the corresponding time interval. Hence, $\vec{a}_{av}$ is not always useful to analyse and understand the variation of velocity of a particle with time. Then, we need to establish the idea of instantaneous acceleration.

If we know the variation of velocity with time as the function $v = f(t)$, we can write $\vec{v_1} = \vec{v}(t)$ and $\vec{v_2} = \vec{v}(t + \Delta t)$. Then the average acceleration is

$$\vec{a}_{av} = \frac{\vec{v_2}-\vec{v_1}}{\Delta t} = \frac{\vec{v}(t + \Delta t) - \vec{v}(t)}{\Delta t}$$

If we want to find the rate of change in velocity over a short time (an instant), we must reduce the time interval gradually to zero, $(\Delta t \to 0)$. In consequence, $\Delta\vec{v}$ tends to zero $(\Delta\vec{v} \to 0)$. Then, the ratio ' $\lim_{\Delta t \to 0} \frac{\Delta\vec{v}_n}{\Delta t}$' attains a limiting magnitude and direction.

We call it *instantaneous* acceleration which is given as

$$\vec{a} = \lim_{\Delta t \to 0} \vec{a}_{av} = \lim_{\Delta t \to 0} \frac{\Delta \vec{v}}{\Delta t},$$

where $\lim_{\Delta t \to 0} \frac{\Delta \vec{v}}{\Delta t} = \frac{d\vec{v}}{dt}$,

$$\Rightarrow \vec{a} = \frac{d\vec{v}}{dt}$$

The instantaneous acceleration (or acceleration at any instant) is defined as the rate of change in velocity with time.

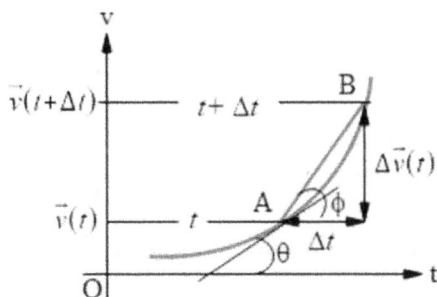

Average acceleration a_{av}= slope of AB ≡ tanφ and slope of the tangent
of v-t graph at the point A = instantaneous acceleration a ≡ tan θ

Example 6 A car moves to the east with a speed of 15 m s^{-1}. It heads due north keeping the speed constant during a time interval of 15 s. Find the (a) change in velocity, (b) average acceleration and (c) change in speed.
Solution
Let us assume that +x- and +y-directions represent east and north, respectively.
  (a) While moving east $\vec{v}_1 = 10\hat{i}$ and when moving north, $\vec{v}_2 = 10\hat{j}$. Then, the change in velocity is

$$\Delta \vec{v} = \vec{v}_2 - \vec{v}_1$$
$$= (15\hat{j} - 15\hat{i})\,\text{ms}^{-1} \text{ Ans.}$$

  (b) The average acceleration is

$$\vec{a}_{av} = \frac{\Delta \vec{v}}{\Delta t}$$
$$= \frac{15(\hat{j} - \hat{i})\text{m s}^{-1}}{15 \text{ s}}$$
$$= (\hat{j} - \hat{i})\text{m s}^{-2} \text{ Ans.}$$

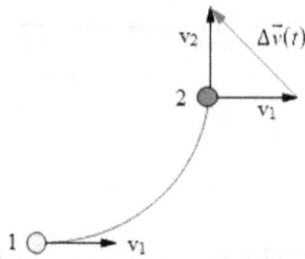

(c) The change in speed is

$$\Delta u = v_2 - v_1 = 15 - 15 = 0 \text{ Ans.}$$

N.B: In the above problem, we can see that $\Delta \vec{v}$ is $15\sqrt{2}$ m s^{-1} towards N–W, $\vec{a}_{av}$ is $\sqrt{2}$ m s^{-2} due N–W and $\Delta u = 0$ but $|\Delta \vec{v}| = 15\sqrt{2} \neq 0$.

2.3.3 General kinematical equations for straight-line motion

When a particle moves in a straight line we call it straight-line motion. It can move in two possible directions in a straight line, +ve and −ve direction. Any direction along the +ve axis is called the +ve direction and along the −ve axis is called the −ve direction. We have three vectors in kinematics, namely position $\vec{r}$ (or displacements), velocity $\vec{v}$ and acceleration $\vec{a}$. Any vector pointing in the +ve direction is positive and one directed in the −ve direction is negative.

Let us assume that a particle moves in the +x-direction having both position vector and acceleration pointing in the +x-direction. Then, the velocity of the particle is

$$v = \frac{dx}{dt}$$

Then, the acceleration is

$$a = \frac{dv}{dt}$$

$$\Rightarrow a = \frac{dv}{dx} \cdot \frac{dx}{dt}$$

$$\Rightarrow a = v\frac{dv}{dx}$$

The acceleration can also be given as

$$a = \frac{dv}{dt} = \frac{d}{dt}\left(\frac{dx}{dt}\right)$$

$$\Rightarrow a = \frac{d^2x}{dt^2}.$$

Example 7 A particle oscillates simple harmonically so that its position changes with time as $x = A \sin \omega t$, where A and ω are constants. Find the:
  (a) velocity,
  (b) acceleration as the function of time t and x.

Solution
  (a) The velocity is

$$v = \frac{dx}{dt} = \frac{d}{dt}(A \sin \omega t) = A\omega \cos \omega t = \omega\sqrt{A^2 - x^2} \text{ Ans.}$$

  (b) The acceleration is

$$a = \frac{dv}{dt} = \frac{d}{dt}(A\omega \cos \omega t) = -A\omega^2 \sin \omega t = -\omega^2 x \text{ Ans.}$$

2.3.4 Uniform accelerated motion

If the average acceleration of a particle remains constant over any time interval, it is said to be moving with uniform acceleration $\vec{a}$. In other words, the instantaneous acceleration of the particle remains constant with both magnitude and direction.

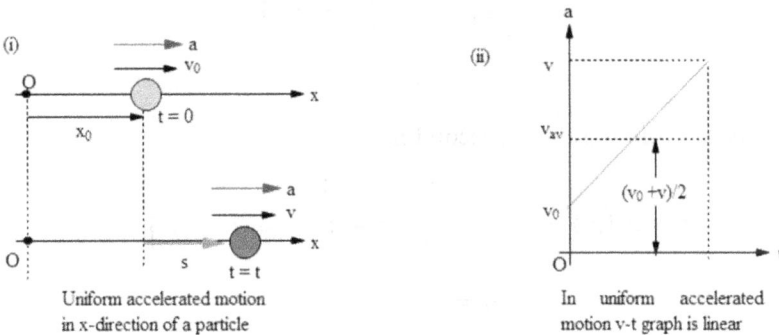

Uniform accelerated motion
in x-direction of a particle

In uniform accelerated
motion v-t graph is linear

Let a particle move along the $+x$-direction with an initial velocity $+v_0$ and a uniform acceleration $+a$. During a time t, let the speed of the particle increase from v_0 to v.

The average acceleration of the particle over a time t is

$$a_{av} = \frac{\Delta v}{\Delta t}$$

$$\Rightarrow a_{av} = \frac{v - v_0}{t}$$

Since, $a_{av} = a$ for uniform acceleration, we have

$$a = \frac{v - v_0}{t}$$

$$\Rightarrow v = v_0 + at$$

The above expression tells us that, in one-dimensional uniform accelerated motion the speed increases linearly with time t.

$$s = v_{av}t$$

$$= \left(\frac{v_0 + v}{2}\right)t \left(\because v_{av} = \frac{v + v_0}{2}\right)$$

$$= \left\{\frac{v_0 + (v_0 + at)}{2}\right\}t \ (\because v = v_0 + at)$$

$$\Rightarrow s = v_0 t + \frac{1}{2}at^2$$

From the above expressions we understand that when acceleration remains constant, velocity varies linearly and the variation of displacement is parabolic with time.

In other words, the a–t graph is a straight line parallel to the t-axis having zero slope. The v–t graph is a slanted straight line of constant slope, that is, $m = \frac{dv}{dt} = a$ and the s–t graph is parabolic having a slope which increases linearly with time.

Substituting, $t = \frac{v - v_0}{a}$, $s = \frac{v + v_0}{2}t$, we have

$$s = \left(\frac{v + v_0}{2}\right)\left(\frac{v - v_0}{a}\right)$$

$$\Rightarrow v^2 = v_0^2 + 2as$$

The displacement during the ith second is

$$s_t = s_t - s_{t-1}$$

$$= \left(v_0 t + \frac{1}{2}at^2\right) - \left\{v_0(t - 1) + \frac{1}{2}a(t - 1)^2\right\}$$

$$\Rightarrow s_t = v_0 + (2t - 1)\frac{a}{2}.$$

Example 8 A car accelerates from rest from A with an acceleration of 2 m s^{-2} for 8 s to reach B, then it applies its brakes to retard uniformly at a rate of 4 m s^{-2}. Find the:
(a) speed at B,
(b) AB,

(c) AC,

(d) total time taken,

(e) average speed over time of motion.

Solution

(a) The speed of the car at B is

$$v_B = v_A + at_{AB}$$
$$= (2)(8)(\because v_A = 0) \ \text{Ans.}$$
$$= 16 \ \text{m} \ \text{s}^{-1}$$

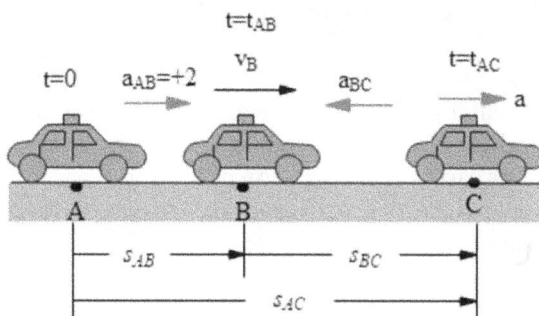

(b) The displacement of the car is

$$s_{AB} = v_A t + \tfrac{1}{2}at_{AB}^2$$
$$= \tfrac{1}{2} \times 2 \times (8)^2 (\because v_A = 0)$$
$$\Rightarrow AB = 64 \ \text{m} \ \text{Ans.}$$

(c) From B to C, we can use the equation

$$v_C^2 = v_B^2 + 2a_{BC}s_{BC}$$

Putting $v_B = 16 \ \text{m} \ \text{s}^{-1}$, $v_C = 0$, $a_{BC} = -4 \ \text{m} \ \text{s}^{-2}$, we have

$$(16)^2 + 2(-4)(s_{BC}) = 0$$

$$\Rightarrow s_{BC} = 32 \ \text{m}.$$

Then $AC = AB + BC = 64 + 32 = 96 \ \text{m. Ans.}$

(d) Then the total time of journey from A to C is

$$t_{AC} = t_{AB} + t_{BC},$$

where $t_{BC} = v_B/a_{BC} = 16/4 = 4$ s and $t_{AB} = 8$ s

$$\Rightarrow t_{AC} = 8 \ \text{s} + 4 \ \text{s} = 12 \ \text{s} \ \text{Ans.}$$

(e) The average speed is

The average velocity of the car over AC is

$$v_{av} = s_{AC}/t_{AC}$$

$$\Rightarrow u_{av} = \frac{AC}{t_{AC}} = \frac{96\ m}{12\ s} = 8\ ms^{-1}\ Ans.$$

2.3.5 Kinematics of freely falling bodies

If you throw a body (a stone, say), ignoring all the forces, it moves under the action of Earth's gravitational field alone. Then, we say that the body is falling freely. The acceleration due to gravity of Earth varies inversely with the square of the distance of the body and the center of Earth. If you take a height $h = 100$ km say, it is much smaller than the radius of Earth ($R_e = 6400$ km). So, it is considered as near Earth's surface. In this case, the acceleration of the body can be taken as a constant quantity. In other words, near Earth's surface the acceleration due to gravity is uniform or a constant which is equal to $g = 9.8$ m s^{-2} pointing vertically downwards.

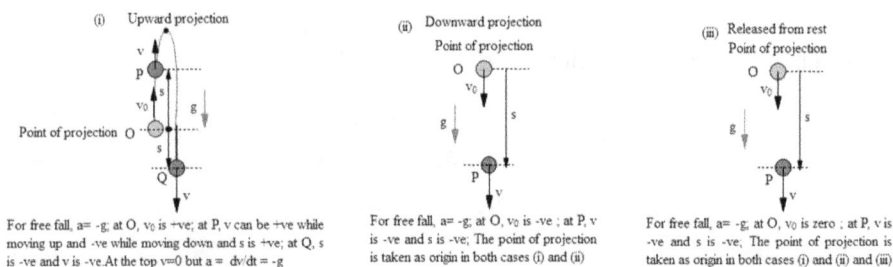

(i) Upward projection

Point of projection O

For free fall, a= -g; at O, v_0 is +ve; at P, v can be +ve while moving up and -ve while moving down and s is +ve; at Q, s is -ve and v is -ve.At the top v=0 but a = dv/dt = -g

(ii) Downward projection
Point of projection

For free fall, a= -g; at O, v_0 is -ve ; at P, v is -ve and s is -ve; The point of projection is taken as origin in both cases (i) and (ii)

(iii) Released from rest
Point of projection

For free fall, a= -g, at O, v_0 is zero ; at P, v is -ve and s is -ve; The point of projection is taken as origin in both cases (i) and (ii) and (iii)

In this section, we will consider the straight-line motion of a particle near Earth's surface. It is experimentally verified that all bodies accelerate towards the center of Earth with an acceleration of 9.8 m s^{-2} 'near Earth's surface'. When viewed near the 'Earth', its surface appears flat. So, any point object appears to accelerate vertically downwards with an acceleration of 9.8 m s^{-2}.

For vertical motion of an object, first of all we choose the point of projection as the origin of the coordinate system. If the particle is projected up, its velocity is considered positive. For downward projection, velocity of projection is assumed as negative. If the final point is above the point of projection (origin or reference point), the displacement is upward; so, we consider s as positive. If the final point is below the point of projection, s is assumed as negative. Since the acceleration of the particle points vertically downwards, $a = -g$.

Example 9 An iron ball is released from rest from the top of a cliff. Find the ratio of the (a) distance covered by the ball during 1, 2, 3 s etc, (b) distance covered by the ball during the first, second and third seconds etc, and (c) velocity of the ball at the first, second and third seconds and so on. (d) Draw the respective graphs considering the magnitudes of displacement, velocity and acceleration.

Solution

(a) The distance covered by the ball is given as

$$s = 4.9t^2$$

Putting $t = 1, 2, 3$ and 4 etc, the distance fallen after 1, 2, 3 and 4 s etc, can be given as; $s_1 = 49$ $s_2 = 4.9 \times 4$, $s_3 = 4.9 \times 9$, $s_4 = 4.9 \times 16$ and so on. The ratio of distance covered is 1:4:9.

(b) The distance covered by the ball in the tth second is given as

$$s = g(2t - 1)/2 = 4.9(2t - 1)$$

Putting $t = 1, 2, 3$ and 4 etc, the distance fallen after 1, 2, 3 and 4 s etc, can be given as; $s_1 = 4.9$ $s_2 = 4.9 \times 3$, $s_3 = 4.9 \times 5$, $s_4 = 4.9 \times 7$ and so on. The ratio of distance covered is 1:3:5.

(c) The velocity of the ball is given as

$$v = ds/dt = 9.8t$$

Putting $t = 1, 2, 3$ and 4 etc, the velocity at 1, 2, 3 and 4 s etc, can be given as; $v_1 = 4.9$ $v_2 = 4.9 \times 2$, $v_3 = 4.9 \times 3$, $v_4 = 4.9 \times 4$ and so on. The ratio of distance covered is 1:2:3.

(d) The s–t, v–t and a–t graphs are given considering their magnitudes only. But, more strictly, all these vectors are negative as they point vertically down; so, their graphs will be inverted.

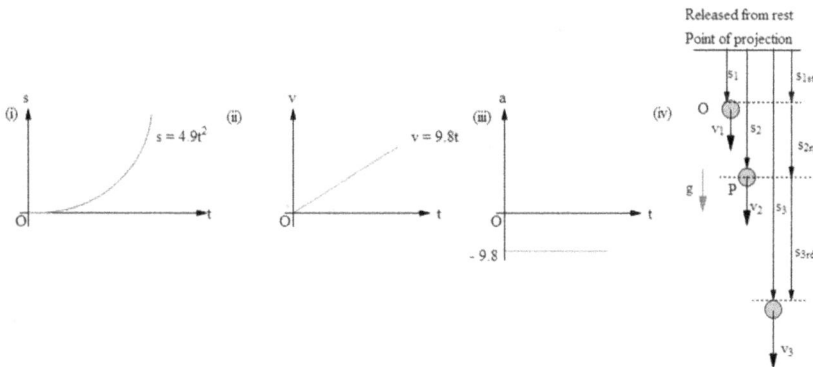

If we plot the s–t graph, we can get a parabola given in part (a),

So, the acceleration of the ball is constant, that is, 9.8 m s^{-2} in a downward direction.

The speed of the ball after a time t can be given as

$$v = \frac{ds}{dt} = 9.8\,t \text{ Ans.}$$

Example 10 When we throw a stone vertically up from the top of a cliff with a speed v_0, it moves up and then comes down. Why? How does the position of the stone from the point of projection change? Draw the y–t graph and snapshots for different positions of the stone. Here, use the sign of the displacement and velocity of the stone while drawing the graphs and trajectory.

Solution

The point of projection O is the origin of the coordinate system.

At the time t, the velocity of the stone is

$$v = u + at$$

For upward projection

$$u = +v_0, \quad a = -g$$

$$\Rightarrow v = v_0 - gt \tag{2.1}$$

At the time t, the displacement of the stone is

$$s = ut + \frac{1}{2}at^2 \tag{2.2}$$

For upward projection $u = +v_0, \quad a = -g$

$$\Rightarrow s = v_0 t - \frac{1}{2}gt^2 \tag{2.3}$$

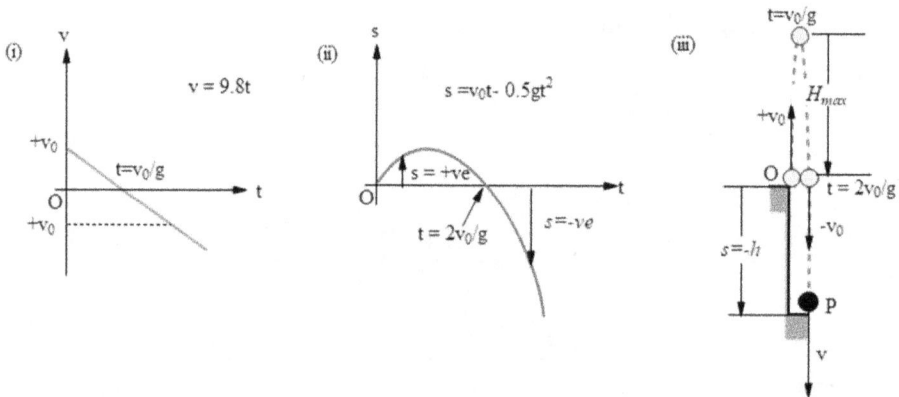

From equation (2.1) we understand that the speed of the stone decreases from v_0 to zero during its ascent. Then it descends with gradually increasing speed,

$$\text{At } t < \frac{v_0}{g}, \quad v \text{ is up}$$

$$\text{At } t = \frac{v_0}{g}, \quad v = 0$$

$$\text{At } t > \frac{v_0}{g}, \quad v \text{ is down}$$

$$\text{At } t = \frac{2v_0}{g}, \quad v = -v(\text{down})$$

From equation(2.2), we understand that the displacement of the stone changes its magnitude and direction as given below:

$$0 < t < \frac{2v_0}{g}, \quad s \text{ is up}$$

$$\text{At } t = \frac{v_0}{g}, \quad s \text{ is maximum}$$

$$\text{At } t = \frac{2v_0}{g}, \quad s \text{ is zero}$$

$$\text{At } t > \frac{2v_0}{g}, \quad s \text{ is down}$$

1. The particle may move up or down but it must accelerate downward with $a = 9.8$ m s^{-2}.
2. For $t < 2v_0/g$ we have two values of $v(+v$ and $-v)$ and for $t > 2v_0/g$ we have one negative value of v.
3. For an upward displacement, we have two values of time solving the quadratic equation $s = v_0 t - gt^2/2$, whereas for downward displacement, we have one value of time. This means the particle has the same upward displacement at two instants: one while going up (ascending) and the other while coming down (descending).

2.3.6 Use of kinematical equations

(a) Use of $a = \frac{dv}{dt}$:

This formula relates acceleration with velocity and time. When acceleration is given as the function of time and velocity, we can use the above formula to find the velocity and displacement at any instant.

Case (i): $a = f(t)$: If the acceleration is the function of time,

$$a = \frac{dv}{dt} = f(t)$$

$$\Rightarrow dv = a \, dt$$

If the velocity changes from v_0 to v during time t integrating both sides, we have

$$\int_{v_0}^{v} dv = \int_{0}^{t} a\, dt$$

$$\Rightarrow v = v_0 + \int_{0}^{t} a\, dt$$

Again, integrating v with respect to time, we have

$$s = \int_{0}^{t} v\, dt = \int_{0}^{t} g(t)dt = h(t), \text{ say.}$$

Case (ii) $a = f(v)$: If the acceleration is the function of time,

$$a = \frac{dv}{dt}$$

$$\Rightarrow dt = \frac{dv}{a}$$

When velocity changes from v_0 to v during time t, integrating both sides, we have

$$\int_{0}^{t} dt = \int_{v_0}^{v} \frac{dv}{a}$$

This yields velocity as the function of time $v = f(t)$.
    Then, integrating velocity, the displacement is

$$s = \int_{0}^{t} v\, dt$$

(b) **Use of** $v\, dv = a\, ds$:

This formula relates acceleration, velocity and displacement. So, this can be used when acceleration is given as the function of distance or velocity.

Case (i); $a = f(s)$:

$$a = \frac{v\, dv}{ds}$$

$$\Rightarrow v\, dv = a\, ds$$

When velocity changes from v_0 to v during a displacement s, integrating both sides, we have

$$\int_{v_0}^{v} v\, dv = \int_{0}^{s} f(s)\, ds$$

This gives the velocity as the function of displacement s

$$v = \sqrt{v_0^2 + 2\int_0^s f(s)\, ds} = f(s)$$

$$\Rightarrow t = \int_0^t dt = \int_0^s \frac{ds}{v}$$

This yields the displacement as the function of time.

Case (ii); $a = f(v)$:

$$\text{Since } a = \frac{v\, dv}{ds},$$

we have $vdv = a\, ds$

$$\Rightarrow ds = \frac{v\, dv}{f(v)}$$

If the velocity of the particle changes from v_0 to v when it undergoes a displacement s, integrating both sides, we have

$$\int_0^s ds = \int_{v_0}^v \frac{v\, dv}{a(v)}$$

This gives velocity as the function of displacement

$$\text{Then, } t = \int_0^s \frac{ds}{v}$$

which gives displacement as the function of time.

Let us tabulate the program to find the displacement from a given force (or acceleration).

Let us use the above ideas in the following problems.

Example 11 A particle slows down from a speed v_0 while moving in a straight line. If the deceleration is directly proportional to the speed of the particle, (a) find its displacement as the function of time (b) draw, s–t, v–t and v–s graphs.

Solution

(a) **Method 1:** $a = \frac{dv}{dt}$, where $a = -\alpha v$

Then, $-\alpha v = \frac{dv}{dt}$

Rearranging the terms, we have $\frac{dv}{v} = -\alpha dt$

When the particle slows down from v_0 to v let it take a time t. Hence, we have

$$\int_{v_0}^v \frac{dv}{v} = -\alpha \int_0^t dt$$

This gives $\ln \frac{v}{v_0} = -\alpha t$. Hence $v = v_0 e^{-\alpha t}$

The displacement $s = \int_0^t v\, dt = v_0 \int_0^t e^{-\alpha t} dt = v_0 \left[\frac{e^{-\alpha t}}{-\alpha} \right]_0^t$

This gives $s = \frac{v_0}{\alpha}(1 - e^{-\alpha t})$ Ans.

Method 2: $v dv = a\, ds$, where $a = -\alpha v$

Then, we have $v\, dv = a\, ds$, where $a = -\alpha v$

This gives $dv = -\alpha ds$

When the particle slows down from v_0 to v, it undergoes a displacement s.

Then, we have $\int_{v_0}^v dv = -\alpha \int_0^s ds$

This yields $v = v_0 - \alpha s$

Since $v = \frac{ds}{dt}$, we have $t = \int_0^s \frac{ds}{v}$. Substituting $v = v_0 - \alpha s$, we have

$$t = \int_{v_0}^s \frac{ds}{v_0 - \alpha s} = -\frac{1}{\alpha}[\ln(v_0 - \alpha s)]_0^s = -\frac{1}{\alpha} \ln\left(\frac{v_0 - \alpha s}{v_0} \right)$$

This gives $(1 - \frac{\alpha s}{v_0}) = e^{-\alpha t}$

So, $s = \frac{v_0}{\alpha}(1 - e^{-\alpha t})$

(b) When t tends to infinite, we can see that s will tend to a maximum value $s_{\max} = v_0/\alpha$, v will tend to a minimum value from a maximum value $v_{\max} = v_0$. However, the velocity increases linearly with distance s. Ans.

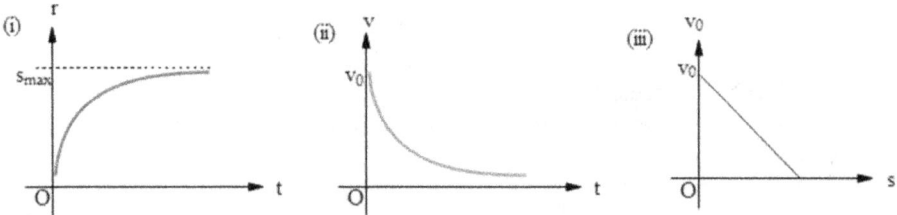

2.3.7 Uniformly accelerated motion

Let us now apply the previously obtained general kinematical equations such as

$$v = u + \int_0^t a\, dt, \quad s = \int_0^t v\, dt, \quad v^2 = u^2 + 2\int_0^s a\, ds$$

for a particle moving with constant acceleration a. Since, a is constant, taking it out of the integral in $v = u + \int_0^t a dt$, we have

$$v = u + at \tag{2.4}$$

Then, substituting v from equation (2.4) in $s = \int v \, dt$, we have

$$s = \int_0^t (u + at)dt$$

$$\Rightarrow s = ut + \frac{1}{2}at^2 \tag{2.5}$$

The average velocity is

$$v_{av} = \frac{s}{t} = \frac{ut + \frac{1}{2}at^2}{t}$$

$$= \frac{2u + at}{2} = \frac{u + (u + at)}{2}$$

$$v_{av} = \frac{u + v}{2} \tag{2.6}$$

Finally, using $v^2 = u + 2\int_0^s a \, ds$ for constant acceleration, we obtain

$$v^2 = u^2 + 2as \tag{2.7}$$

Example 12 A particle moves with an initial velocity 10 m s^{-1} in the negative x-direction. If the acceleration of the particle is $a = 2$ m s^{-2} in the positive x-direction, find the (i) velocity of the particle at $t = 10$ s, (ii) displacement of the particle in the first ten seconds, (iii) average velocity of the particle over first 5 s and (iv) distance first 10 s and average speed over first 10 s.

Solution

(i) As the particle moves in the $-$ve x-direction put $u = -10$; as the acceleration points in the $+$ve x-direction, put $a = +2$. So, $v = u + at = (-10) + (+2)(10) = +10$ m s^{-1} Ans.

(ii) The displacement is $s = ut + \frac{1}{2}at^2 = (-10)(10) + \frac{1}{2}(2)(10)^2 = 0$ Ans.

(iii) The average velocity is $v_{av} = \frac{u + v}{2} = \frac{-10 + 10}{2} = 0$ m s^{-1} Ans. You can also write $v_{av} = \frac{s}{t} = \frac{0}{10} = 0$.

(iv) The displacement is maximum when velocity is zero momentarily. Putting $v = 0$, $a = +2$ and $u = -10$ in $v^2 = u^2 + 2as$ $s_{max} = \frac{100}{2} = 50$ m. As the particle returns back to its initial position to undergo zero displacement and the total distance covered is d $= 2s_{max} = 2 \times 50 = 100$ m. Then, the average speed is $u_{av} = \frac{d}{t} = \frac{100}{10} = 10$ for the first 10 s.

2.3.8 Graphs and graphical solutions

Introduction: There are four main parameters in rectilinear kinematics such as displacement (s), velocity (v), acceleration (a) and time (t). One can be plotted as the function of the other. We can plot them in six possible ways, i.e., $s-t$, $v-t$, $a-t$, $v-s$, v^2-s and $a-s$.

The time derivative of position vector or displacement gives velocity; $v = \frac{dr}{dt} = \frac{ds}{dt}$. The time derivative of velocity gives acceleration; $a = \frac{dv}{dt}$. Following the reverse process, integration of acceleration with respect to time gives us change in velocity; $\Delta v = \int a \, dt$. Similarly, integration velocity with respect to time we find the displacement (change in position vector);

$$s = \Delta r = \int v \, dt$$

Sometimes, the velocity is given as the function of displacement. In that case, we find acceleration by using the formula

$$a = \frac{v \, dv}{ds} = \frac{1}{2} \frac{dv^2}{ds}$$

This may also be written as

$$\frac{dv}{ds} = \frac{a}{v}.$$

That means, derivative of v with displacement s yields the ratio of acceleration and speed.

When acceleration is given as the function of displacement, integrating acceleration with respect to displacement, we have

$$\int \vec{a} \; d\vec{s} = \int_{v_0}^{v} v \, dv = \frac{1}{2}(v^2 - v_0^2)$$

From the above explanations we understand that, sometimes the derivative of one parameter leads to the other and in some other cases, integration of one parameter results in the other. As we know, derivative of a function refers to its slope and its integration gives the area bounded by the function. Let us discuss the slope and area bounded by the kinematical functions. We will use the graphs in solving the kinematical problems.

(a) **Displacement–time graph**

   (i) **Average velocity:** As derived earlier, average velocity of a particle over a time interval Δt can be given as $v_{av} = \frac{\Delta s}{\Delta t} = \tan \phi$

$$\vec{v}_{av} = \frac{\Delta \vec{r}}{\Delta t}$$

slope of the line joining the points in the s–t graph

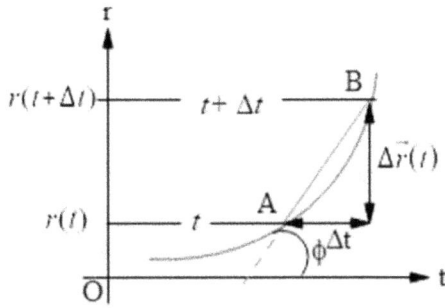

Average velocity v_{av}= slope of AB $\equiv \tan\phi$

From the above analysis, we conclude the following points:

1. The slope of the straight line joining any two points in an s–t graph gives the average velocity between these points of time. Symbolically, $v_{av} = \frac{\Delta s}{\Delta t}$=slope of the line joining the points in the s–t graph. If the slope is positive, v_{av} is positive; if the slope is negative, a_{av} is negative; if the slope is zero, v_{av} is zero.

2. In the process of a graphical solution, we should not mistake the displacement–time graph the equation of trajectory.

   (ii) **Instantaneous velocity:**

      Since the instantaneous velocity is given as

$$v = \frac{ds}{dt} = \tan\theta$$

we have, $v =$ slope of an s–t graph.

   If the slope is positive, v is positive (the particle moves in the positive direction) and when the slope is negative, v is negative (the particle moves in the negative direction). It is a general rule or convention that *up* refers to +ve and *down* is −ve; right is +ve and left refers to −ve; outward is +ve and inward is −ve.

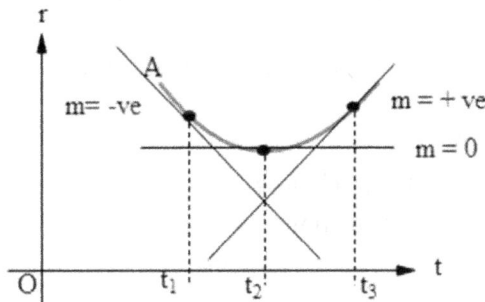

The slope m of r - t graph = velovity ;$m = dr/dt$=v ; if m is +ve, v is +ve;if m is -ve, v is -ve ; if m=0, $v = 0$

Example 13 A particle moves rectilinearly such that, after some time t_0 from starting, say let its (instantaneous) velocity be equal to its average velocity over that time. Referring to the graph as shown in the figure below for the motion of the cockroach, find the time t_0 and the average velocity of the cockroach over the time t_0.

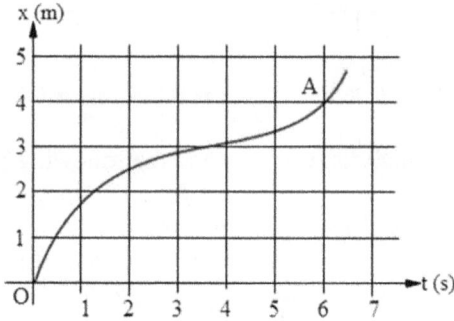

Solution

Since $a_{av} = v$ (given), the 'slope of AB = slope of the tangent AC'. Hence, ABC is a straight line.

We can see that at point 'B' of the x–t graph, both $\vec{v}_{av}$ and $\vec{v}$ are equal. The average velocity during time interval $AC = t_0$ is given by the slope of the line OAB.

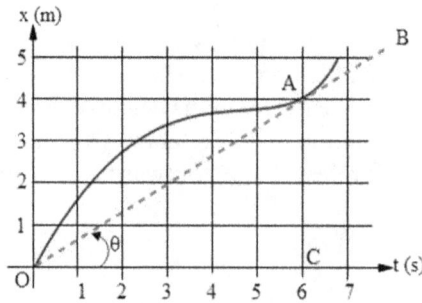

The slope of OA = slope of AB = slope of OAB = tanθ = AC/OC=4/6=2/3.

That means, $v_{av} = \dfrac{AC}{t_0}$ by referring to the x–t graph. On the other hand, the velocity v of the particle at the time t is equal to the slope of the x–t graph at that instant (point P); $\vec{v} = \dfrac{d\vec{s}}{dt} \equiv$ slope of the tangent AB at point A.

Since $\vec{v}_{av}$ and $\vec{v}$ are equal, the straight line OAB must be tangential at A. Hence, the point B has coordinates $(3, 3)$.

It gives $v_{av} = v = \dfrac{\Delta s}{\Delta t} = \dfrac{4}{6} = \dfrac{2 \text{ m}}{3 \text{ s}} = 0.67 \text{ m s}^{-1}$. Ans.

(b) Velocity–time graph

From a v–t graph, we can find

(i) **Displacement:** When the velocity of a particle is given as the function of time, i.e., $v = f(t)$, we find the total displacement $\vec{s}$ by summing up (integrating) the elementary displacements $\vec{ds}$; $\vec{s} = \int \vec{v} \, dt$.

The elementary displacement $\vec{ds} = \vec{v} \, dt$ can be given as the area of the shaded elementary strip. Summing up all the elementary areas between $t = t_1$ and $t = t_2$, we can find the total area under the v–t graph enclosed within the time interval which gives the total displacement $\vec{s}$ of the particle during the time interval $\Delta t(=t_2 - t_1)$. So, we can write

$$\vec{s} = \int \vec{v} \, dt = A(=\text{Area under } v\text{–}t \text{ graph})$$

When the area lies above the time axis, it is considered as positive; if it lies below the time axis, it is said to be negative. A positive area reveals a positive displacement (displacement directed along positive directions of the coordinate system) and negative area signifies a negative displacement (displacement pointing in negative directions).

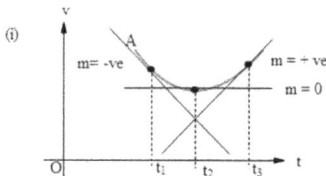

(i) The slope of v-t graph = acceleration; m = dv/dt = a;if m is +ve, v is +ve;if m is -ve, v will be -ve ; if m=0, v=0.

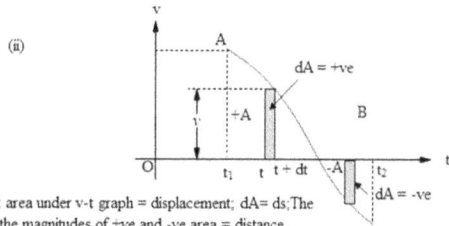

(ii) The net area under v-t graph = displacement; dA= ds;The sum of the magnitudes of +ve and -ve area = distance.

(ii) **Distance:** Since, the distance covered D is given as $D = \int |\vec{v}| \, dt$ it can be given graphically as the sum of the magnitude of positive and negative areas under the v–t graph.

$$D = \int |\vec{v}| \, dt = |A_+| + |A_-|$$

Example 14 A particle moving in the x-axis is given. Find the distance and displacement of the particle during 2 s from starting.

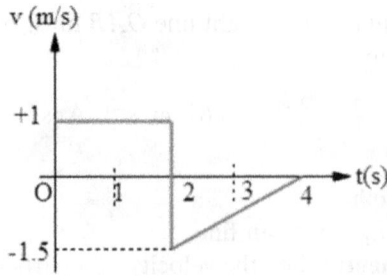

Solution

Th displacement is

$$s = A_+ - A_-, \quad \text{where} \quad A_+ = (+1)(2) = +2 \quad \text{and} \quad A_- = -(1.5)(2)/2 = -1.5.$$

$$\Rightarrow s = 0.5 \ \text{m}.$$

The particle moves a distance of 3 m to the right (+x-direction) and then moves to the left by a distance of 0.5 m (-x-direction); so, the net displacement is +0.5 m. Ans.

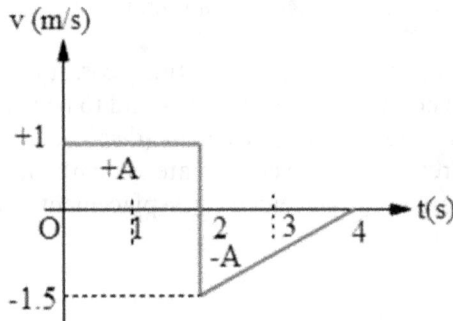

The distance is

$$D = |A_+| + |A_-|$$
$$= |+2| + |-1.5|$$
$$= 3.5 \ \text{m}$$

The sum of magnitude of displacement to the right (2 m) and left (1.5 m) is the total distance of 3.5 m. Ans.

(iii) **Average acceleration:** As derived earlier, the average acceleration

$$\vec{a}_{av} = \frac{\overrightarrow{\Delta v}}{\Delta t} = \tan \phi$$

Then, the average acceleration can be given as

$\vec{a_{av}}$ = slope of the line joining two points in $v-t$ graph.

(iv) **Instantaneous acceleration:** The instantaneous acceleration is given as

$$\vec{a} = \frac{d\vec{v}}{dt} = \tan \theta$$

$$\Rightarrow \vec{a} = \text{ slope of } v - t \text{ graph}$$

If the slope of the $v–t$ graph is positive at any point of time on the graph, the acceleration of the particle is positive at that time. If the slope of the $v–t$ graph is zero, it signifies that the acceleration is zero. When the slope of the $v–t$ graph is negative, it means a negative acceleration (or retardation).

Example 15 A Diwali rocket moves vertically up with a constant acceleration $a_1 = \frac{20}{3}$ m s^{-2}. After some time its fuel gets exhausted and then it falls freely with an acceleration $a_2 = 10$ ms^{-2}. (a) Draw a $v–t$ graph. (b) If the maximum height attained by the Diwali rocket is $h = 50$ m, find its speed v when its fuel just gets exhausted.
  Solution
  (a) During its ascent with the fuel its motion is upward accelerating. So, the slope of the $v–t$ graph is positive during time t_1. After its fuel gets exhausted, it accelerates down with $a = g = 9.8$ m s^{-2} and then the slope of the $v–t$ graph is negative.
  The total displacement $= s =$ area under the $v–t$ graph

$$= \frac{v}{2}(t_1 + t_2) \tag{2.8}$$

Since the slope of the $v–t$ graph $=$ acceleration,

$$a_1 = \frac{v}{t_1} \tag{2.9}$$

and

$$a_2 = \frac{v}{t_2} \tag{2.10}$$

Using equations (2.8), (2.9) and (2.10), we have

$$v = \sqrt{\frac{2a_1a_2h}{a_1 + a_2}} = \sqrt{\frac{2 \times \frac{20}{3} \times 10 \times 50}{\frac{20}{3} + 10}}$$

or $v = 20$ m s^{-1} Ans.

(b) **Acceleration–time graph**

Let the acceleration be given as the function of time, i.e., $a = f(t)$. Then the elementary change in velocity during a time dt is:

$$\vec{dv} = \vec{a}\,dt$$

We can observe that in an a–t graph, $\vec{a}\,dt$ is equal to the area of the shaded elementary strip (rectangle of sides a and dt). Summing up all the elementary areas dA, we have the total area

$$A = \int dA = \int \vec{a}\,dt$$

Since $\displaystyle\int_{t_1}^{t_2} \vec{a}\,dt = \int_{v_1}^{v_2} d\vec{v} = \Delta\vec{v}$,

we have $\Delta\vec{v} = \int \vec{a}\,dt = A =$ area under the a–t graph.

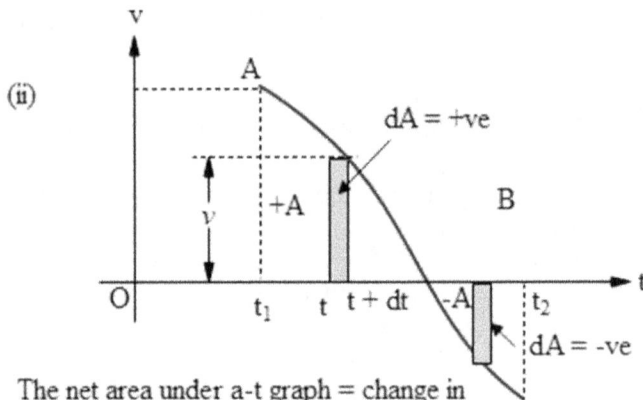

(ii)

The net area under a-t graph = change in velocity vector; Net area $A = A_+ + A_-$

This means that the change in velocity $\Delta\vec{v}$ over any time interval is represented by the algebraic sum of all positive and negative areas which we call 'area under the a–t graph' over that time interval; $A = \Sigma A_x + \Sigma A_-$. If the area is negative, $\Delta\vec{v}$ is negative; $\Delta\vec{v}$ is pointing in the negative direction of the coordinate axes. When area is zero, $\Delta\vec{v} = 0$, which means no change in velocity. When $\Delta\vec{v}$ is positive, you should not immediately accept it as $\Delta\,|\vec{v}| > 0$, because $\Delta\vec{v}$ is a vector quantity. So, positive area means a positive $\Delta\vec{v}$ but not positive $\Delta\,|\vec{v}|(\neq\Delta v)$. In other words, the change in velocity is a vector, whereas the change in speed (magnitude of velocity) is a scalar quantity.

Example 16 A particle moves vertically with an upward initial speed $v_0 = 5$ m s^{-1}. If its acceleration varies with time as shown in the a–t graph in the figure, find the velocity of the particle at $t = 4$ s.

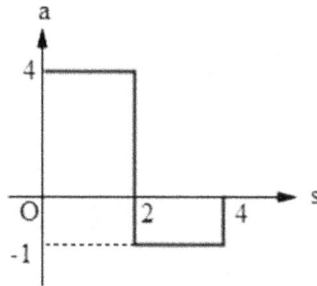

Solution

Let us write the vector equations without using vector notation. The velocity at $t = 4$ s is,

$$v = v_0 + \Delta v$$
$$= v_0 + A, \quad \text{where } A = A_+ + A_-$$
$$= (4 \times 2) + (-1 \times 2) = 6$$

and

$$v_0 = +5$$

Then, $v = 5 + 6 = 11$ m s^{-1}.

(c) **Velocity–displacement graph**

When v is given as the function of s, we can find the acceleration at any position by using the v–s graph as described below.

Since the acceleration is given as

$$a = \frac{v \, dv}{ds},$$

we have $\frac{dv}{ds} = \frac{a}{v}$

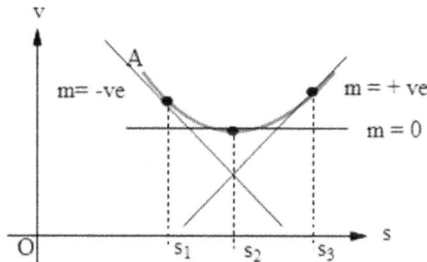

The slope m of v - s graph $=a/v$; if m is +ve,
a is +ve;if m is -ve, a is -ve ; if m=0, a=0

2-33

So, the slope of $v - s$ graph $= \dfrac{\text{acceleration}}{\text{velocity}}$

Recasting the expression

$$a = \frac{v\,dv}{ds} = \frac{1}{2}\frac{d}{ds}(v^2),$$

we have $a = \frac{1}{2}\frac{d}{ds}(v^2)$. This tells us that the slope of v^2-s graph $= 2$ (acceleration)

Summarizing the above facts, we conclude the following points:

1. Acceleration $=$ speed X slope of the v-s graph $= \frac{1}{2} \times$ slope of v^2-s graph.
2. The slopes can be positive, negative or zero which signify positive, negative or zero accelerations (acceleration pointing in positive, negative directions and zero acceleration vector, respectively).

Example 17 The v-s and $v^2 - s$ graphs are given for two particles. Find the accelerations of the particles *at $s = 0$*.

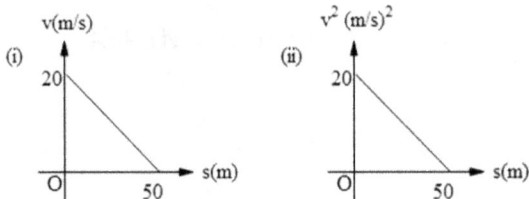

Solution

In figure (i), the acceleration is $a_1 = \frac{v\,dv}{ds}$, where $v = 20$ (at $s = 0$) and $\frac{dv}{ds} = -\frac{2}{5}$
Then, $a_1 = -8$ ms^{-2} Ans.

In figure (ii), the acceleration is

$$a_2 = \frac{1}{2}\frac{d(v^2)}{ds}, \quad \text{where} \quad \frac{d(v^2)}{ds} = -\frac{2}{5}$$

$$\Rightarrow a_2 = -0.2 \text{ ms}^{-2} \text{ Ans.}$$

(d) **Acceleration–displacement graph**

By using the formula $v\,dv = \vec{a}\cdot\vec{ds}$ and integrating both sides, we have $\int_{v_0}^{v} v\,dv = \int_{s_1}^{s_2} \vec{a}\cdot\vec{ds}$. It gives the speed

$$v^2 = v_0^2 + 2\int_{s_1}^{s_2} \vec{a}\cdot\vec{ds}$$

As $\int_{s_1}^{s_2} \vec{a}\cdot\vec{ds}$ represents the area under the a–s graph, we have

$v = \sqrt{v_0^2 + 2(\text{area under } a - s \text{ graph})}$

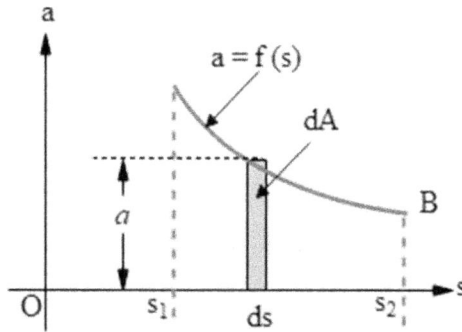

The area under a-s graph $=2$ (change
in kinetic energy) per unit mass

Example 18 A particle moves in a straight line with an initial speed of 5 m s^{-1} when
$s = 0$. Referring to the acceleration–displacement graph for the speeding boat, find
its speed when it pases a raft at a distance of 40 m from the starting point.

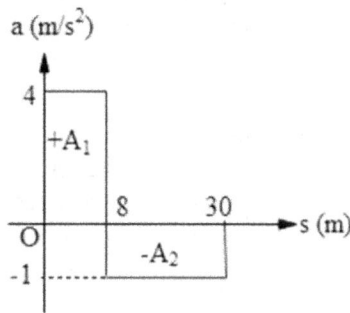

Solution
The velocity of the particle is

$$v = \sqrt{v_0^2 + 2(A)}$$

where,

$$A = 8 \times 4 - 22 \times 1 = 10$$

and

$$v_0 = 5.$$

This gives $v = \sqrt{v_0^2 + 2(A)} = \sqrt{5^2 + 2(10)} = 3\sqrt{5}$ Ans.

Problem 1 The velocity of a particle varies with time as shown in the following
diagrams (graphs).

Give the comments on the above graphs.

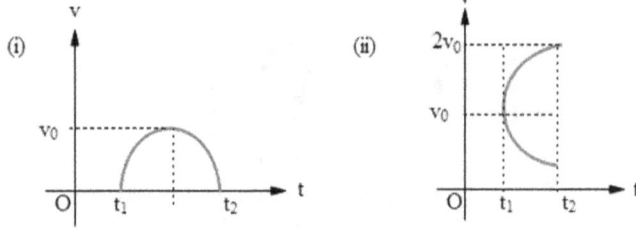

Solution

In graph (i), at two times the particle can have the same velocity v except at the top. Hence, this graph is valid and the area under this v–t graph is equal to the displacement of the particle.

In graph (ii), at any time t we have two velocities which is not possible. Hence, this graph is invalid. We cannot have two velocities at an instant. This means that a particle cannot move in two different directions with two different speeds simultaneously. However, we can have the same velocity at two different instants, i.e., an oscillating particle can have the same velocity at each time interval equal to the time period of oscillation.

Problem 2 If you walk 10 m due east during 20 seconds, then 50 m 37° south of west during 40 seconds, finally climb a pillar of height 20 m during 2 minutes, find your (a) displacement, (b) average velocity, (c) distance covered, (d) average speed, over the total journey of 3 minutes. Assume that the ground is in the x–z plane and x-direction represents east.

Solution

(a) Your net or total displacement is

$$\vec{s} = \vec{s_1} + \vec{s_2} + \vec{s_3},$$

where $\vec{s_1} = 10\hat{i}$ m, $\vec{s_2} = (-50\cos 37°\,\hat{i} + 50\sin 37°\,\hat{k})$ m $= (-40\hat{i} + 30\hat{k})$ m and $\vec{s_3} = 20\hat{j}$ m.

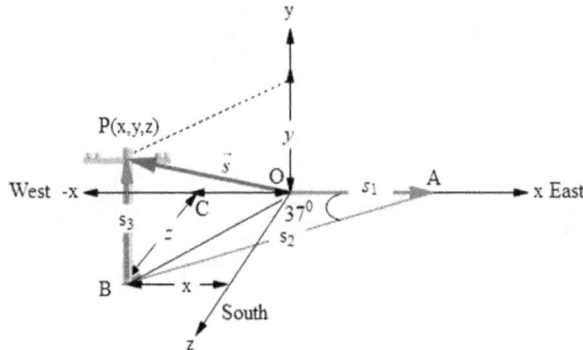

Substituting $\vec{s}_1$, $\vec{s}_2$ and $\vec{s}_3$, we have

$$\vec{s} = \left[-30\hat{i} + 20\hat{j} + 30\hat{k}\right] \text{m Ans.}$$

(b) The total time interval is 3 minutes or 180 seconds. So, the average velocity is

$$\vec{v}_{av} = \frac{\vec{s}}{t} = \frac{-30\hat{i} + 20\hat{j} + 30\hat{k}}{3 \times 60} = \left(-\frac{\hat{i}}{6} + \frac{\hat{j}}{9} + 30\frac{\hat{k}}{6}\right) \text{m s}^{-1} \text{ Ans.}$$

(c) The total distance is

$$D = OC + CB + BP = \left(30 + 20 + 30\right) = 80 \text{ m Ans.}$$

(d) So, the average speed is

$$u_{av} = \frac{D}{t} = \frac{80}{180} = \frac{4}{9} \text{ m s}^{-1} \text{ Ans.}$$

Problem 3

(a) A particle moves in a straight line in such a way that the average velocity over a time t from starting is equal to its instantaneous velocity. How does the particle move?

(b) If $x = be^{at}$, find the ratio of the average velocity over a time interval t measured from starting and instantaneous velocity at any point of time t.

(c) If $x = A \sin\omega t$, find the ratio of the average velocity over a time interval t measured from starting and instantaneous velocity at any point of time t.

Solution

(a) It is given that

$$u_{av} = v$$

$$\Rightarrow \frac{x}{t} = \frac{dx}{dt}$$

Separating the variables and integrating,

$$\int_{x_0}^{x} \frac{dx}{x} = \int_{t_0}^{t} \frac{dt}{t}$$

Evaluating the integrals,

$$\ln\frac{x}{x_0} = \ln\frac{t}{t_0}$$

So, the equation of motion is given as

$$x = \frac{x_0 t}{t_0} + C,$$

where C is a constant.

Then, the particle must move with a constant velocity

$$v = \frac{dx}{dt} = \frac{x_0}{t_0}$$

It must be a one-directional uniform motion. Ans.

(b) If $x = be^{at}$, the instantaneous velocity is

$$v = dx/dt = abe^{at}$$

The average velocity is

$$v_{av} = x/t = \frac{be^{at}}{t}$$

The ratio of the two velocities at any point of time t, we have

$$\frac{v_{av}}{v} = \frac{\frac{be^{at}}{t}}{abe^{at}} = \frac{1}{a}$$

Then the required time is

$$t = 1/a \text{ Ans.}$$

(c) If $x = A \sin\omega t$, the instantaneous velocity is

$$v = dx/dt = A\omega \cos\omega t$$

The average velocity is

$$v_{av} = x/t = \frac{A \sin\omega t}{t}$$

The ratio of the two velocities at any point of time t, we have

$$\frac{v_{av}}{v} = \frac{\frac{A \sin\omega t}{t}}{\omega A \cos\omega t} = \frac{\tan(\omega t)}{\omega t} \text{ Ans.}$$

Problem 4 In (a) a square cut, (b) straight drive, let the speed of a cricket ball change from 30 to 40 m s^{-1} during the time of its contact $\Delta t = 0.01$ s with the bat. Find the magnitude of average acceleration of the ball during the shot

Solution

(a) In the square cut, the ball deflects by a right angle (90-degrees). The average acceleration is

$$a = \frac{|\Delta\vec{v}|}{\Delta t} = \frac{\sqrt{v_1^2 + v_2^2 + 2v_1 v_2 \cos\theta}}{\Delta t}$$

$$= \frac{\sqrt{(30)^2 + (40)^2 + 2(30)(40)\cos 90°}}{0.01}$$

$$= 50 \times 10^2 \text{ ms}^{-2} \text{ Ans.}$$

(b) In the straight drive, the ball deflects by 180 degrees, reversing its velocity during the shot. Then, the net change in velocity

$$= (-40) - (+30) = -70 \text{ ms}^{-1}$$

So, the average acceleration is (display eqn)

$$a = \frac{70}{0.01} = 7000 \text{ ms}^{-2} \text{ Ans.}$$

Problem 5 A particle retards from a velocity v_0 while moving in a straight line. If the magnitude of deceleration is directly proportional to the square root of the speed of the particle, find its average for the total time of its motion.

Solution

The average velocity

$$v_{av} = \frac{s}{t} = \frac{\text{Total displacement}}{\text{Total time}}$$

Use of $v\, dv = a\, ds$: Let us calculate the displacement by substituting $a = -\alpha\sqrt{v}$ (for retardation) in $v\, dv = a \cdot ds$. Then, we have

$$v\, dv = (-\alpha\sqrt{v})ds.$$

$$\Rightarrow \sqrt{v}\, dv = -\alpha\, ds$$

When the particle slows down from v_0 to 0, it covers a distance s. So, we can write

$$\int_{v_0}^{0} \sqrt{v}\, dv = -\int_{0}^{s} \alpha\, ds$$

$$\Rightarrow s = \frac{2v_0^{3/2}}{3\alpha}$$

Use of $a = \frac{dv}{dt}$; By substituting $a = -\alpha\sqrt{v}$, we have

$$-\alpha\sqrt{v} = \frac{dv}{dt}$$

$$\Rightarrow \frac{dv}{\sqrt{v}} = -\alpha dt$$

If the particle takes a time t to stop, integrating both sides, we have

$$\int_{v_0}^{0} \frac{dv}{\sqrt{v}} = -\alpha \int_{0}^{t} dt$$

$$\Rightarrow t = \frac{\sqrt{v_0}}{\alpha}$$

Finally, substituting s and t in $v_{av} = \frac{s}{t}$, we have

$$v_{av} = \frac{2v_0}{3} \text{ Ans.}$$

Problem 6 A particle moves along the x-axis with acceleration $a = 6(t - 1)$ where t is in seconds. If the particle is initially at the origin and it moves along the positive x-axis with $v_0 = 2$ m s^{-1}, discuss the nature of motion of the particle by drawing x–t, v–t and $a - t$ graphs.

Solution

Assuming $t > 1$ we find a is +ve which is given by $a = +6(t - 1)$.

Let the particle have velocity +ve as shown in the figure below.

Here, $s = +x$, $v = +v$, $a = +6(t - 1)$.

Putting these values in

$$a = \frac{dv}{dt},$$

we have $\dfrac{dv}{dt} = 6(t - 1)$

When velocity changes from v_0 to v during time t, integrating both sides with respect to time, we have

$$\int_{v_0}^{v} dv = \int_{0}^{t} 6(t - 1)dt = 3t^2 - 6t$$

Substituting $v_0 = 2$ m s^{-1}, we have

$$v = \frac{dx}{dt} = 3t^2 - 6t + 2$$

Again, integrating both sides, we have

$$\int_{0}^{x} dx = \int_{0}^{t} (3t^2 - 6t + 2)dt$$

$$\Rightarrow x = t^3 - 3t^2 + 2t = t(t - 1)(t - 2)$$

(i) x_1

$-x_2$ $1-1/\sqrt{3}$ $s=-ve$

(ii) $1+1/\sqrt{3}$

$+2$

-1 $s=-ve$

(iii)

$+1$

-1

(iv)

$t=0$ v

$t=1+1/\sqrt{3}$ $t=1-1/\sqrt{3}$ x

$t=1$

$t=2$

The particle moves along the x-axis

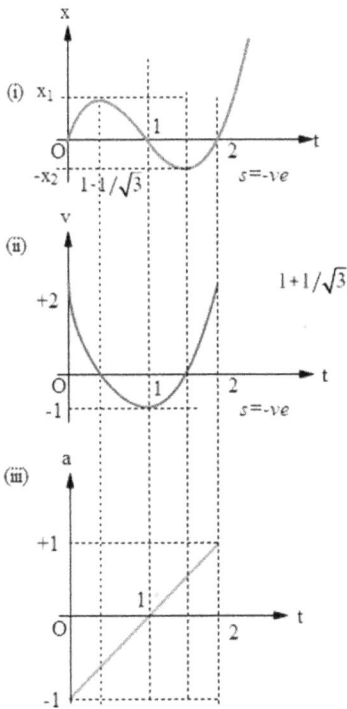

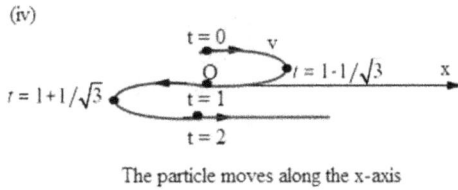

We can see that $x = 0$ at $t = 0,1$ and 2.; $dx/dt = v$ is zero when $t = 1 - 1/\sqrt{3}, 1 + 1/\sqrt{3}$. The velocity $v + 2$ m s^{-1} at $t = 0$, this means that the particle moves to the right at $t = 0$ and stops instantaneously at time $t = t = 1 - 1/\sqrt{3}$ and then moves back to the left. At $t = 1$ s, the velocity becomes maximum which is equal to -1 m s^{-1}. At $t = 1 + 1/\sqrt{3}$, it will stop momentarily and then move to the right increasing its speed. It passes through the origin at $t = 2$ s and thereafter it will move to the right, as shown in the above graphs. Ans.

Problem 7 A driver applies the brakes of his car while making with a speed of 72 km h^{-1} on a straight road through a distance of 48 m. (a) For what time does the driver apply the brakes? Assume the retardation due to braking is 4 m s^{-2}. (b) Find the average velocity of the car during the braking.

Solution

(a) Let $t =$ the time of braking. Putting $s = +48$ m $a = -4$ m s^{-2} and $u = 72$ km h$^{-1} = 20$ m s^{-1} in the formula $s = ut + \frac{1}{2}at^2$, we have

$$48 = 20t + \frac{1}{2}(-4)t^2$$

$$\Rightarrow t^2 - 10t + 24 = 0$$

Solving the quadratic equation, we have

$$t = \frac{(10) \pm \sqrt{(-10)^2 - 4(24)(1)}}{2} \text{ s}$$

$$\Rightarrow t = 4, 6 \text{ s}$$

(b) Using the expression $v = u + at$ and putting $v = 0$, $u = 20$ m s^{-1} and $a = -4$ m s^{-2}, we have $t = 5$ s. After this time the car will remain stationary. So, we should disregard the value of time $t = 6$ s. Then, the average velocity is $v_{av} = (20 + 0)/2 = 10$ m s^{-1}. Ans.

Problem 8 A motorcyclist at $t = 0$, situated at a distance $s = 12$ m behind a car moves with a velocity $v = 8$ m s^{-1}. If the car starts accelerating from rest with $a = 2$ m s^{-2} at $t = 0$ when do they meet? Justify your answer.

Solution

Let them meet after a time t. During this time the motorcycle covers a distance x_m, say.

$$x_m = vt \tag{2.11}$$

During the time t, the car moves through a distance

$$x_c = \frac{1}{2}at^2 \tag{2.12}$$

From the figure, we have

$$x_m - x_c = 12 \tag{2.13}$$

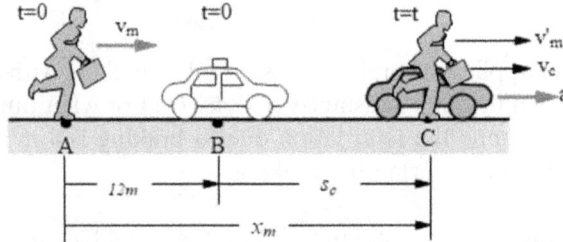

Substituting x_m from equation (2.11) and x_c from equation (2.12) in equation (2.13), we have

$$vt - \frac{1}{2}at^2 = 12$$

Then substituting $v = 8$ m s^{-1} and $a = 2$ m s^{-2}, we have

$$t^2 - 8t + 12 = 0$$

This yields two real values of time, i.e., $t = 2$ and 6 s. Ans.

At $t = 2$ a motorcycle overtakes the car and at $t = 6$, the car overtakes the motorcycle. During first two seconds the motorcycle approaches the car. At $t = 2$ s, the motorcycle overtakes the car. Then it moves away from the car till the car acquires a velocity equal to that of the motorcycle at $t = 4$ s. Then the distance of their separation will gradually decrease as the car moves faster than the motorcycle after 4 s. In consequence, at $t = 6$ s the car overtakes the motorcycle. After that, the car will take a lead leaving the motorcycle behind it.

Problem 9 A particle is projected vertically up with a speed of 10 m s^{-1}. Find the time after which it will pass through a point (a) 3.2 m above (b) 15 m below the point of projection.

Solution

(a) You use $s = ut + \frac{1}{2}at^2$, where $s = +3.2$, $u = +10$

$$\Rightarrow 3.2 = 10t + \frac{1}{2}(-10)t^2$$

$$\Rightarrow 5t^2 - 10t + 3.2 = 0$$

$$\Rightarrow t = 0.4 \text{ and } 1.8 \text{ s Ans.}$$

(b) You use $s = ut + \frac{1}{2}at^2$, where $s = -15$, $u = +10$ and $a = -10$

$$\Rightarrow -15 = 10t + \frac{1}{2}(-10)t^2$$

$$\Rightarrow t^2 - 2t - 3 = 0$$

$$\Rightarrow t = 3 \text{ s Ans.}$$

N.B: 1. If the given point is above the point of projection the particle passes the point twice, once during ascent and again during descent.
2. If the given point is below the point of projection the particle passes the point once.

Problem 10 A ball is projected vertically up such that it passes through a fixed point after time t_1 and t_2. Find the:

    (a) height at which the point is located with respect to the point of projection and speed of projection of the ball,

    (b) velocity of the ball at the time of passing through the point P,

    (c) maximum height reached by the ball relative to the:

        (i) point of projection O,

        (ii) point P under consideration,

    (d) average velocity of the ball during the motion from O to P for the time t_1 and t_2, respectively,

    (e) average speed of the ball during the motion from O to P for the time t_1 and t_2, respectively.

Solution:

Method 1

    (a) Let the ball be projected up with a velocity v_0. It passes through the point at $t = t_1$ during its ascent and at $t = t_2$ during its descent.

        For the motion of the ball from O to P

$$s = +h, \quad v_0 = +v_0, \quad a = -g \quad \text{and} \quad t = t_1 \text{ and } t_2.$$

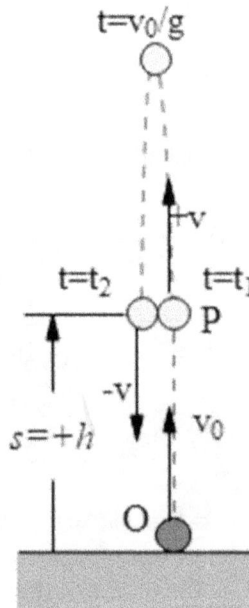

Substituting the above values in

$$s = v_0 t + \frac{1}{2}at^2, \quad \text{where } h = vt_1 - \frac{1}{2}gt_1^2 = vt_2 - \frac{1}{2}gt_2^2$$

Solving the above equation, we have

$$h = \frac{gt_1t_2}{2} \text{ and } v = \frac{g(t_1 + t_2)}{2} \text{ Ans.}$$

Method 2
For the motion of the ball from O to P

$$s = +h, \ v_0 = v, \ a = -g \text{ and } t = t_1 \text{ and } t_2.$$

Substituting the above values in

$$s = v_0t + \frac{1}{2}at^2,$$

we have

$$h = vt - \frac{1}{2}gt^2$$

This yields a quadratic equation $t^2 - \frac{2v}{g} + \frac{2h}{g} = 0$. If this equation is satisfied for two roots t_1 and t_2, we have $t_1t_2 = \frac{2h}{g}$ and $t_1 + t_2 = \frac{2v}{g}$ (because in the quadratic equation $Ax^2 + Bx + C = 0$, $x_1x_2 = \frac{C}{A}$ and $x_1 + x_2 = \frac{-B}{A}$). Ans.

(b) The velocity of the stone at P is

$$v = v_0 - gt = g(t_1 + t_2)/2 - gt_1 = g(t_1 - t_2)/2 \text{ Ans.}$$

(c) (i) The maximum height from O is

$$h_{max} = v_0t - \frac{1}{2}gt^2 = g\frac{(t_1 + t_2)}{2}\frac{(t_1 + t_2)}{2} - \frac{1}{2}g\left\{\frac{(t_1 + t_2)}{2}\right\}^2$$
$$= \frac{g}{8}(t_1 + t_2)^2 \text{ Ans.}$$

(ii) The maximum height from P is

$$h'_{max} = vt - \frac{1}{2}gt^2 = g\left\{\frac{(t_1 - t_2)}{2}\right\}\frac{(t_1 - t_2)}{2} - \frac{1}{2}g\left\{\frac{(t_1 - t_2)}{2}\right\}^2$$
$$= \frac{g}{8}(t_1 - t_2)^2 \text{ Ans.}$$

(d) (i) The average velocity of the ball during the motion from O to P for the time t_1 is
$$v_{av} = h/t_1 = \frac{gt_1t_2}{2t_1} = \frac{gt_2}{2} \text{ (up) Ans.}$$

(ii) The average velocity of the ball during the motion from O to P for the time t_1 is

$$v_{av} = h/t_2 = \frac{g t_1 t_2}{2 t_2} = \frac{g t_1}{2} \text{ (up) Ans.}$$

(e) (i) The average speed of the ball during the motion from O to P for the time t_1 is

$$v_{av} = d_1/t_1 = \frac{g t_1 t_2}{2 t_1} = \frac{g t_2}{2} \text{ Ans.}$$

(ii) The average speed of the ball during the motion from O to P for the time t_1 is

$$v_{av} = d_2/t_2 = (h + 2h'_{max})/t_2$$

Putting $h = \frac{g}{8}(t_1 + t_2)^2$ and $h' = \frac{g}{8}(t_1 - t_2)^2$, we have

$$v_{av} = \frac{g(t_1^2 + t_2^2)}{4 t_2} \text{ Ans.}$$

Problem 11 An elevator of height h moves up with an upward acceleration a. At certain instant $t = t'$ during the motion of the elevator, a bolt loses contact with the ceiling of the elevator. Find the (a) time of fall of the bolt till it strikes the base of the elevator, (b) velocity of the bolt when the bolt strikes the base of the elevator, and (c) velocity of the bolt and elevator when the bolt strikes the base of the elevator.

Solution

(a) Let us assume that, the elevator has a velocity v when the bolt loses contact with the elevator at $t = t_1$. Let the bolt strike the base of the elevator after a time $t = T$, below the point of projection of the bolt. So, by substituting

$$s_b = -y, \ v_0 = v, \ a = -g \text{ and } t = T \text{ in } s = v_0 t + \frac{1}{2} a t^2,$$

we have

$$y = \frac{1}{2} g T^2 - vT \tag{2.14}$$

Likewise, for the elevator, substituting $s_e = +y' = h - y$,

$$v_0 = v, \ a = +a \text{ and } t = T$$

In the equation

$$s = v_0 t + \frac{1}{2} a t^2,$$

we have

$$h - y = vT + \frac{1}{2} a T^2 \qquad (2.15)$$

Adding equations (2.14) and (2.15), we have

$$h = \frac{1}{2}(g + a)T^2$$

$$\Rightarrow T = \sqrt{\frac{2h}{g + a}} \quad \text{Ans.}$$

(b) The velocity of the bolt when the bolt strikes the base of the elevator is

$$v_b = v' - gT = at' - gT = at' - g\sqrt{\frac{2h}{g + a}} \quad \text{Ans.}$$

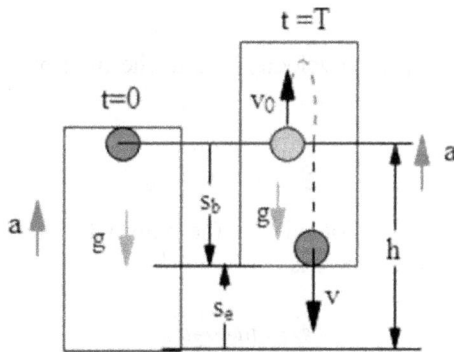

(c) The velocity of the elevator when the bolt strikes the base of the elevator is

$$v_e = aT = a\sqrt{\frac{2h}{g + a}} \quad \text{Ans.}$$

N.B: 1. The time of fall of the bolt does not depend upon the velocity of the elevator at the time of releasing the bolt, given as $T = \sqrt{\frac{2h}{g+a}}$.

2. If the elevator accelerates down, the time of fall of the bolt can be given as $T = \sqrt{\frac{2h}{|g-a|}}$.

Problem 12 A man hangs from a balloon which starts ascending with constant upwards acceleration of $a_0 = 2$ m s^{-2} from the ground. After $t_0 = 4$ s, the man jumps (releases) from the ascending balloon. Find the time measured from the instant of release after which the man touches the ground.

Solution

The height at which the man is released, is

$$h = \tfrac{1}{2}a_0 t_0^2$$

$$= \tfrac{1}{2} \times 2 \times (4)^2 = 16 \text{ m}.$$

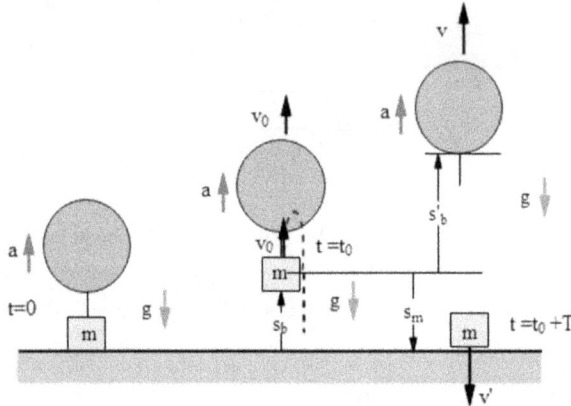

Let the man have an upward velocity v_0 and the time of release given as

$$v_0 = at_0$$

$$= 2 \times 4 = 8 \text{ m s}^{-1}$$

After a time t from the instant of release, the man will reach the ground.

The displacement during time t is

$$\vec{s} = \vec{u}t + \frac{1}{2}\vec{a}t^2$$

Putting $u = +8$, $a = -10$, $s = -16$, we have

$$-16 = 8t + \frac{1}{2} \times (-10)t^2$$

$$\Rightarrow 5t^2 - 8t - 16 = 0$$

$$\Rightarrow t = \frac{8 \pm \sqrt{64 + 320}}{10} = 2.75. \text{ Ans.}$$

N.B.: Just after release, the acceleration of the man is $g\downarrow$ but not $(g + a)\downarrow$ and its velocity is equal to the velocity of the balloon at that instant.

Problem 13 A car moves in a straight line from rest with constant acceleration α for some time then moves with a constant velocity v and constant retardation β. (a) If the total time and distance of journey are T and s, respectively, prove that

$$T = \frac{s}{v} + \frac{v}{2}\left(\frac{1}{\alpha} + \frac{1}{\beta}\right)$$

Solution

Let the car acceleration for a time t_1, move uniformly for a time t_2 and then retard for a time t_3 as shown in the $v - t$ graph.

Then, the total time of motion of the car is

$$T = t_1 + t_2 + t_3 \tag{2.16}$$

The slope of the $v - t$ graph gives

$$\frac{v}{t_1} = \alpha \text{ and } \frac{-v}{t_3} = -\beta \tag{2.17}$$

$$\Rightarrow t_1 + t_3 = \frac{v}{\alpha} + \frac{v}{\beta}$$

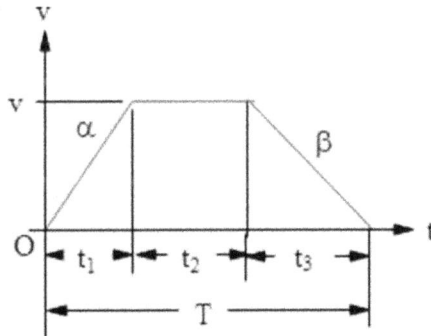

Area under v–t graph gives the total displacement

$$s = \text{Area of the trapezium}$$
$$= \frac{v}{2}(t_2 + T) \tag{2.18}$$

$$\Rightarrow t_2 = \frac{2s}{v} - T$$

Substituting $t_1 + t_3$ from equation(2.17) and t_2 from equation(2.19) in equation(2.16), we have

$$T = \frac{v}{\alpha} + \frac{v}{\beta} - \frac{2s}{v} - T$$

$$\Rightarrow T = \frac{s}{v} + \frac{v}{2}\left(\frac{1}{\alpha} + \frac{1}{\beta}\right) \text{ Ans.}$$

Problem 14 A car moves in a straight line from rest with constant acceleration α for some time then moves with a constant velocity v and constant retardation β. If the total time and distance of journey are T and s, respectively, find the (a) maximum velocity and (b) average velocity of the car.

Solution

(a) Referring to the answer of the last problem, we have

$$T = \frac{s}{v} + \frac{v}{2}\left(\frac{1}{\alpha} + \frac{1}{\beta}\right)$$

Putting $\alpha = \beta = a$, we have

$$T = \frac{s}{v} + \frac{v}{2}\left(\frac{1}{a} + \frac{1}{a}\right) = \frac{s}{v} + \frac{v}{a}$$

$$\Rightarrow T = \frac{s}{v} + \frac{v}{a}$$

$$\Rightarrow avT = v^2 + as$$

$$\Rightarrow v^2 - avT + as = 0$$

$$\Rightarrow v = \frac{aT \pm \sqrt{(aT)^2 - 4as}}{2} \text{ Ans.}$$

(b) The average velocity is

$$v_{av} = \frac{v_{max}}{2} = \frac{v}{2} = \frac{aT \pm \sqrt{(aT)^2 - 4as}}{4}$$

$$= \frac{aT}{4}\left\{1 \pm \sqrt{1 - 4as/(aT)^2}\right\} \text{ Ans.}$$

Problem 15 The velocity–displacement graph for a jet plane on a straight runaway is shown in the figure below. Find the (a) speed and acceleration of the jet plane at $s = 150$ m, (b) $a = f(s)$ and draw the a–v graph.

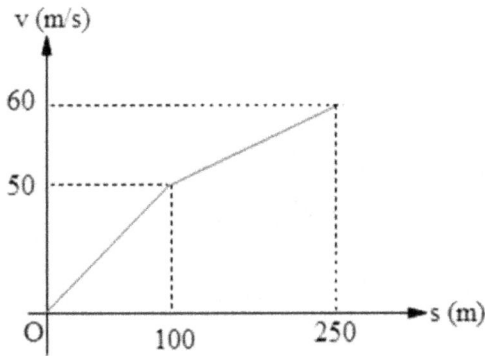

Solution

(a) Referring to the v-s graph, its slope for $100 \leqslant s \leqslant 250$ is equal to

$$\frac{60 - 50}{250 - 100} = \frac{1}{15}$$

The equation of the above graph is given by

$$\frac{v - 50}{s - 100} = \frac{1}{15}$$

$$\Rightarrow v = \frac{650 + s}{15}; \ 100 \leqslant s \leqslant 250$$

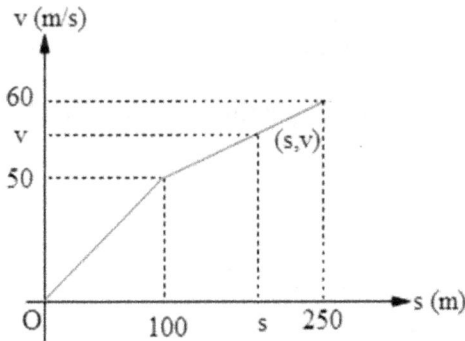

Substituting $s = 150$, we have $v = \frac{160}{3}$ ms^{-1} Ans.

Since,

$$a = \frac{v\,dv}{ds} \quad \text{and} \quad \frac{dv}{ds} = \frac{1}{15} \text{ for } 100 \leqslant s \leqslant 250,$$

we have $a = \frac{32}{9}$ ms^{-2} Ans.

(b) Putting $\frac{dv}{ds} = \frac{1}{2}$ in $a = \frac{v\,dv}{ds}$, we have $a = \frac{v}{2}$; $0 \leqslant s \leqslant 100$

Putting $v = \frac{s}{2}$, we have $a = \frac{s}{4}$; $0 \leqslant s \leqslant 100$

Similarly, putting

$$\frac{dv}{ds} = \frac{1}{15} \text{ in } a = \frac{vdv}{ds},$$

we have $a = \frac{v}{15}$; $100 \leqslant s \leqslant 250$.

Putting $v = \frac{650+s}{15}$; $100 \leqslant s \leqslant 250$, we have

$$a = \frac{650 + s}{225}; \ 100 \leqslant s \leqslant 250 \text{ Ans.}$$

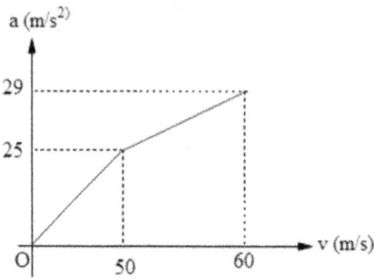

Problem 16 A body slows down such that v^2 is varying linearly with displacement s as shown in the figure below. Assuming rectilinear motion, find the:

(a) acceleration of the body,
(b) speed of the body when it just crosses 100 m,
(c) v–s graph,
(d) time taken to undergo a displacement from A to B.

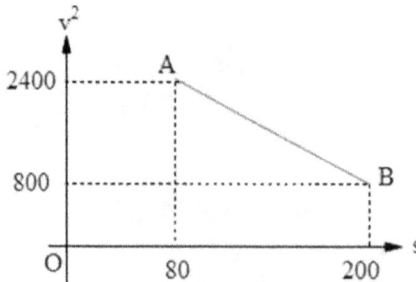

Solution

(a) The slope of v^2–s graph, that is $\frac{d(v^2)}{ds}$ is equal to $\frac{800 - 2400}{200 - 80} = -\frac{40}{3}\text{ms}^{-2}$, obtained from the given as shown in the figure.

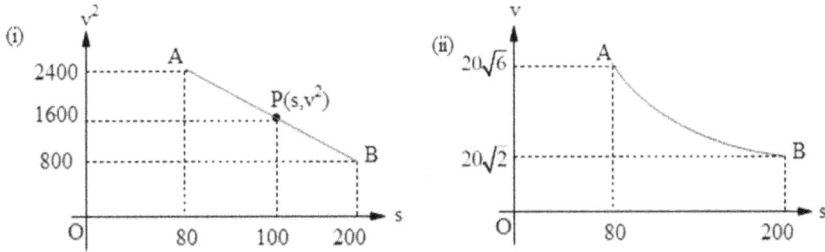

Since, $\frac{d(v^2)}{ds} = 2a$ and $\frac{d(v^2)}{ds} = -\frac{40}{3}$, we have $a = \frac{-20}{3}$ m s^{-2} Ans.

(b) From the graph, $\frac{v^2 - 2400}{200 - 80} = \frac{-40}{3}$. This yields $v = 20\sqrt{2}$ m s^{-1} Ans.

(c) Considering a point $P(s, v^2)$ on the v^2-s graph, the equation of AB is given by $\frac{v^2 - 2400}{s - 80} = -\frac{40}{3}$. It yields $v = \sqrt{\frac{40}{3}(260 - s)}$ as plotted in the above figure.

(d) Putting $v = \sqrt{\frac{40}{3}(260 - s)} = \frac{ds}{dt}$, we have

$$\Rightarrow \sqrt{\frac{40}{3}}\, dt = \frac{ds}{\sqrt{(260 - s)}}$$

$$\Rightarrow \sqrt{\frac{40}{3}} \int_0^t dt = \int_{80}^{200} \frac{ds}{\sqrt{(260 - s)}}$$

$$\Rightarrow \sqrt{\frac{40}{3}}\, t = -2\left(\sqrt{(260 - 200)} - \sqrt{(260 - 80)}\right)$$

$$\Rightarrow \sqrt{\frac{40}{3}}\, t = 2(\sqrt{180} - \sqrt{60})$$

$$\Rightarrow t = 3\left(\sqrt{3/2} - \sqrt{1/2}\right) \text{ s Ans.}$$

Problem 17 Two bodies 1 and 2 are projected simultaneously with velocities v_1 and v_2, respectively. The body 1 is projected vertically up from the top of a cliff of height h and the body 2 is projected vertically up from the bottom of the cliff. Find the time of meeting of the bodies.

Solution

Let the particles meet after a time t. First of all, we choose the point of collision above the top of the cliff.

For the first particle, $s = -s_1$, $v_0 = v_1$, $a = -g$

Then, applying the equation $s = ut + \frac{at^2}{2}$, we have

$$s_1 = -v_1 t - \frac{1}{2}gt^2 \qquad (2.19)$$

For the second particle, $s = s_2$, $v_0 = v_2$, $a = -g$

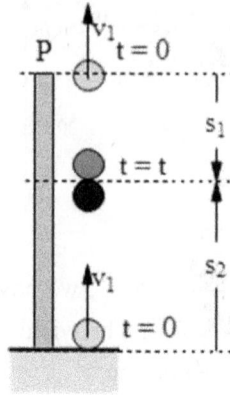

Then, applying the equation $s = ut + \frac{at^2}{2}$, we have

$$s_2 = v_2t - \frac{1}{2}gt^2 \qquad (2.20)$$

From the last two equations, we have

$$s_1 + s_2 = -v_1t + \frac{1}{2}gt^2 + v_2t - \frac{1}{2}gt^2$$

$$\Rightarrow h = (v_2 - v_1)t$$

$$\Rightarrow t = h/(v_2 - v_1) \text{ Ans.}$$

N.B: Depending upon the value of time, the bodies will meet at different points. For simultaneous projection, the time of meeting is, $t =$ relative distance (h)/initial relative velocity.

Problem 18 A stone is thrown vertically up from a height $h = 40$ m with a velocity of 20 m s^{-1}. Find the time taken to cover half of the total (i) distance and (ii) displacement.

   Solution

   (a) Let the stone strike the ground after a time t. If the extra height attained by the stone from the point of projection is y, the total distance covered till the stone hits the ground is $d = 2h + H$. The half of the distance is $d/2 = h + H/2$ from the ground P. Then the position of the stone from the point of projection O is given as

$$s = H - d/2 = H/2 - h = 60/2 - h \qquad (2.21)$$

the value of h is given as

$$h = v_0^2/2g = 20^2/2(10) = 20 \text{ m} \qquad (2.22)$$

Using the last two equations, we have

$s = 30 - 20 = 10 \text{ m from } 0 \text{ or } 10 = 20 = 30 \text{ m from the highest}$

position Q. Use the equation

$$s = v_0 t + (1/2)at^2$$

$$\Rightarrow -10 = 20t + (1/2)(-10)t^2$$

$$\Rightarrow t^2 - 4t - 2 = 0$$

Solving this quadratic equation, we have

$$t = 2 + \sqrt{6} \text{ s Ans.}$$

(b) Half of the total displacement will be given as $s = -H/2 = -60/2 = -30$ m.

Use the equation

$$s = v_0 t + (1/2)at^2$$

$$\Rightarrow -30 = 20t + (1/2)(-10)t^2$$

$$\Rightarrow t^2 - 4t - 6 = 0$$

Solving this quadratic equation, we have

$$t = 2 + \sqrt{10} \text{ s Ans.}$$

Problem 19
(a) A particle moving in a straight line in one direction, moves with velocity u for 40% of its total time and with a velocity v for rest of time. Find its average velocity over total time.
(b) A particle moving in a straight line in one direction, moves with velocity u for 40% of its total distance and with a velocity v for rest of the distance. Find its average velocity over total distance.

Solution
Let $T = $ total time of journey of the particle. The total distance covered by it during time T is

$$D = u(2T/5) + v(3T/5)$$

Then, the average velocity is

$$v_{av} = D/T = \frac{u(2T/5) + v(3T/5)}{T} = \frac{2u + 3v}{5} \quad \text{Ans.}$$

Let D = total distance of journey of the particle. The total time elapsed by it is

$$T = \frac{2D/5}{u} + \frac{3D/5}{v} = \left(\frac{2}{u} + \frac{3}{v}\right)\frac{D}{5}$$

Then, the average velocity is

$$v'_{av} = D/T = \frac{D}{\left(\frac{2}{u} + \frac{3}{v}\right)\frac{D}{5}} = \frac{5uv}{(3u + 3v)} \quad \text{Ans.}$$

Problem 20 A ball is projected up with velocity v_0 from a horizontal floor. On each collision the velocity decreases by a factor η. Draw the velocity–time graph till the ball strops after so many collisions. Find the total distance covered by the ball (c) Find the time of motion of the ball.

Solution

(a) Just after first, second and third collisions and so on, the velocities of the ball are $\eta v_0, \eta^2 v_0, \eta^3 v_0$ etc. Then, the times of flight before the first collision, between the first and second collisions, second and third collisions and so on, are $t_1 = 2v_0/g$, $t_2 = 2\eta v_0/g$, $t_3 = 2\eta^2 v_0/g$, respectively. Then, the total time of motion is

$$T = t_1 + t_2 + t_3 + \ldots = v_0/g + 2v_0/g + 2\eta v_0/g + 2\eta^2 v_0/g + \cdots.$$

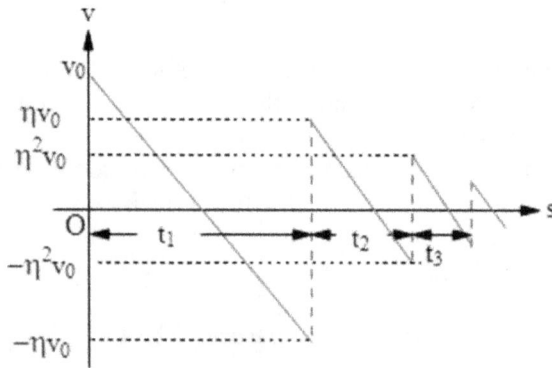

$$\Rightarrow T = v_0/g + 2v_0/g + 2\eta v_0/g + 2\eta^2 v_0/g + \cdots.$$

$$\Rightarrow T = 2v_0/g(1 + \eta + \eta^2 + \ldots.) = 2v_0/\{(1 - \eta)g\} \quad \text{Ans.}$$

(b) Taking the sum of magnitude of area of the v-t graph, the total distance covered by the ball is

$$D = v_0 t_1/2 + \eta v_0 t_2/2 + \eta^2 v_0 t_3/2 + \cdots.$$

$$\Rightarrow D = v_0(t_1 + \eta t_2 + \eta^2 t_3 + \cdots)/2$$

$$\Rightarrow D = v_0(2v_0/g + \eta(2\eta v_0/g) + \eta^2(2\eta^2 v_0/g) + \cdots)/2$$

$$\Rightarrow D = (v_0^2/g)(1 + \eta^2 + \eta^4 + \cdots)$$

$$\Rightarrow D = v_0^2/\{(1 - \eta^2)g\} \text{ Ans.}$$

Problem 21 n water drops are falling from a water tap with negligible speed from a height h above ground level. When the first drop touches the ground, the nth drop just leaves the tap. Find the distance between the first and the $(n-1)$th drop when the nth drop is about to leave the tap.

Solution

Since the first drop touches the ground when the nth drop just leaves the tap, the time interval between two consecutive drops is equal to $T/(n-1)$, where T = time of fall of each drop = $(2\,h/g)^{1/2}$. The distance covered by the first drop = h, as the second drop gets a time $T/(n-1)$, the distance covered by the second drop is

$$s_2 = gt'^2/2$$

Putting $t' = T/(n-1)$, we have $s_2 = gT^2/(n-1)^2$. Putting T $T^2 = 2h/g$, we have

$$s_2 = h/(n-1)^2$$

Then, the distance between the $(n-1)$th drop and the nth drop is given as

$$s_2 = h - s_2 = h - h/(n-1)^2 = h\{1 - 1/(n-1)^2\} \text{ Ans.}$$

IOP Publishing

Problems and Solutions in Particle Mechanics

Pradeep Kumar Sharma

Chapter 3

Motion in two and three dimensions

3.1 Introduction

In the last chapter, you learned about the motion of a particle in a straight line, for example, an oscillating block connected with a spring and the straight-line motion of a body released from rest under gravity etc.

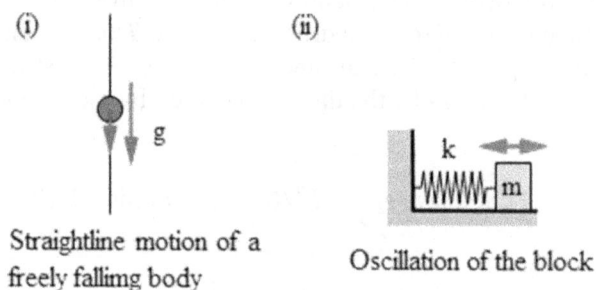

(i)

Straightline motion of a freely fallimg body

(ii)

Oscillation of the block

When a particle moves in a curve in a plane (2D) or in any arbitrary path (curve) in three-dimensional space (3D), it is known as curvilinear motion. Circular motion, projectile motion, elliptical motion etc, are two-dimensional motion.

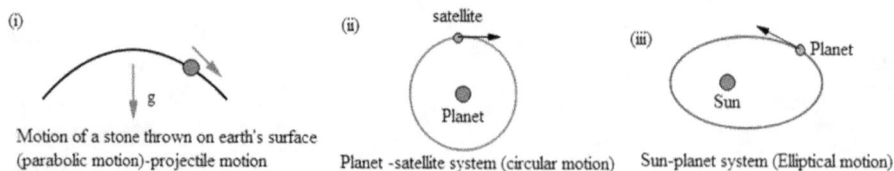

(i)

Motion of a stone thrown on earth's surface
(parabolic motion)-projectile motion

(ii) satellite

Planet

Planet -satellite system (circular motion)

(iii) Planet

Sun

Sun-planet system (Elliptical motion)

doi:10.1088/978-0-7503-6442-3ch3

The motion of birds, insects, dust particles in air, of a charged particle in a magnetic field etc, are a few examples of three-dimensional motion.

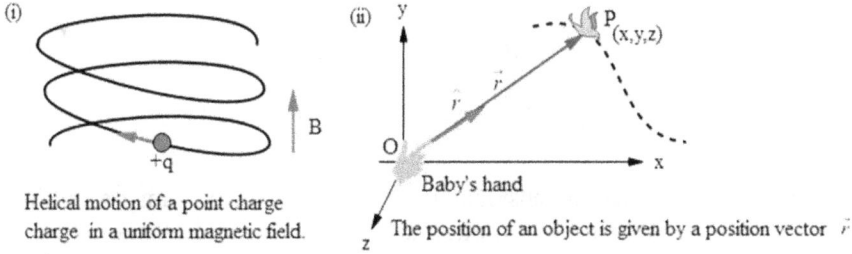

(i)

B

+q

Helical motion of a point charge
charge in a uniform magnetic field.

(ii)

$P_{(x,y,z)}$

$\hat{r}$

$\vec{r}$

O

x

z

Baby's hand

The position of an object is given by a position vector $\vec{r}$

Generally, we think of curvilinear motion as the superposition of rectilinear (one-dimensional) motions of a particle in x-, y- and z-directions. Using the idea of vector addition of displacement, velocity and acceleration we will discuss mainly projectile motion, circular motion and relative motion in this chapter.

3.2 Position, velocity and acceleration

As discussed earlier, when a particle moves, its position vector changes. In other words, the coordinates of the particle change with time. So, at any instant, the position, displacement, average and instantaneous velocity and acceleration of the particle moving in an arbitrary path can be given as the superposition (combination or resultant) of the respective vector components as follows:

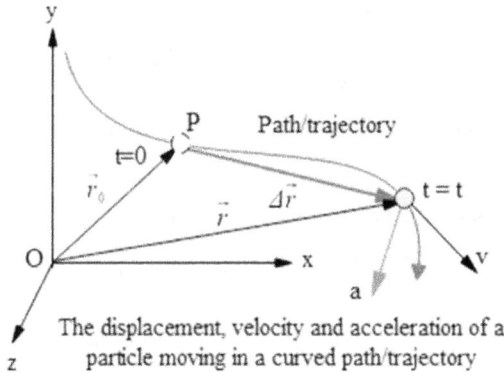

P

Path/trajectory

t=0

$\vec{r_0}$

$\vec{r}$

$\Delta\vec{r}$

t = t

O

x

a

v

z

The displacement, velocity and acceleration of a
particle moving in a curved path/trajectory

$$\vec{r} = x\hat{i} + y\hat{j} + z\hat{k} \tag{3.1}$$

$$\Delta\vec{r} = \Delta x\hat{i} + \Delta y\hat{j} + \Delta z\hat{k} \tag{3.2}$$

$$\vec{v}_{av} = \frac{\Delta x}{\Delta t}\hat{i} + \frac{\Delta y}{\Delta t}\hat{j} + \frac{\Delta z}{\Delta t}\hat{k} \tag{3.3}$$

$$\vec{v} = \frac{dx}{dt}\hat{i} + \frac{dy}{dt}\hat{j} + \frac{dz}{dt}\hat{k} \tag{3.4}$$

$$\vec{a}_{av} = \frac{\Delta v_x}{\Delta t}\hat{i} + \frac{\Delta v_y}{\Delta t}\hat{j} + \frac{\Delta v_z}{\Delta t}\hat{k} \tag{3.5}$$

$$\vec{a} = \frac{dv_x}{dt}\hat{i} + \frac{dv_y}{dt}\hat{j} + \frac{dv_z}{dt}\hat{k} \tag{3.6}$$

N.B: The two- and three-dimensional motion is the vector sum or resultant or superposition of one-dimensional motion along the x-, y- and z-axes.

Example 1 A particle moves so that its coordinates vary with time as $x = a \sin \omega t$, $y = a \cos \omega t$ and $z = bt^3$. Find the initial:
   (a) position,
   (b) velocity,
   (c) acceleration of the particle.

Solution

   (a) The initial position is

$$\vec{r_0} = [a \sin \omega t \hat{i} + a \cos \omega t \hat{j} + bt^3 \hat{k}]_{t=0} = a\hat{j} \text{ Ans.}$$

   (b) The speed of the particle at $t = 0$ is

$$v_0 = \left[\frac{dx}{dt}\hat{i} + \frac{dy}{dt}\hat{j} + \frac{dz}{dt}\hat{k} \right]_{t=0}$$
$$= [a\omega \cos \omega t \hat{i} - a\omega \sin \omega t \hat{k} + 3bt^2 \hat{k}]_{t=0} = a\omega \text{ Ans.}$$

   (c) The initial acceleration is

$$\vec{a_0} = \left[\frac{dv_x}{dt}\hat{i} + \frac{dv_y}{dt}\hat{j} + \frac{dv_z}{dt}\hat{k} \right]_{t=0}$$
$$= [-a\omega^2 \sin \omega t \hat{i} - a\omega^2 \cos \omega t \hat{j} + 6bt\hat{k}]_{t=0} \text{ Ans.}$$
$$= -a\omega^2 \hat{j} .$$

3.3 Tractory and its equations

The path traced by a particle is called trajectory. Let the coordinates of a particle moving in a plane vary with time as $x = f(t)$ and $y = g(t)$. By eliminating time t between these equations, we can express y as the function of x, that is, $y = \phi(x)$ which is known as the equation of trajectory.

Example 2 The position of a particle is given as

$$\vec{r} = at\,\hat{i} + bt^2\hat{j}$$

Find the equation of trajectory of the particle.
Solution
The coordinates of the particle are

$$x = at \tag{3.7}$$

$$y = bt^2 \tag{3.8}$$

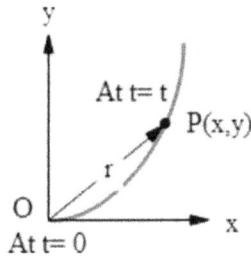

Substituting $t = \frac{x}{a}$ from equation (3.7) in equation (3.8), we have

$$y = b\left(\frac{x}{a}\right)^2$$

$$\Rightarrow y = \frac{bx^2}{a^2}$$

Since this equation is a parabola, the trajectory is parabolic.

3.4 Uniform accelerated motion in a plane

If a particle moves in the x–y-plane with a constant acceleration $\vec{a}$, for the superposition of two one-dimensional motions along the x- and y-axes, we can write the following expressions:

$$v_x = v_{0x} + a_x t$$

$$v_y = v_{0y} + a_y t$$

$$\Delta x = v_{0_x} t + \frac{1}{2}a_x t^2$$

$$\Delta y = v_{0_y} t + \frac{1}{2} a_y t^2$$

$$v_x^2 = v_{0x}^2 + 2a_x \Delta x$$

$$v_y^2 = v_{0y}^2 + 2a_y \Delta y$$

Alternately, putting $\vec{v} = v_x \hat{i} + v_y \hat{j}$, $\vec{v}_0 = v_{0x}\hat{i} + v_{0y}\hat{j}$, $\vec{a} = a_x\hat{i} + a_y\hat{j}$ and $\vec{s} = \Delta\vec{r} = \Delta x\hat{i} + \Delta y\hat{j}$ in the formulae, $\vec{v} = \vec{v}_0 + \vec{a}\,t$, $\vec{s} = \vec{v}_0 t + \frac{1}{2}\vec{a}\,t^2$ and $v^2 = v_0^2 + 2\,\vec{a}.\,\Delta\vec{r}$ and separating the variables we can get the above six equations.

Example 3 A ball is projected horizontally in air with a velocity v_0 so that it moves with a constant horizontal acceleration due to air flow. Taking the gravitational acceleration into account, find the (a) velocity and (b) displacement as the function of time till it strikes the ground.

Solution

(a) It is given that $a_x = a$, $a_y = -g$ and $v_{0x} = v_0$ and $v_{0y} = 0$. The x-component of velocity components is

$$v_x = v_{0x} + a_x t$$

$$\Rightarrow v_x = v_0 + at$$

The y-component of velocity components is

$$v_y = v_{0y} + a_y t$$

$$\Rightarrow v_y = 0 + (-g)t$$

$$\Rightarrow v_y = -gt$$

Then, the velocity of the particle is

$$\vec{v} = v_x\hat{i} + v_y\hat{j} = (v_0 + at)\hat{i} - gt\hat{j} \text{ Ans.}$$

(b) The x-component of displacement is given as

$$\Delta x = x = v_{0x}t + \frac{1}{2}a_x t^2$$

$$\Rightarrow x = v_0 t + \frac{1}{2}at^2.$$

The y-component of displacement is given as

$$\Delta y = y = v_{0y}t + \frac{1}{2}a_y t^2$$

$$\Rightarrow y = -\frac{1}{2}gt^2.$$

Then, the net displacement is

$$\vec{s} = \vec{r} = \left(v_0 t + \frac{1}{2}at^2\right)\hat{i} - \frac{1}{2}gt^2 \quad \text{Ans.}$$

3.5 Projectile motion

3.5.1 Projection onto horizontal plane

3.5.1.1 Velocity as the function of time

When we throw an object near Earth's surface, the acceleration due to gravity remains uniform.

Here we need to neglect the viscosity and buoyancy of air on the objects. Also, the flow of air is ignored. So, the only force acting on the body is gravity and each object falls freely with a constant downward acceleration $g = 9.8 \text{ m s}^{-2}$.

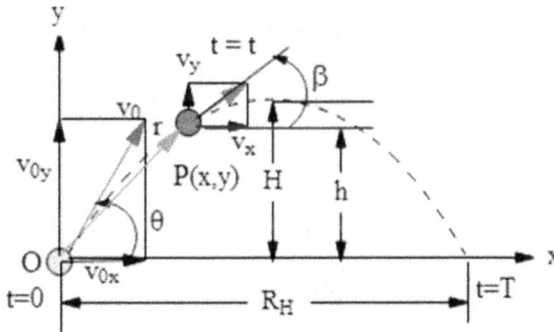

Let the object be thrown from O with a velocity $\vec{v_0}$ at an angle θ with the horizontal. Then, using the component method, the velocity of projection can be given as

$$\vec{v_0} = \vec{v}_{0x} + \vec{v}_{0y} = v_0 \cos\theta\hat{i} + v_0 \sin\theta\hat{j}$$

$$\Rightarrow \vec{v_0} = v_0(\cos\theta\hat{i} + \sin\theta\hat{j}) \tag{3.9}$$

Since there is zero horizontal force, $a_x = 0$ (no air drag), the total acceleration is equal to the vertically downward gravitational acceleration, which is given as

$$\vec{a} = \vec{a_y} = -g\hat{j} \tag{3.10}$$

The velocity of the object after a time $t = t$ is

$$\vec{v} = \vec{v_0} + \vec{a}\,t \tag{3.11}$$

Using the last three equations, we have

$$\vec{v} = v_0(\cos\theta\hat{i} + \sin\theta\hat{j}) - gt\hat{j}$$

$$\Rightarrow \vec{v} = v_x\hat{i} + v_y\hat{j} = v_0\cos\theta\hat{i} + (v_0\sin\theta - gt)\hat{j} \tag{3.12}$$

So, the horizontal and vertical velocities are given as

$$v_x = v_0\cos\theta_0 \tag{3.13}$$

$$v_y = v_0\sin\theta_0 - gt \tag{3.14}$$

So, the magnitude of velocity is given as

$$v = \sqrt{(v_0\cos\theta_0)^2 + (v_0\sin\theta_0 - gt)^2}$$

$$v = \sqrt{v_0^2 - 2v_0gt\sin\theta_0 + g^2t^2} \tag{3.15}$$

The angle made by the velocity vector with horizontal is given as

$$\phi = \tan^{-1}(v_y/v_x) = \tan^{-1}\left(\frac{v_0\sin\theta_0 - gt}{v_0\cos\theta_0}\right). \tag{3.16}$$

3.5.1.2 Displacement
The displacement of the object after a time $t = t$ is

$$\vec{s} = \vec{v_0}t + \frac{1}{2}\vec{a}\,t^2 \tag{3.17}$$

Using equations (3.9), (3.10) and (3.17), we have

$$\vec{s} = v_0t(\cos\theta\hat{i} + \sin\theta\hat{j})t - \frac{1}{2}gt^2\hat{j}$$

$$\vec{s} = x\hat{i} + y\hat{j} = v_0t\cos\theta\hat{i} + \left(v_0t\sin\theta - \frac{1}{2}gt^2\right)\hat{j} \tag{3.18}$$

So, the displacements along the x- and y-axes are given as follows:

$$x = v_{0x}t = (v_0\cos\theta_0)t \tag{3.19}$$

$$y = v_0 t \sin \theta_0 - \frac{1}{2} g t^2 \tag{3.20}$$

The above equations tell us that projectile motion is the superposition of uniform motion along the horizontal and uniformly accelerated motion (free fall) along the vertical.

3.5.1.3 Velocities are given as the function of y
We know that for uniform accelerated motion,

$$v^2 = v_0^2 + 2\vec{a} \cdot \vec{s} \tag{3.21}$$

Using equations (3.10), (3.16) and (3.19), we have

$$v^2 = v_0^2 + 2(-g\hat{j}) \cdot (x\hat{i} + y\hat{j})$$

$$v = \sqrt{v_0^2 - 2gy} \tag{3.22}$$

However, the horizontal and vertical velocities are given as the function of y as follows

$$v_x = v_0 \cos \theta_0 \tag{3.23}$$

$$v_y^2 = (v_0 \sin \theta_0)^2 + 2(-g)(y)$$

$$\Rightarrow v_y = \sqrt{v_0^2 \sin^2 \theta_0 - 2gy}. \tag{3.24}$$

3.5.1.4 Time of flight
Putting $y = 0$ when $t = T$ in equation (3.18), we have

$$T = \frac{2v_0 \sin \theta_0}{g} \tag{3.25}$$

Maximum height: Putting $y = H$ and the value of T in equation (3.12) or putting $v_y = 0$ in equation (3.27), we obtain

$$H = \frac{v_0^2 \sin^2 \theta_0}{2g}. \tag{3.26}$$

3.5.1.5 Horizontal range
Putting $T = \frac{2v_0 \sin \theta_0}{g}$ in equation (3.11), for $x = R$, we have

$$R = (v_0 \cos \theta_0)\left(\frac{2v_0 \sin \theta_0}{g}\right)$$

$$\Rightarrow R = \frac{v_0^2 \sin 2\theta_0}{g}. \tag{3.27}$$

3.5.1.6 Equation of trajectory

Putting $t = \frac{x}{v_0 \cos \theta_0}$ from equation (3.11) in equation (3.13), we have

$$y = (v_0 \sin \theta t)\left(\frac{x}{v_0 \cos \theta_0}\right) - \frac{1}{2}g\left(\frac{x}{v_0 \cos \theta_0}\right)^2$$

$$\Rightarrow y = x \tan \theta_0 - \frac{gx^2}{2v_0^2 \cos^2 \theta_0}. \qquad (3.28)$$

As this equation is a parabola, the trajectory of the projectile is parabolic.

Example 4 A particle is projected with a velocity $\vec{v_0} = (3\hat{i} + 4\hat{j})m$ onto a horizontal plane, find the:
  (a) time of flight,
  (b) maximum height,
  (c) horizontal range,
  (d) velocity at $t = \frac{1}{2}$s,
  (e) equation of trajectory.

Solution

  (a) It is given that $v_{0x} = 3$, $v_{0y} = 4$
      The time of flight is

$$T = \frac{2v_{0y}}{g} = \frac{2 \times 4}{10} = 0.8 \text{ s Ans.}$$

  (b) The maximum height is

$$H = \frac{v_0^2 y}{2g} = \frac{(4)^2}{2(10)} = 0.8 \text{ m Ans.}$$

  (c) The horizontal range is

$$R = \frac{2(v_0 \sin \theta)(v_0 \cos \theta)}{g} = \frac{2v_{0x}v_{0y}}{g}$$

$$= \frac{2(3)(4)}{10} = 2.4 \text{ m Ans.}$$

  (d) The components of velocities are

$$v_x = v_{0x} = 3 \text{ and } v_y = v_{0y} - gt = 4 - 10 \times \frac{1}{2} = -1 \text{ Ans.}$$

(e) Then, the velocity of the particle is

$$\vec{v} = v_x\hat{i} + v_y\hat{j} = 3\hat{i} - \hat{j} \text{ m s}^{-1} \text{ Ans.}$$

(f) The equation of trajectory is

$$y = x(\tan\theta_0) - \frac{1}{2}g\frac{x^2}{(v_0\cos\theta)^2},$$

where $\tan\theta_0 = \frac{v_{y0}}{v_{x0}} = \frac{4}{3}$, $v_0\cos\theta_0 = 3$

$$\Rightarrow y = \frac{4}{3}x - \frac{1}{2} \times 10\frac{x^2}{(3)^2}$$

$$\Rightarrow y = \frac{4}{3}x - \frac{5}{9}x^2 \text{ Ans.}$$

3.5.2 Projection onto an inclined plane

3.5.2.1 Range and time of flight

Let us project a particle with a speed v_0 at an angle θ with the inclined plane of angle of inclination β. Let us take x-axis parallel to the inclined plane (in the line of greatest slope) and y-axis is perpendicular to the inclined plane. Let the object strike the inclined plane at a distance R from the point of projection after a time t.

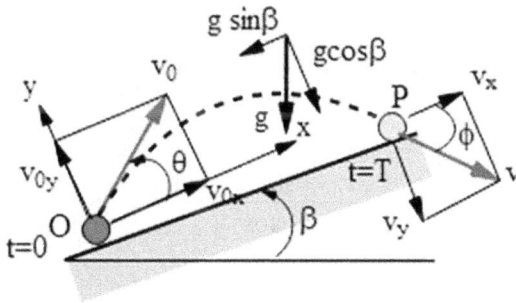

The body is projected up on to the inclined plane

In this case, resolving $\vec{v_0}$, $\vec{a} = \vec{g}$, $\vec{s}$ and $\vec{v}$ along the x- and y-axes, we have

$$v_{0x} = v_0\cos\theta, \quad v_{0y} = v_0\sin\theta, \quad a_x = -g\sin\beta, \quad a_y = -g\cos\beta$$

Then we can write

$$v_x = v_{0x} + a_x t$$

$$v_x = v_0\cos\theta - (g\sin\beta)T \tag{3.29}$$

$$v_y = v_{0y} + a_y t$$

$$v_y = v_0 \sin\theta - (g \cos\beta)T \tag{3.30}$$

$$x = R = v_{0x}t + \frac{1}{2}a_x t^2$$

$$\Rightarrow x = (v_0 \cos\theta_0)T - \frac{1}{2}(g \sin\beta)T^2 \tag{3.31}$$

$$y = 0 = v_{0y}t + \frac{1}{2}a_y t^2$$

$$\Rightarrow y = (v_0 \sin\theta_0)T - \frac{1}{2}(g \cos\beta)T^2 \tag{3.32}$$

$$\Rightarrow T = \frac{2v_0 \sin\theta}{g \cos\beta} \tag{3.33}$$

Putting T in equations (3.29) and (3.30) and simplifying, we have

$$v_x = v_0 \cos\theta(1 - 2\tan\theta\tan\beta) \tag{3.34}$$

$$v_y = -v_0 \sin\theta \tag{3.35}$$

So, the initial and final components of velocity normal to the inclined plain are equal in magnitude. Substituting $t = \frac{2v_0 \sin\theta}{g \cos\beta}$ in equation (3.31), we have

$$R = v_0\left(\frac{2v_0 \sin\theta}{g \cos\beta}\right)\cos\theta - \frac{g}{2}\left(\frac{2v_0 \sin\theta}{g \cos\beta}\right)^2 \sin\beta$$

$$\Rightarrow R = \frac{2v_0^2 \cos(\theta + \beta)\sin\phi}{g \cos^2\beta}. \tag{3.36}$$

3.5.2.2 Maximum range
Recasting the last equation we have

$$R = \frac{2v_0^2 \cos(\theta + \beta)\sin\phi}{g \cos^2\beta}$$

$$= \frac{v_0^2}{g \cos^2\beta}\{\sin(2\theta + \beta) - \sin\beta\}$$

R is maximum when $\sin(2\theta + \beta) = 1$

$$\Rightarrow \theta = \frac{\pi}{4} - \frac{\beta}{2} \tag{3.37}$$

So, the maximum range up the inclined plane is

$$R_{max} = \frac{v_0^2}{g(1 + \sin \beta)} \tag{3.38}$$

For downward projection put ' $- \beta$' instead of β, the maximum range down the plane is

$$R_{max} = \frac{v_0^2}{g(1 - \sin \beta)} \tag{3.39}$$

Example 5 A stone is projected perpendicular to the inclined plane with a speed v_0. If the angel of inclination is θ, find the:
  (a) time of flight,
  (b) speed of striking the inclined plane,
  (c) range along the inclined plane.

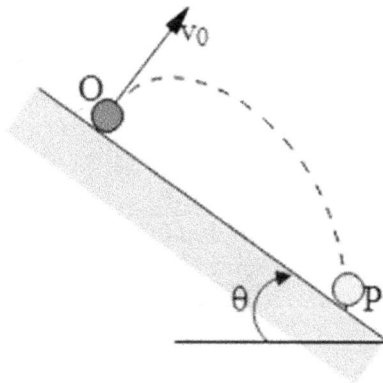

Solution

(a) Here $v_{0x} = 0$, $v_{0y} = v_0$, $a_x = g \sin \theta$, $a_y = -g \cos \theta$. Along the x-axis, the velocity just before striking the inclined plane at time t is

$$v_x = v_{0x} + a_x t = 0 + (g \sin \theta)t \tag{3.40}$$

The time of flight is

$$t = (-v_0 - v_0)/-a_y = 2v_0/a_y = =2v_0/g \cos \theta \text{ Ans.}$$

(b) Putting the value of time t in equation (3.40),

$$v_x = (g \sin \theta)(2v_0/g \cos \theta) = 2v_0 \tan \theta \tag{3.41}$$

Furthermore,

$$v_y = -v_0 \qquad (3.42)$$

Then, the speed of striking is

$$v = \sqrt{v_x^2 + v_y^2} \qquad (3.43)$$

Putting the values of v_x and v_x from equations (3.41) and (3.42) in equation (3.43), we have

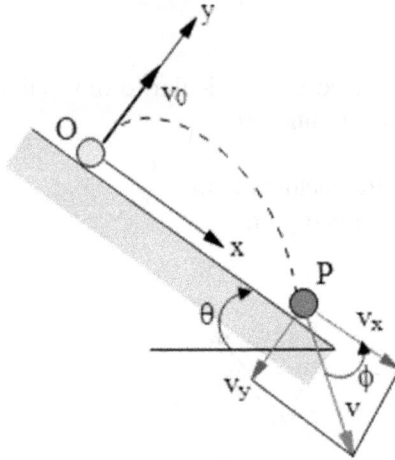

$$\Rightarrow v = \sqrt{(2v_0 \tan \theta)^2 + (-v_0)^2}$$

$$\Rightarrow v = v_0\sqrt{4 \tan^2 \theta + 1} \quad \text{Ans.}$$

(c) The range of the particle OP along the inclined plane is

$$OP = v_{0x}t + \frac{1}{2}a_x t^2 = \frac{1}{2}(g \sin \theta)t^2 \text{ (because } v_{0x} = 0, a_x = g \sin \theta)$$

$$\Rightarrow OP = \frac{1}{2}(g \sin \theta)\{2v_0/g \cos \theta\}^2$$

$$\Rightarrow OP = \frac{2v_0^2 \sin \theta}{g \cos^2 \theta} \quad \text{Ans.}$$

3.6 Relative motion

Suppose that you are going to Delhi in a train. If you ask your co-passenger, 'sir, when will the New Delhi Railway station (NDLS) come?' Your co-passenger would answer 'NDLS will come in five minutes'. This answer seems to be half right and half wrong. It is partially wrong because NDLS is not moving anywhere relative to the ground; rather, you are going towards Delhi as the train moves relative to the ground. However, the question posed by the passenger can also be correct relative to the train; so, the answer given by the co-passenger can be correct relative to the train.

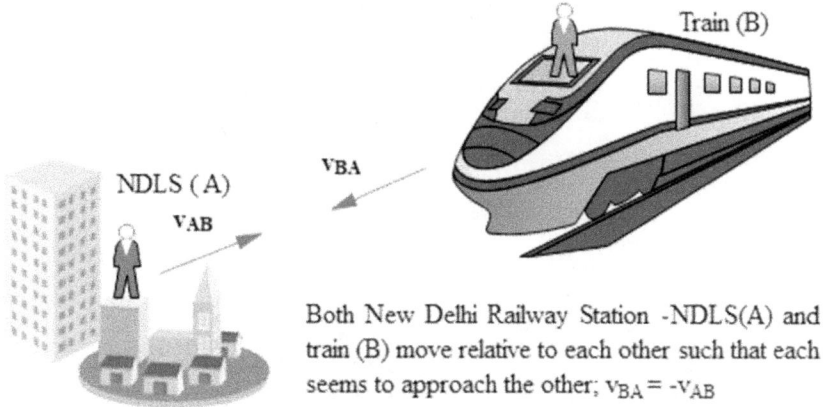

Both New Delhi Railway Station -NDLS(A) and train (B) move relative to each other such that each seems to approach the other; $v_{BA} = -v_{AB}$

Let us discuss the relative motion between two points, A and B, say. Let us assume point A as an 'object' and point B as an 'observer'. When the position vector of A relative to B, that is, $\vec{r}_{AB}$ changes, we say that, 'A moves relative to B'. The rate at which the position vector changes, that is, $\frac{d\,\vec{r}_{AB}}{dt}$ is defined as velocity of A relative to B which is denoted by $\vec{v}_{AB}$. Then, we can write

$$\vec{v}_{AB} = \frac{d\,\vec{r}_{AB}}{dt}$$

A seems to move with a velocity $\vec{v}_{AB}$ relative to B and B seems to move with a velocity $\vec{v}_{BA}$ relative to A such that $\vec{v}_{AB} = -\vec{v}_{BA}$. This signifies the fact that each seems to approach the other with the velocity of equal magnitude.

3.6.1 Galilean transformation of velocities and accelerations

When you drop a coin in a uniformly moving train, it appears to move vertically down relative to the train, whereas a stone moves in a parabola relative to the ground. So, the nature of motion of a particle depends upon the reference frame. Relative to different observers (reference frames) the motion of a particle can be different.

Let us assume that observers B and C are fixed with reference frames S' and S, respectively and observe the motion of an object A. Let the observers B and C measure the position vectors of A as $\vec{r}_{AB}$ and $\vec{r}_{AC}$, respectively. If the position vector of B relative to C is $\vec{r}_{BC}$, following the triangle law of vectors, we have,

$$\vec{r_{AC}} = \vec{r_{AB}} + \vec{r_{BC}}$$

Differentiating the above expression with time, we have,

$$\frac{d\vec{r_{AC}}}{dt} = \frac{d\vec{r_{AB}}}{dt} + \frac{d\vec{r_{BC}}}{dt}$$

$$\Rightarrow \vec{v_{AC}} = \vec{v_{AB}} + \vec{v_{BC}}$$

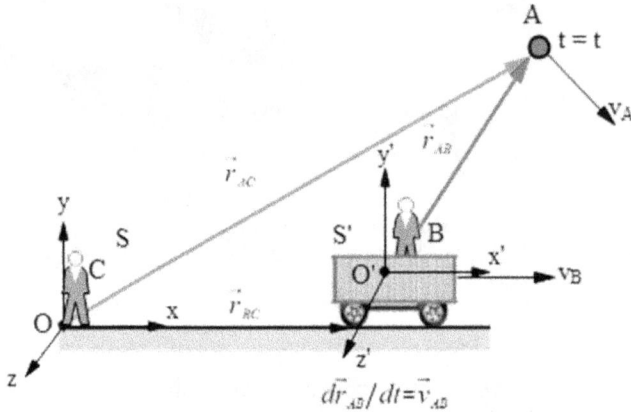

The velocity of A relative to the reference frame S (ground observer C) is v_{AC} but relative to the reference frame S' (observer B fixed with the trolley car) is v_{AB}

Differentiating both sides with time,

$$\vec{a_{AC}} = \vec{a_{AB}} + \vec{a_{BC}}$$

The above expression tells us that different observers measures different velocities of the same object, if they move relative to each other.

1. The motion of an object A relative to different observers is different ($v_{AB} \neq v_{AC}$) so long as the observers B and C move relative to each other ($v_{BC} \neq 0$)
2. $\vec{v_{AB}} = \vec{v_{AC}} - \vec{v_{BC}}$; the velocity of A relative to B is equal to velocity of A relative to C *minus* velocity of B relative to C; C may be in any reference frame including the ground frame.

Example 6 *(Man in the rain)*

To a person standing on the ground the rain falls vertically with a speed of 3 m s^{-1}. When the man moves to the right with a speed of 2 m s^{-1}, how does the rain appear to fall? Explain by using Galilean transformation of velocities.

Solution

The rain appears to fall with velocity $\vec{v_{rg}}$ relative to the reference frame S (ground). The velocity of rain relative to the moving man (reference frame S')

is $\vec{v}_{rm}$. These two velocities are different because there is a relative velocity between S' (man) and S (ground).

According to Galilean transformation of velocities,

$$\vec{v}_{rg} = \vec{v}_{rm} + \vec{v}_{mg}$$

$$\Rightarrow \vec{v}_{rm} = \vec{v}_{rg} - \vec{v}_{mg} = -3\hat{j} - 2\hat{j}$$

So, the rain appears to fall with a velocity $v_{rg} = \sqrt{13}$ m s^{-1} at angle of $\theta = \tan^{-1}\frac{2}{3}$ with the vertical.

Ans.

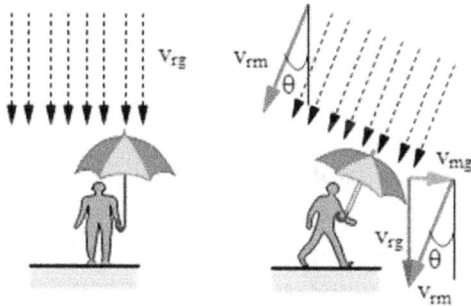

The rain falls vertically down relative to ground, but falls backward at an angle $\theta = \tan^{-1}(v_{mg}/v_{rg})$ with vertical

Example 7 *(Boat in water)*

Let us assume that a boat can move with a speed v_{bg} relative to water. If water flows with a velocity v_{wg} relative to the ground, using Galilean transformation of velocities explain the motion of the boat relative to both reference frames.

Solution

Man 1, who is standing on a raft (water frame), observes that the raft heads at an angle θ (with the direction of water current $\vec{v}_{wg}$) with a velocity $\vec{v}_{bw}$. Man 2 standing on the ground frame observes the raft moving with a velocity $\vec{v}_{bg}$ at an angle ϕ with the direction of flow. We can call θ and ϕ 'angle of heading' and 'angle of motion', respectively.

According to Galilean transformation of velocities, the boat seems to move with different velocities $\vec{v}_{gw}$ and $\vec{v}_{wg}$ relative to water and the ground frame because the water frame moves relative to the ground frame with a velocity $\vec{v}_{wg}$;

$$\vec{v}_{bg} = \vec{v}_{bw} + \vec{v}_{wg}$$

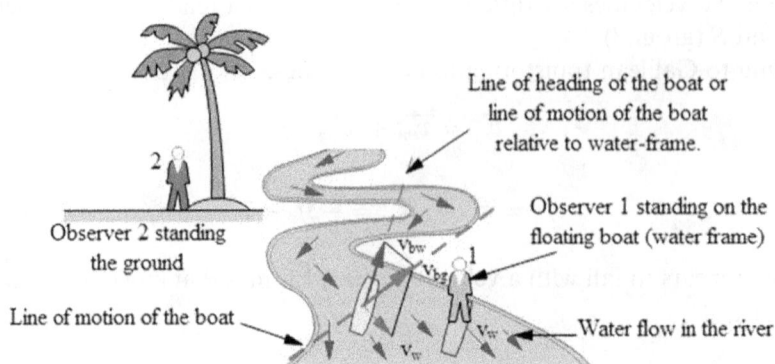

Line of heading of the boat or
line of motion of the boat
relative to water-frame.

Observer 1 standing on the
floating boat (water frame)

Observer 2 standing
the ground

2

Line of motion of the boat

Water flow in the river

The motion of the boat relative to ground-frame and water-frame. The observer fixed with the water frame(standing on the floating boat) osbervers that the boat is moving with a velocity v_{bw} along the blue dotted line;but, the observer 2 fixed with ground sees that the boat is moving with a velocity v_{bg} along the red dotted line.

3.7 Kinematics of circular motion

When a particle or point moves along the perimeter of a circle it is called circular motion. Orbiting of a satellite around a planet, motion of a pendulum bob of a simple pendulum, motion of a point P, say, on a rotating wheel are some familiar examples of circular motion. Let us discuss the angular parameters or variables used in circular motion.

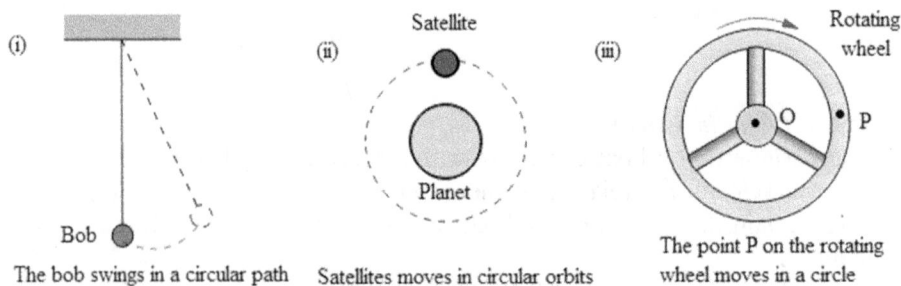

(i)

Bob

The bob swings in a circular path

(ii)

Satellite

Planet

Satellites moves in circular orbits

(iii)

Rotating wheel

O

P

The point P on the rotating wheel moves in a circle

(i) **Angular position**

The position of a particle P at any given instant may be described by angle θ, where OP makes the reference line OX (x-axis). We call θ the angular position of the particle.

(ii) **Angular velocity**

Suppose the particle goes to a nearby point P' in a time interval Δt so that θ increases to $\theta + \Delta\theta$. The rate of change of angular position is called angular velocity. Thus, the instantaneous angular velocity is

$$\omega = \lim_{\Delta t \to 0} \frac{\Delta\theta}{\Delta t} = \frac{d\theta}{dt}.$$

(iii) **Angular acceleration**

The rate of change of angular velocity is called angular acceleration, given as

$$\alpha = \frac{d\omega}{dt} = \frac{d^2\theta}{dt^2}.$$

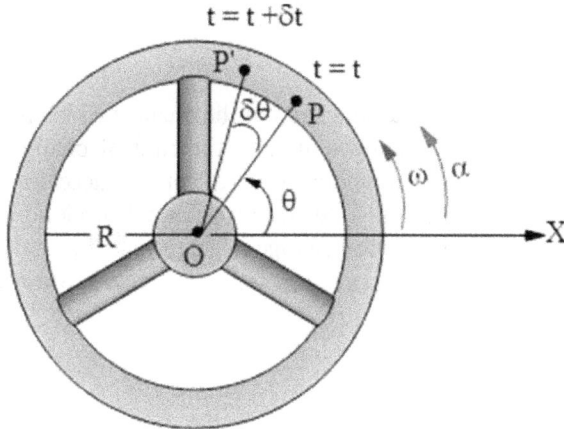

The angular position, angular displacement, angular velocity and angular acceleration of the point P on the rotating wheel

If θ increases with time, ω and θ have the same direction; if ω increases with time, α and ω have the same direction; otherwise, these vectors oppose each other.

3.7.1 Constant angular acceleration

If the angular acceleration α is constant, without derivation we have the following equations:

$$\theta = \omega_0 t + \frac{1}{2}\alpha t^2 \tag{3.44}$$

$$\omega = \omega_0 + \alpha t \tag{3.45}$$

$$\omega^2 = \omega_0^2 + 2\alpha\theta \tag{3.46}$$

where ω_0 and ω are the angular velocities at $t = 0$ and at time t and θ is the angular position at time 't'.

A particle has traveled the linear distance PP' in time δt, so,

$$\delta s = r\delta\theta$$

Dividing both sides by time interval, we have

$$\frac{\delta s}{\delta t} = r\frac{\delta\theta}{\delta t}$$

$$\Rightarrow v = r\omega,$$

where v is the linear speed of the particle. Differentiating equation (3.4) with respect to time, the rate of change of speed is

$$a_t = \frac{dv}{dt} = r\frac{d\omega}{dt} = r\alpha$$

$$a_t = r\alpha$$

Here, a_t is the tangential acceleration of the particle which is responsible for change in speed of the particle (but not equal to the rate of change of velocity). So, it is a component of the net acceleration. Tangential acceleration cannot change the direction of motion. Normal acceleration or radial acceleration (also called centripetal acceleration) is responsible for change in direction of motion of the particle.

Let us now derive the expression for radial acceleration of a particle moving on the circular path.

3.7.2 Radial acceleration

Let the velocity of a fixed mark (black dot) at the position P on the perimeter of the wheel be $\vec{v_1} = \vec{v}(t)$ at $t = t_1$. After a very small time δt, at $t = t + \delta t$, let the position of the mark be P' and the velocity becomes $\vec{v_2} = \vec{v}(t + \delta t)$. Let us assume that $\vec{v_1}$ and $\vec{v_2}$ have the same magnitude; this means that the mark has uniform speed. So, the velocities differ only in the direction as their magnitudes are the same $(v_1 = v_2 = v)$.

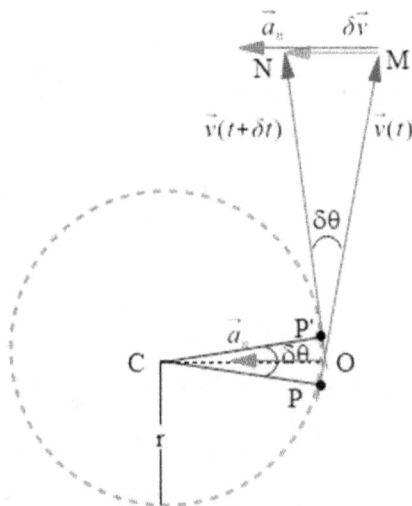

In uniform circular motion, the net acceleration points towards the centre C of the circle

The elementary change in velocity $d\vec{v}$ of the mark is given by joining the heads M and N of these two velocity vectors $\vec{v_1} = \vec{v}(t)$ and $\vec{v_2} = \vec{v}(t + \delta t)$. As the velocities are nearly vertical, we can see that $d\vec{v}$ is perpendicular to the velocity vector $\vec{v_1} = \vec{v}(t)$; so it is horizontal pointing radially inward. From the geometry of the figure, we have

$$|d\vec{v}| = MN = (OM)\delta\theta = v\delta\theta$$

So, the magnitude of instantaneous acceleration can be given as

$$a_n = a_{cp} = \left|\frac{\delta\vec{v}}{\delta t}\right|$$

Using the last two equations, we have

$$a_c = \frac{v\delta\theta}{\delta t} = v\omega$$

Putting $\omega = v/r$, we have

$$a_C = \frac{v^2}{r}$$

Putting $v = \omega r$, we have

$$a_C = r\omega^2,$$

where a_C is called centripetal or radial because it is pointing towards the center of the circle.

If the speed of the particle changes, the motion is said to be non-uniform circular motion. Now, we have a tangential acceleration due to the change in magnitude of the velocity, that is, speed of the particle. So, the total acceleration of a particle in circular motion can be written as

$$\vec{a} = \vec{a_t} + \vec{a_C},$$

where $a_t = \frac{dv}{dt} = \alpha r$, and $a_c = \frac{v^2}{r} = \omega^2 r$

So, the magnitude of acceleration is

$$a = \sqrt{a_t^2 + a_C^2}$$

$$\Rightarrow a = \sqrt{\alpha^2 r^2 + \left(\frac{v^2}{r}\right)^2}$$

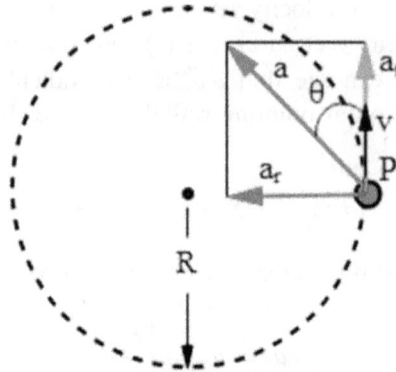

The total acceleration is equal to the vector
sum of tangential and radial accelerations

3.8 Curvilinear motion

Let us consider the motion of a particle along a curved path, where the velocity changes both in direction and magnitude, as shown in the figure below.

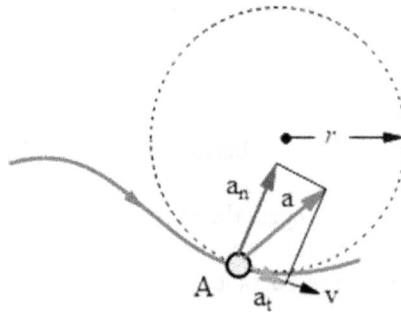

$a_t = dv/dt; \ a_n = v^2/r; \ \vec{a}_t + \vec{a}_n = \vec{a}$

Let us consider a point A on the curve. We can draw a circle (dotted) that fits the curve at the given point. The radius 'r' of the model circle is the radius of curvature of the path at that point A.

Now we can resolve the total or net acceleration in two components, such as tangential (a_t) and normal (a_n), as shown in figure. Tangential acceleration is responsible for change in magnitude of the velocity, whereas the normal acceleration is responsible for change in direction of velocity. Then, putting

$$a_t = \frac{dv}{dt} \text{ and } a_n = \frac{v^2}{r}$$

in the expression,

$$\vec{a} = \vec{a_t} + \vec{a_N},$$

the magnitude of net acceleration is

$$|\vec{a}| = \sqrt{|a_t|^2 + |a_n|^2}$$

The radius of curvature at point A is given as

$$r = \frac{v^2}{a_n}$$

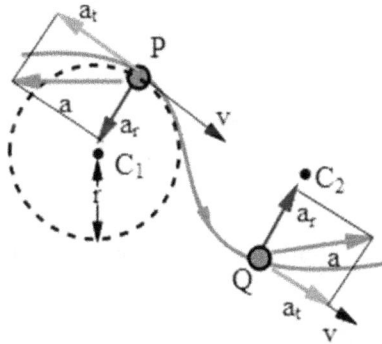

If v decreases, a_t opposes v and if v
increases, a_t and v have same direction

We can see that if the velocity opposes the tangential acceleration a_t, speed decreases with time or the particle decelerates or retards; if the particle speeds up, the tangential acceleration a_t favors the velocity vector. If the particle moves with uniform speed, the tangential acceleration will be zero; so, the net acceleration will be centripetal.

3.9 Relative angular velocity

When an object A moves with a velocity $\vec{v_A}$, its velocity relative to another object B (that is moving with velocity $\vec{v_B}$) is given by:

$$\vec{v_{AB}} = \vec{v_A} - \vec{v_B}.$$

Let us now resolve the velocity $\vec{v_{AB}}$ parallel (along) OP and perpendicular (normal) to BA along r- and θ- directions. We will get $(v_{AB})_r$ and $(v_{AB})_\theta$, respectively. This means that A goes away from B with $(v_{AB})_r$ and it turns around B with a velocity $(v_{AB})_\theta$. The angular velocity of turning of A relative to B (or around B) is called 'angular velocity' of A relative to B denoted as ω_{AB}.

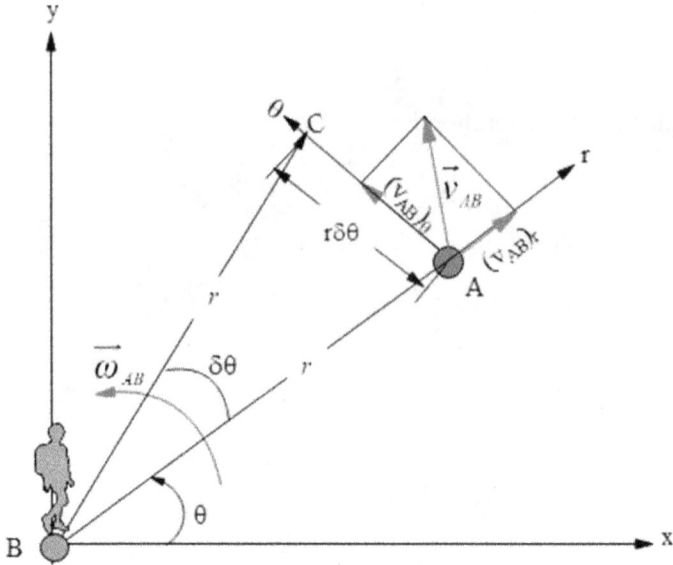

Angular velocity of A relative to B = $\omega_{AB} = (v_{AB})_n / r = (v_{AB})_\theta / r$

In other words, particle/point A turns around point B with an angular velocity

$$\omega_{AB} = \frac{d\theta}{dt}$$

$$\Rightarrow \omega_{PO} = \frac{r_{AB}}{r_{AB}} \frac{d\theta}{dt} = \frac{r}{r} \frac{d\theta}{dt} = \left(\frac{AC}{r_{AB}}\right) \frac{1}{dt}$$

$$= \left(\frac{(v_{AB})_n \, dt}{r \, dt}\right) (\because \ AC = (v_{AB})_n \, dt)$$

$$\Rightarrow \omega_{AB} = \frac{(v_{AB})_n}{r} = \frac{(v_{AB})_\theta}{r}.$$

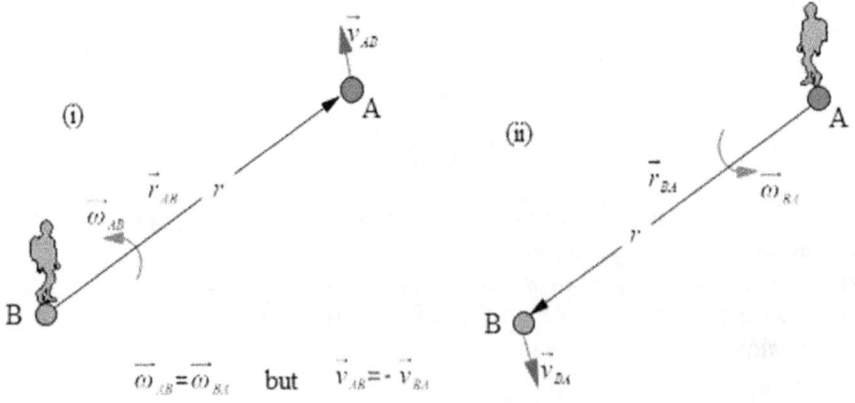

$$\vec{\omega}_{AB} = \vec{\omega}_{BA} \quad \text{but} \quad \vec{v}_{AB} = -\vec{v}_{BA}$$

We can see that $\vec{\omega}_{AB} = \vec{\omega}_{BA}$, but $\vec{v}_{PO} = -\vec{v}_{OP}$. This means that two points turn around each other with the same angular velocity but their relative velocities are equal and opposite.

Example 8 Two particles 1 and 2 move with velocities v_1 and v_2, respectively, as shown in the figure. If the distance between them is l, find their relative angular velocity.

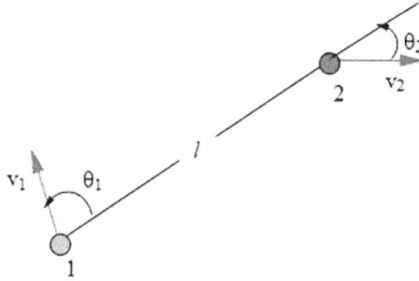

Solution
The relative angular velocity is

$$\omega_{rel} = |\vec{v}_{12}|_n / l$$

$$= \frac{|\vec{v}_{1\,n} - v_{2\,n}|}{l}$$

$$= \frac{v_1 \sin \theta_1 - (-v_2 \sin \theta_2)}{l}$$

$$= \frac{v_1 \sin \theta_1 + v_2 \sin \theta_2}{l} \quad \text{Ans.}$$

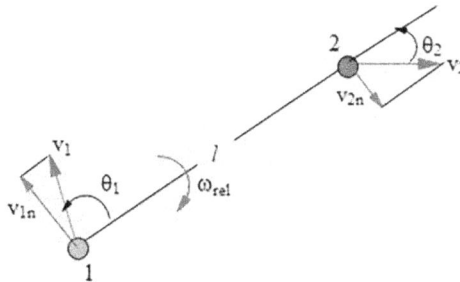

3.10 Equation of trajectory of one point relative to the other

When we observe A from B, A seems to move with a velocity of magnitude $v_{AB}(=v$ say) tracing an arbitrary curved path, as shown in the figure below. In other words, A approaches B with a velocity of magnitude $v = v_r$. Let the distance r of separation reduce by dr during an elementary time dt. So, we can write

$$-dr = v_r dt \tag{3.47}$$

Furthermore, A turns about/around B by an elementary angle $d\theta$ during the time dt covering an elementary distance $rd\theta$ perpendicular to the line joining A and B. If $\vec{v}_n$ is the component of $\vec{v}$ perpendicular to the line joining the particles, we have

$$rd\theta = v_n dt = v_\theta dt \qquad\qquad (3.48)$$

Dividing equation (3.47) by equation (3.48), we have

$$-\frac{dr}{r} = \frac{v_r}{v_\theta}d\theta$$

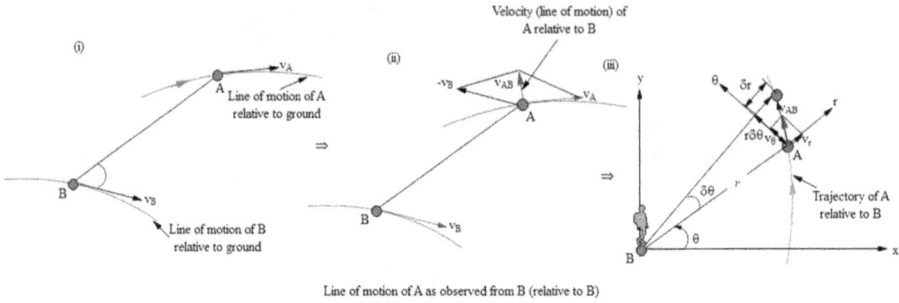

Since, at $t = 0$, $r = r_0$; by integrating both sides we have,

$$-\int_{r_0}^{r}\frac{dr}{r} = \int\frac{v_r}{v_\theta}d\theta$$

$$\Rightarrow \ln\frac{r_0}{r} = \int\frac{v_r}{v_\theta}d\theta$$

$$\Rightarrow r = r_0 e^{-\int\frac{v_r}{v_\theta}d\theta},$$

where r is the distance of separation between A and B and θ is the angle of orientation of $\vec{r}_{AB}$.

The above equation gives us the path traced by A relative to B.

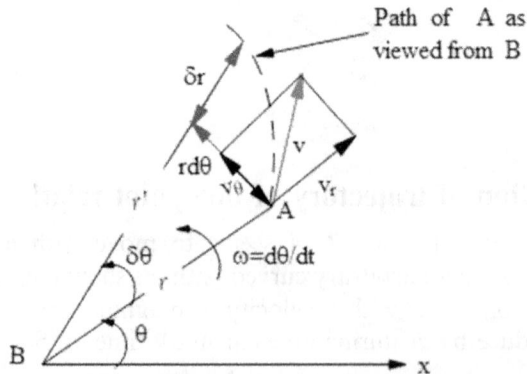

At time t, the path traced by the particle A as viewed from the particle B

3.10.1 Time of meeting of particles

If two particles have velocities v_1 and v_2, they approach each other with a velocity $v_{app} = |\vec{v}_{1r} - \vec{v}_{2r}|$.

Since, their relative separation r decreases,

$$v_{app} = -\frac{dr}{dt}$$

Then, the time of meeting can be given as;

$$t = -\int_{r_0}^{0} \frac{dr}{v_{approach}},$$

where r_0 = initial separation.

Example 9 Three snails A, B and C are situated at the vertices of an equilateral triangle of side 1. Snail A heads towards B, B towards C, C towards A with constant speeds v such that they always remain at the vertices of an equilateral triangle. Find the (a) time after which the distance between the snails will be equal x, (b) time of meeting, (c) total distance overed by each snail, (d) average velocity, and (e) average speed till the snails meet, (f) equation of path traced by one snail relative to the other, (g) equation of path traced by one snail relative to the ground.

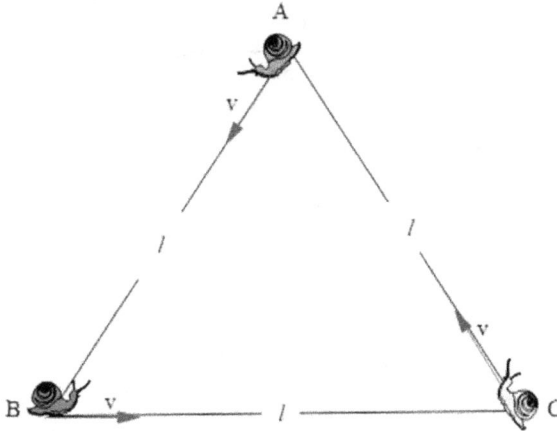

Solution

(a) After a time t, let the distance of separation between the snails be r. The velocity of approach (component of relative velocity v_{rel} between the snails along the line of their separation) is

$$v_{app} = v + v \cos 60° = \frac{3v}{2}$$

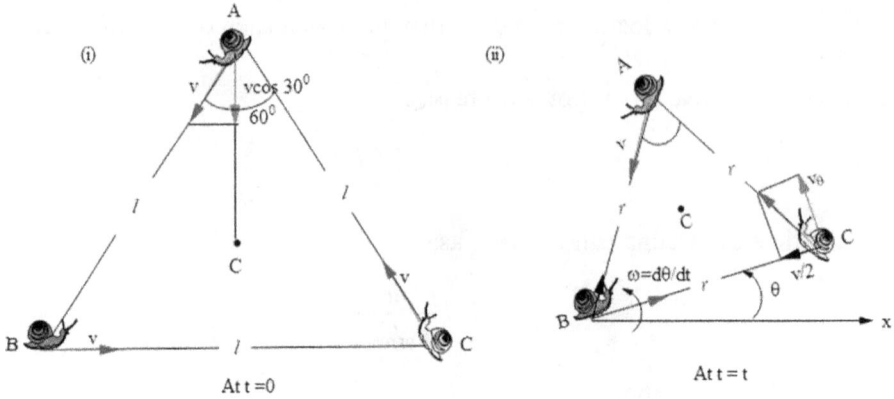

Since, $v_{app}\,dt = -dr$, substituting $v_{app} = \frac{3v}{2}$, we obtain

$$\frac{3v}{2}dt = -dr$$

As, $r = l$ at $t = 0$ and $r = 0$ at $t = \tau$ (time of meeting), integrating both sides, we have,

$$\frac{3v}{2}\int_0^\tau dt = -\int_l^x dr$$

$$\Rightarrow t = \frac{2(l - x)}{3v} \quad \text{Ans.}$$

(b) Putting $x = 0$ in the last expression, we have
$$t = \frac{2(l - 0)}{3v} = \frac{2l}{3v}$$

(c) The distance covered by each snail is
$$d = vt = \frac{2l}{3} \quad \text{Ans.}$$

(d) The average velocity is
$$v_{av} = \frac{s}{t} = \frac{AC}{t}$$

where s = displacement of each insect = $AC = l/\sqrt{3}$

$$\Rightarrow v_{av} = \frac{l/\sqrt{3}}{2l/3v} = \frac{\sqrt{3}}{2}v \quad \text{Ans.}$$

(e) As the speed of each snail remains the same, the average speed is
$$v_{av} = v \quad \text{Ans.}$$

(f) Referring to the last equation of section 3.10, the equation of path traced by one snail relative to the other is given as
$$r = r_0 e^{-\int \frac{v_r}{v_\theta} d\theta},$$
where r is the distance of separation between A and B and θ is the angle of orientation of $\vec{r}_{AB} = \vec{r}$. Putting $v_r = v \cos 60° + v = 3v/2$ and $v_\theta = v \cos 30° = \sqrt{3}v/2$, we have
$$r = le^{-\int \frac{3v/2}{\sqrt{3}v/2} d\theta} = le^{-\int \sqrt{3} d\theta} = le^{-\sqrt{3}\theta} \quad \text{Ans.}$$

(g) Let x = distance between the center C and vertex A of the triangle. Taking 'C' as the reference point, after a time t, let the sides of the triangle rotate through an angle θ. The snail at A approaches C with a velocity
$$v' = v \cos 30° = \sqrt{3}v/2$$
which is equal to the rate of decrease of the separation x; so, we can write
$$-\frac{dx}{dt} = \sqrt{3}v/2$$

Furthermore, the snail at the vertex A of the shrinking equilateral triangle turns around the center C with an angular velocity
$$\frac{d\theta}{dt} = (v/2)/x = v/2x$$

Eliminating dt between the last two equations, we have
$$-\frac{dx}{d\theta} = \sqrt{3}x$$

Separating the variables, we have
$$-\frac{dx}{x} = \sqrt{3} d\theta$$

Integrating both sides,
$$-\int_{x_0}^{x} \frac{dx}{x} = \sqrt{3} \int_{0}^{\theta} d\theta$$

Evaluating the integration and simplifying the expressions, we have
$$x = x_0 e^{-\sqrt{3}\theta}$$
$$\Rightarrow \frac{x}{x_0} = \frac{r}{l} = e^{-\sqrt{3}\theta} \quad \text{Ans.}$$

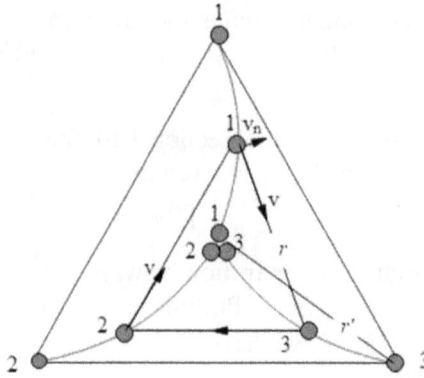

N.B.: As each snail aims at the other in the same sense (either clockwise or anticlockwise) the path of the snails relative to the ground will be spiral and eventually they all meet at the center of the triangle. Since the speed of each is equal to v and each will meet at the center of the equilateral triangle, we can say that each snail is approaching the center of the triangle with a velocity of

$$v' = v \cos 30° = \sqrt{3}v/2$$

in spiral paths relative to the ground so also relative to the others following the same function as shown in the above figure.

Problem 1 Find the average velocity of a projectile projected with a velocity $\vec{v_0}$ during the time equal to twice the time of ascent, measured from the instant of projection.

Solution

The required time interval is $t = 2t'$, By putting $t' =$ time of ascent $= \frac{v_0 \sin \theta_0}{g}$, we have

$$t = \frac{2v_0 \sin \theta_0}{g}$$

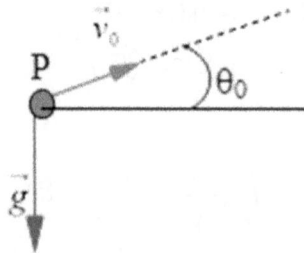

Substituting the above value of time in the equation

$$\vec{v_{av}} = \vec{v_0} + \frac{\vec{g}t}{2}$$

as derived earlier, we have,

$$\vec{v}_{av} = \vec{v}_0 + (v_0 \sin \theta_0)\vec{g} \qquad (3.49)$$

To eliminate $\sin \theta_0$ in equation (3.49), we write

$$\vec{v}_0 \cdot \vec{g} = v_0 g \cos(90° + \theta_0) = -v_0 g \sin \theta_0$$

$$\Rightarrow \sin \theta_0 = -\frac{\vec{v}_0 \cdot \vec{g}}{v_0 g}$$

Now substituting $\sin \theta_0$ in equation (3.49), we have,

$$\vec{v}_{av} = \vec{v}_0 - \left(\frac{\vec{v}_0 \cdot \vec{g}}{g}\right)\vec{g} \quad \text{Ans.}$$

Problem 2 A particle is projected from a point O such that it passes through a point P and again passes through a point Q at same level of projections. If $\angle POQ = \alpha = 37°$ and $\angle PQO = \beta = 45°$, find the angle of projection.
   Solution
   By substituting $\dfrac{v_0^2}{g} = \dfrac{R}{2 \sin \theta_0 \cos \theta_0}$ in the trajectory equation

$$y = x \tan \theta_0 - \frac{gx^2}{2v_0^2 \cos^2 \theta_0}$$

we have,

$$\tan \theta_0 = \frac{Ry}{x(R - x)} = \left(\frac{y}{R - x} + \frac{y}{x}\right)$$

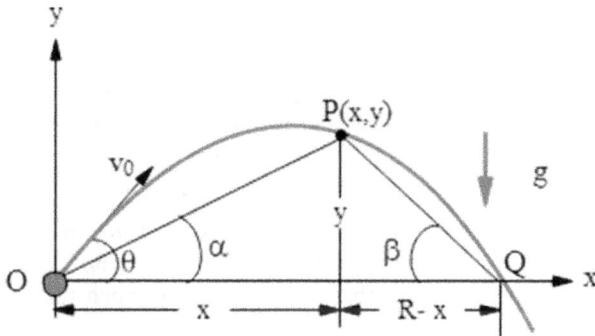

Then, substituting $y/x = \tan \alpha$ and $y/(R - x) = \tan \beta$, we have

$$\tan \theta_0 = \tan \alpha + \tan \beta$$

$$\Rightarrow \tan \theta_0 = \tan 37° + \tan 45° = 3/4 + 1 = 7/4$$

$$\theta_0 = \tan^{-1}(7/4) = 60.25° \quad \text{Ans.}$$

Problem 3 Two water jets are projected from the same spot with the same speed v with angles of projection $\theta_1 = 30°$ and $\theta_2 = 60°$, respectively. Find the position where they meet.

Solution

Let the jets meet at $P(x_1, y_1)$. Then these coordinates must satisfy the trajectory equations of both jets at the point of intersection. Substituting x_1, y_1 in the corresponding trajectory equations, we have

$$y_1 = x_1 \tan \theta_1 - \frac{gx_1^2}{2v \cos^2 \theta_1} \qquad (3.50)$$

$$y_1 = x_1 \tan \theta_2 - \frac{gx_1^2}{2v \cos^2 \theta_2} \qquad (3.51)$$

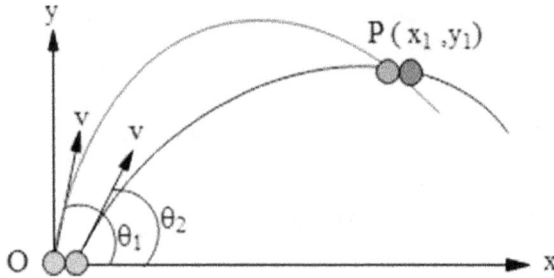

Solving equations (3.50) and (3.51), we have,

$$x_1 = \frac{2v^2 \cos \theta_1 \cos \theta_2 \sin(\theta_1 - \theta_2)}{g(\cos^2 \theta_2 - \cos^2 \theta_1)}$$

$$x_1 = \frac{2v^2 \cos 60° \cos 30° \sin(60° - 30°)}{g(\cos^2 60° - \cos^2 30°)} = \frac{\sqrt{3}v^2}{10g} \quad \text{Ans.}$$

Problem 4 Two balls are projected one after the other from the same point with the same speed v with angles of projection $\theta_1 = 30°$ and $\theta_2 = 60°$, respectively, so that they meet above the ground. Find the time interval between their projection.

Solution

Method 1

Let the balls meet at $P(x_1, y_1)$. Then these coordinates must satisfy the trajectory equations of both jets at the point of intersection. Substituting x_1, y_1 in the corresponding trajectory equations, we have

$$y_1 = x_1 \tan \theta_1 - \frac{gx_1^2}{2v \cos^2 \theta_1} \qquad (3.52)$$

3-31

$$y_1 = x_1 \tan \theta_2 - \frac{g x_1^2}{2v \cos^2 \theta_2} \tag{3.53}$$

Solving equations (3.52) and (3.53), we have,

$$x_1 = \frac{2v^2 \cos \theta_1 \cos \theta_2 \sin(\theta_1 - \theta_2)}{g(\cos^2 \theta_2 - \cos^2 \theta_1)} \tag{3.54}$$

The time taken by ball 1 to traverse a horizontal distance x is

$$t_1 = \frac{x_1}{v \cos \theta_1} \tag{3.55}$$

The time taken by ball 2 to traverse a horizontal distance x is

$$t_2 = \frac{x_1}{v \cos \theta_2} \tag{3.56}$$

Then, the time interval between the projection of the two balls is

$$\Delta t = t_1 - t_2 = \frac{x_1}{v \cos \theta_1} - \frac{x_1}{v \cos \theta_2}$$

$$= \frac{x_1}{v}\left(\frac{1}{\cos \theta_1} - \frac{1}{\cos \theta_2}\right) = \frac{x_1}{v}\left(\frac{\cos \theta_2 - \cos \theta_1}{\cos \theta_1 \cos \theta_2}\right)$$

$$\Rightarrow \Delta t = \frac{x_1}{v}\left(\frac{\cos \theta_2 - \cos \theta_1}{\cos \theta_1 \cos \theta_2}\right) \tag{3.57}$$

By putting the value of x_1 from equation (3.54) in (3.56), we have

$$\Rightarrow \Delta t = \frac{2v^2 \cos \theta_1 \cos \theta_2 \sin(\theta_1 - \theta_2)}{vg(\cos^2 \theta_2 - \cos^2 \theta_1)}\left(\frac{\cos \theta_2 - \cos \theta_1}{\cos \theta_1 \cos \theta_2}\right)$$

$$\Rightarrow \Delta t = \frac{2v \sin(\theta_1 - \theta_2)}{g(\cos \theta_1 + \cos \theta_2)} \quad \text{Ans.}$$

Method 2

Let us use the equation

$$y = vt \sin \theta - \frac{g t^2}{2}$$

For the two balls 1 and 2, if the times of flight of the balls are t_1 and t_2 till they meet, their equations can be given as follows:

$$y = vt_1 \sin \theta_1 - \frac{g t_1^2}{2}$$

$$y = vt_2 \sin \theta_2 - \frac{g t_2^2}{2}$$

Eliminating 'y' from both equations, we have

$$vt_1 \sin \theta_1 - \frac{gt_1^2}{2} = vt_2 \sin \theta_2 - \frac{gt_2^2}{2}$$

$$\Rightarrow v(t_1 \sin \theta_1 - t_2 \sin \theta_2) = \frac{g(t_1^2 - t_2^2)}{2}$$

$$\Rightarrow v(t_1 \sin \theta_1 - t_2 \sin \theta_2) = \frac{g(t_1 + t_2)}{2}(t_1 - t_2)$$

$$\Rightarrow (t_1 - t_2) = \frac{2v(t_1 \sin \theta_1 - t_2 \sin \theta_2)}{g(t_1 + t_2)}$$

$$\Rightarrow (t_1 - t_2) = \frac{2v\{(t_1/t_2)\sin \theta_1 - \sin \theta_2\}}{g(t_1/t_2 + 1)} \tag{3.58}$$

From equation (3.55) in (3.56), we have

$$t_1/t_2 = \frac{\cos \theta_2}{\cos \theta_1} \tag{3.59}$$

Putting t_1/t_2 from equation (3.59) in equation (3.58), we have

$$(t_1 - t_2) = \frac{2v\left\{\left(\frac{\cos \theta_2}{\cos \theta_1}\right)\sin \theta_1 - \sin \theta_2\right\}}{g\left\{\left(\frac{\cos \theta_2}{\cos \theta_1}\right) + 1\right\}}$$

$$= \frac{2v(\sin \theta_1 \cos \theta_2 - \sin \theta_2 \cos \theta_1)}{g(\cos \theta_1 + \cos \theta_2)}$$

$$= \frac{2v \sin(\theta_1 - \theta_2)}{g(\cos \theta_1 + \cos \theta_2)} \quad \text{Ans.}$$

Problem 5 A particle is projected from the ground with an angle θ_0 with the horizontal such that its horizontal range is equal to its maximum height. (a) Find θ_0. (b) When it moves through a horizontal distance equal to one third of the horizontal range, find the
   (i) slope of the trajectory,
   (ii) tangential and radial (normal) acceleration of the particle, at that instant.

Solution

(a) We can find the angle of projection θ_0 from the given condition

$$R_H = y_{\max} \tag{3.60}$$

The horizontal range is

$$R_H = \frac{2v_0^2 \sin \theta_0 \cos \theta_0}{g} \tag{3.61}$$

The maximum height is

$$y_{max} = \frac{v_0^2 \sin^2 \theta_0}{2g} \tag{3.62}$$

Using the last three equations,

$$\tan \theta_0 = 4$$

$$\theta_0 = \tan^{-1} 4 \simeq 76° \text{ Ans.}$$

(b) (i) It is given that $x = R/3$, where

$$R = \frac{2v_0^2 \sin \theta_0 \cos \theta_0}{g}$$

$$\Rightarrow x = R/3 = \frac{2v_0^2 \sin \theta_0 \cos \theta_0}{3g} \tag{3.63}$$

The slope of the trajectory is given as

$$\tan \theta = \frac{dy}{dx} = \tan \theta_0 - \frac{gx}{v_0^2 \cos^2 \theta_0} \tag{3.64}$$

Using the last two equations, we have

$$\tan \theta = \tan \theta_0 - \frac{g\left(\dfrac{2v_0^2 \sin \theta_0 \cos \theta_0}{3g}\right)}{v_0^2 \cos^2 \theta_0}$$

$$\Rightarrow \tan \theta = \frac{\tan \theta_0}{3} \tag{3.65}$$

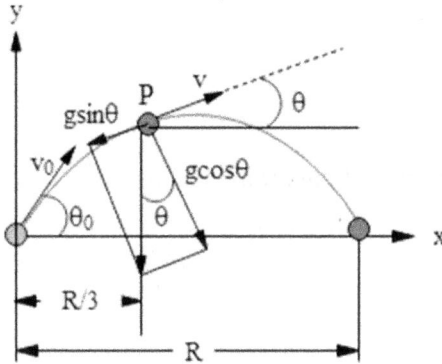

Substituting $\tan \theta_0 = 4$ in equation (3.65), we have $\tan \theta = 4/3$. Ans.

(ii) Substituting $\sin \theta = \frac{4}{5}$ and $\cos \theta = \frac{3}{5}$ in $|a_t| = g \sin \theta$ and $a_r = g \cos \theta$, we obtain

$$|a_t| = \frac{4g}{5} \quad \text{and} \quad a_r = \frac{3g}{5} \quad \text{Ans.}$$

Problem 6 A particle is projected with a velocity $\vec{v} = a\hat{i} + b\hat{j}$. Find the radius of curvature of the trajectory of the particle at the (i) point of projection (ii) highest point.

Solution

(i) Let the angle of projection be θ. At the point of projection P, $a_n = g \cos \theta_0$. Hence, the radius of curvature at P is

$$\rho_p = \frac{v_p^2}{a_n} = \frac{v_0^2}{g \cos \theta_0}.$$

Since, $\tan \theta_0 = \frac{b}{a}$, $\cos \theta_0 = \frac{a}{\sqrt{a^2 + b^2}}$, substituting $v_0 = \sqrt{a^2 + b^2}$ and $\cos \theta_0 = \frac{a}{\sqrt{a^2 + b^2}}$, we have,

$$\rho_p = \frac{(a^2 + b^2)^{3/2}}{ga} \quad \text{Ans.}$$

(ii) At the highest position Q, the velocity of the particle is $v_Q = v_0 \cos \theta_0$. Since, it moves horizontally at the highest point Q, $\vec{a}_n = \vec{g}(\perp \vec{v})$.

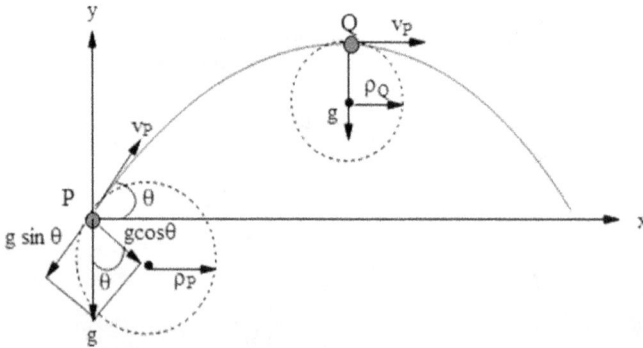

Hence, the radius of curvature at Q is

$$\rho_Q = \frac{v_Q^2}{a_n} = \frac{v_0^2 \cos^2 \theta}{g},$$

where, $v_0 \cos \theta = v_x = a$ (given)

$$\Rightarrow \rho_Q = \frac{a^2}{g} \quad \text{Ans.}$$

Problem 7 A particle is projected with a velocity v_0 so as to cover a maximum horizontal range. Find the ratio of (a) centripetal accelerations, (b) radii of curvature, at the point of projection and highest point of the path traced by the particle.

Solution

(a) Let the angle of projection be θ. At the point of projection P,

$$a_n = g \cos \theta_0$$

At the highest position Q, the velocity of the particle is $v_Q = v_0 \cos \theta_0$. Since, it moves horizontally at the highest point Q,

$$\vec{a}_n = \vec{g}\left(\perp \vec{v}\right)$$

The ratio of the centripetal accelerations is

$$g \cos \theta_0 / g = \cos \theta_0 = 1/\sqrt{2} \text{ Ans.}$$

(b) Hence, the radius of curvature at P is

$$\rho_p = \frac{v_p^2}{a_n} = \frac{v_0^2}{g \cos \theta_0}$$

Hence, the radius of curvature at Q is

$$\rho_Q = \frac{v_Q^2}{a_n} = \frac{v_0^2 \cos^2 \theta}{g}$$

So, the ratio of radii of curvature is

$$\rho_p / \rho_Q = \left(\frac{v_0^2}{g \cos \theta_0}\right) \bigg/ \left(\frac{v_0^2 \cos^2 \theta}{g}\right) = \sec^3 \theta$$

$$= (\sqrt{2})^3 = 2\sqrt{2} \text{ Ans.}$$

Problem 8 A stone is thrown from the ground such that it sweeps the top of two poles of heights b and a situated at distances a and b, respectively, from the point of projection. Find the (a) angle of projection, (b) horizontal range of the stone, and (c) maximum height attained by the stone.

Solution

(a) The stone passes through two points $P \equiv (a, b)$ and $Q \equiv (b, a)$.
Then, putting $x_1 = a_1$, $x_2 = b$; $y_1 = b$, $y_2 = a$ in the expression

$$\tan \theta = \frac{y_1 x_2^2 - y_2 x_1^2}{x_1 x_2^2 - x_2 x_1^2},$$

we have,

$$\tan \theta = \frac{b \cdot b^2 - a \cdot a^2}{ab^2 - ba^2} = \frac{a^2 + ab + b^2}{ab} \quad \text{Ans.}$$

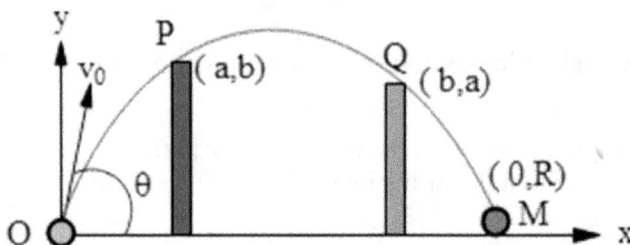

(b) The equation of trajectory can also be written as

$$y = x \tan \theta \left(1 - \frac{x}{R} \right)$$

Now substituting $x = a$, $y = b$ and $\tan \theta = \frac{a^2 + ab + b^2}{ab}$, we have,

$$R = \frac{a^2 + ab + b^2}{a + b} \quad \text{Ans.}$$

(c) The maximum height is

$$H = \frac{v_0^2 \sin^2 \theta_0}{2g}$$

The horizontal range is

$$R = \frac{2v_0^2 \sin \theta_0 \cos \theta_0}{g}$$

Using the last two equations, we have

$$\Rightarrow H/R_H = \left(\frac{v_0^2 \sin^2 \theta_0}{2g} \right) \bigg/ \left(\frac{2v_0^2 \sin \theta_0 \cos \theta_0}{g} \right)$$

$$\Rightarrow H = R \left(\frac{\tan \theta_0}{4} \right)$$

Putting the obtained value of R and $\tan \theta_0$, we have

$$H = \left(\frac{a^2 + ab + b^2}{a + b}\right)\left(\frac{\frac{a^2 + ab + b^2}{ab}}{4}\right) = \frac{(a^2 + ab + b^2)^2}{4(a + b)ab} \quad \text{Ans.}$$

N.B.: You can prove that the angle of projection is greater than $\tan^{-1} 3$.

Problem 9 Find the critical speed of projection to hit a point $P(x,y)$.
Solution
Assuming all standard notations, the locus equation is given as

$$y = x \tan \theta_0 - \frac{gx^2}{2v_0^2 \cos^2 \theta_0}$$

$$\Rightarrow y = x \tan \theta_0 - \frac{gx^2}{2v_0^2}(1 + \tan^2 \theta_0)$$

$$\Rightarrow \tan^2 \theta_0 - \frac{2v_0^2}{gx} \tan \theta_0 + \left(\frac{2v_0^2}{gx^2}y + 1\right) = 0.$$

For the above quadratic equation, for the real roots, the discriminant is

$$\Delta = \left(-\frac{2v_0^2}{gx}\right)^2 - 4\left(\frac{2v_0^2}{gx^2}y + 1\right) \geqslant 0$$

$$\Rightarrow \Delta = \left(-\frac{v_0^2}{gx}\right)^2 - \left(\frac{2v_0^2}{gx^2}y + 1\right) \geqslant 0$$

$$\Rightarrow \frac{v_0^4}{g^2x^2} - \frac{2v_0^2}{gx^2}y - 1 \geqslant 0$$

$$\Rightarrow v_0^4 - 2v_0^2 gy - g^2x^2 \geqslant 0$$

$$\Rightarrow v_0^2 \geqslant \frac{2gy \pm \sqrt{(2gy)^2 + 4(g^2x^2)}}{2}$$

$$\Rightarrow v_0^2 \geqslant g\left(y + \sqrt{x^2 + y^2}\right)$$

$$\Rightarrow v_0 \geqslant \sqrt{g\left(y + \sqrt{x^2 + y^2}\right)}.$$

Problem 10 A shell is fired from the foot of a hill along the line of the greatest slope of the slant portion of the hill such that it strikes a target situated at a distance b from the point of projection and at a height of h from the base of the hill. Find the ratio of minimum possible speed of projection to the speed of its striking the target.

Solution

Referring to the last problem, the minimum velocity of projection to hit the target P is given as

$$v_0 = \sqrt{g\left(\sqrt{x^2 + y^2} + y\right)}$$

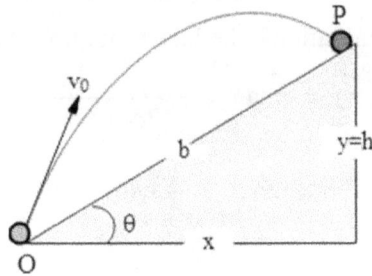

Substituting $\sqrt{x^2 + y^2} = b$ and $x = h$ in the above expression, we have

$$v_0 = \sqrt{g(b + h)} \tag{3.66}$$

The speed of striking is

$$v = \sqrt{v_0^2 - 2gy} \tag{3.67}$$

Substituting $y = h$ and $v_0 = \sqrt{g(b + h)}$ from equation (3.66) in equation (3.67), we have,

$$v = \sqrt{g(b - h)}$$

Hence, the ratio of v_0 and v is

$$\frac{v_0}{v} = \sqrt{\frac{b + h}{b - h}}$$

Problem 11 A shotput thrower of height h can throw a shot with a speed v_0. (a) Find the maximum horizontal distance that the shotputter can throw the shot. (b) What is the critical angle of projection for maximum horizontal distance?

Solution

(a) Let the shot hit the ground at point $P(x, y)$.
   P is situated below the point of projection (collision).

Substituting $x_{\max} = R$ and $y = -h$ in the expression

$$v_0 \geqslant \sqrt{g\left(y + \sqrt{x^2 + y^2}\right)},$$

we have

$$v_0 \geqslant \sqrt{g\left(\sqrt{R^2 + h^2} - h\right)}$$

$$\Rightarrow R \leqslant \frac{v_0\sqrt{v_0^2 + 2gh}}{g}$$

$$\Rightarrow R_{\max} = \frac{v_0\sqrt{v_0^2 + 2gh}}{g} \quad \text{Ans.}$$

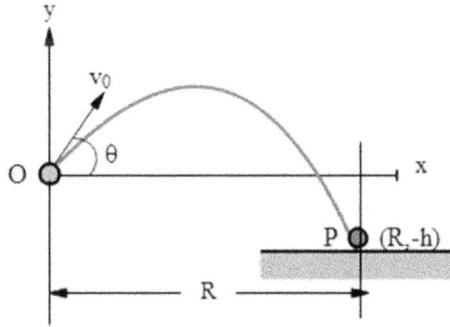

(b) The equation of trajectory is

$$\Rightarrow y = x\tan\theta - \frac{gx^2}{2v_0^2}(1 + \tan^2\theta) \tag{3.68}$$

For a critical θ, $\frac{dy}{d(\tan\theta)} = 0$ or $\frac{dy}{d\theta} = 0$

$$\Rightarrow \frac{dy}{d(\tan\theta)} = x - \frac{gx^2}{2v_0^2}(2\tan\theta) = 0$$

$$\Rightarrow \tan\theta = \frac{v_0^2}{gx} \tag{3.69}$$

Putting $x = R_{\max} = \frac{v_0\sqrt{v_0^2 + 2gh}}{g}$, we have

$$\tan\theta = \frac{v_0^2}{g\left(\dfrac{v_0\sqrt{v_0^2 + 2gh}}{g}\right)} = \frac{v_0}{\sqrt{v_0^2 + 2gh}}$$

$$\theta = \tan^{-1}\left(\frac{v_0}{\sqrt{v_0^2 + 2gh}}\right) \quad \text{Ans.}$$

Alternative method:

Putting $\tan\theta = \frac{v_0^2}{gx}$ from equation (3.69) in the locus equation (3.68), we have

$$y = x\left(\frac{v_0^2}{gx}\right) - \frac{gx^2}{2v_0^2}\left\{1 + \left(\frac{v_0^2}{gx}\right)^2\right\}$$

$$\Rightarrow y = \frac{v_0^2}{g} - \frac{gx^2}{2v_0^2}\left\{1 + \frac{v_0^4}{g^2x^2}\right\} = \frac{v_0^2}{2g} - \frac{gx^2}{2v_0^2}$$

$$\Rightarrow y = \frac{v_0^2}{2g} - \frac{gx^2}{2v_0^2}$$

$$\Rightarrow v_0^4 - 2v_0^2 gy - g^2x^2 \geqslant 0$$

$$\Rightarrow v_0^2 \geqslant \frac{2gy \pm \sqrt{(2gy)^2 + 4(g^2x^2)}}{2}$$

$$\Rightarrow v_0^2 \geqslant g(y + \sqrt{x^2 + y^2})$$

$$\Rightarrow v_0 \geqslant \sqrt{g(y + \sqrt{x^2 + y^2})}$$

Putting $y = -h$ and $x = R$ in the last expression, we can also get the answer.

Problem 12 A body has maximum range R_1 when projected up the inclined plane. The same body when projected down the inclined plane has maximum range R_2. Find (a) the maximum horizontal range, (b) the ratio of times of projection if $\beta = 30°$. Assume the equal speed of projection in each case and the body is projected onto the inclined plane in the line of the greatest slope.

Solution

(a) As derived earlier (theory of section 3.5), for upward projection,

$$R_{\max} = \frac{v_0^2}{g(1 + \sin\beta)} = R_1 \tag{3.70}$$

For downward projection,

$$R_{max} = \frac{v_0^2}{g(1 - \sin \beta)} = R_2 \qquad (3.71)$$

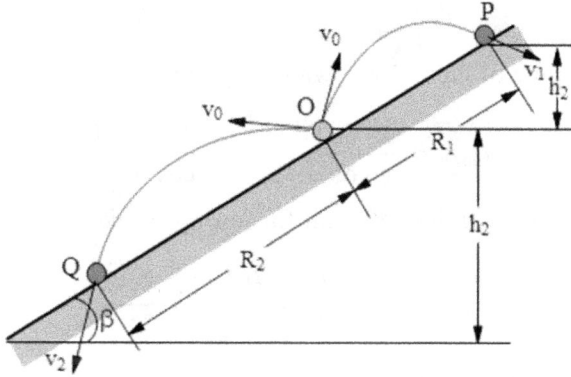

For a projection on horizontal surface substituting $\beta = 0$,
Then, we have

$$R_{max} = \frac{v_0^2}{g} = R \qquad (3.72)$$

To establish a relation between R, R_1 and R_2, we need to eliminate $\sin \beta$.
Adding $\frac{1}{R_1}$ from equation (3.70) with $\frac{1}{R_2}$ from equation (3.71), we have

$$\frac{2}{R} = \frac{1}{R_1} + \frac{1}{R_2}$$

$$\Rightarrow R = \frac{2R_1 R_2}{R_1 + R_2} \quad \text{Ans.}$$

(b) The angle of projection with horizontal for downward projection is

$$\theta = \frac{\pi}{4} + \frac{\beta}{2}$$

The angle of projection with the inclined plane for downward projection is

$$\theta_1 = \frac{\pi}{4} + \frac{\beta}{2} - \beta = \frac{\pi}{4} - \frac{\beta}{2}$$

The angle of projection with horizontal for upward projection is

$$\theta = \frac{\pi}{4} - \frac{\beta}{2}$$

The angle of projection with the inclined plane for upward projection is

$$\theta_1 = \frac{\pi}{4} - \frac{\beta}{2} + \beta = \frac{\pi}{4} + \frac{\beta}{2}$$

Then the time of flight for downward projection is

$$t_{\text{down}} = \frac{2v \sin \theta_1}{g \cos \beta} = \frac{2v \sin\left(\frac{\pi}{4} - \frac{\beta}{2}\right)}{g \cos \beta}$$

Similarly, then the time of flight for upward projection is

$$t_{\text{up}} = \frac{2v \sin \theta_2}{g \cos \beta} = \frac{2v \sin\left(\frac{\pi}{4} + \frac{\beta}{2}\right)}{g \cos \beta}$$

Then, the ratio of times is

$$\frac{t_{\text{down}}}{t_{\text{up}}} = \frac{\sin\left(\frac{\pi}{4} - \frac{\beta}{2}\right)}{\sin\left(\frac{\pi}{4} - \frac{\beta}{2}\right)} = \frac{\sin\left(45° - \frac{30°}{2}\right)}{\sin\left(45° + \frac{30°}{2}\right)}$$

$$= \frac{\sin 30°}{\sin 60°} = \frac{1}{\sqrt{3}} \quad \text{Ans.}$$

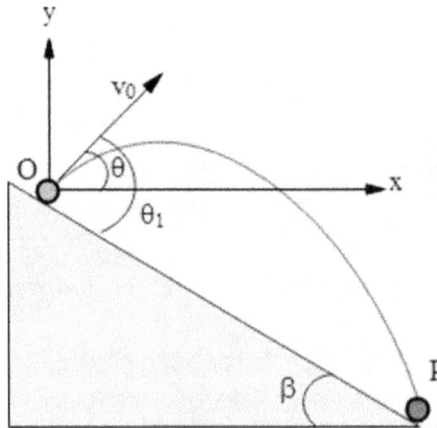

Problem 13 In the previous problem, (a) find the velocities with which the body will strike the inclined plane in terms of the angle of inclination β of the inclined plane and speed of projection v_0; (b) if the body strikes the inclined plane with speeds v_1

and v_2 for upward and downward projections onto the inclined plane, respectively, find the speed of projection of the body.

Solution

(a) Referring to the last problem, for upward projection,

$$v_0^2 = gR_1(1 + \sin \beta) \tag{3.73}$$

Then, the velocity of striking the inclined plane at P is

$$v_1^2 = v_0^2 - 2gh_1 = gR_1(1 + \sin \beta) - 2gh_1$$

$$\Rightarrow v_1^2 = gR_1(1 + \sin \beta) - 2gR_1 \sin \beta = gR_1(1 - \sin \beta)$$

$$\Rightarrow v_1^2 = gR_1(1 - \sin \beta) \tag{3.74}$$

Using equations (3.73) and (3.74), we have

$$\Rightarrow v_1 = \sqrt{\frac{(1 - \sin \beta)}{(1 + \sin \beta)}} v_0 \text{ Ans.}$$

For downward projection,

$$v_0^2 = gR_2(1 - \sin \beta) \tag{3.75}$$

Then, the velocity of striking the inclined plane at P is

$$v_2^2 = v_0^2 + 2gh_2 = gR_2(1 - \sin \beta) + 2gh_2$$

$$\Rightarrow v_2^2 = gR_2(1 - \sin \beta) + 2gR_2 \sin \beta$$

$$\Rightarrow v_2^2 = gR_2(1 + \sin \beta) \tag{3.76}$$

Using equations (3.75) and (3.76), we have

$$\Rightarrow v_2 = \sqrt{\frac{(1 + \sin \beta)}{(1 - \sin \beta)}} v_0 \text{ Ans.}$$

(b) From the last two results, we can see that

$$v_1 v_2 = \sqrt{\frac{(1 - \sin \beta)}{(1 + \sin \beta)}} v_0 \sqrt{\frac{(1 + \sin \beta)}{(1 - \sin \beta)}} v_0 = v_0^2$$

$$v_0 = \sqrt{v_1 v_2} \text{ Ans.}$$

Problem 14 (a) A large trolley car moves down an inclined plane of angle of inclination β with an acceleration a, find the range of a shell along the inclined plane, which is projected up the inclined plane at an angle θ with a trolley car with a velocity u relative to the sledge at $t = 0$.

(b) If the trolley car is smooth, find the:

(i) range of the projectile (shell) relative to the trolley car.

(ii) maximum height attained by the shell relative to the point of projection;

(iii) maximum height attained by the shell relative to the trolley car.

Solution

(a) Let us fix the coordinate x-axis with the trolley car. Then we write

$$\vec{s}_{SC} = R\hat{i} + 0\hat{j},$$

$$\vec{u}_{sc} = u\cos\theta\hat{i} + u\sin\theta\hat{j} \quad \text{(given)},$$

and

$$\vec{a}_{sc} = \vec{a}_s - \vec{a}_c,$$

where $\vec{a}_s = -g\sin\beta\hat{i} - g\cos\beta\hat{j}$ and $\vec{a}_c = -a\hat{i}$.

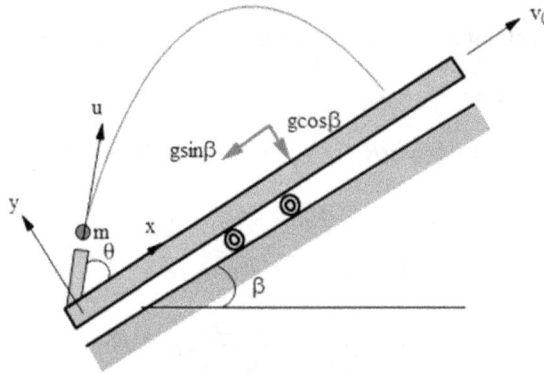

Substituting all the above quantities in $\vec{s}_{sc} = \vec{u}_{sc}t + \frac{1}{2}\vec{a}_{sc}t^2$, we have the following two equations:

$$t = \frac{2u\sin\theta}{g\cos\beta} \tag{3.77}$$

$$R = ut\cos\theta - \frac{1}{2}(g\sin\beta - a)t^2 \tag{3.78}$$

Substituting the value of t from equation (3.77) in equation (3.78), we have

$$R = \frac{2u^2 \sin\theta \cos\theta}{g \cos\beta} - \frac{2u^2 \sin^2\theta}{g^2 \cos^2\beta}(g \sin\beta - a) \text{ Ans.}$$

(b) (i) If the sledge is smooth, the acceleration of the trolley car is

$$a = g \sin\beta$$

So, the range of the shell is

$$R = \frac{2u^2 \sin\theta \cos\theta}{g \cos\beta} - \frac{2u^2 \sin^2\theta}{g^2 \cos^2\beta}(g \sin\beta - g \sin\beta)$$

$$= \frac{2u^2 \sin\theta \cos\theta}{g \cos\beta} = \frac{u^2 \sin 2\theta}{g \cos\beta} \text{ Ans.}$$

(ii) The vertical component of the velocity of the shell is

$$v_{\text{vertical}} = u \sin(\theta + \beta) + v_0 \sin\beta$$

The maximum height attained by the shell relative to the point of projection (ground frame) is given as

$$H = \frac{u^2_{\text{vertical}}}{2g} = \frac{\{u \sin(\theta + \beta) + v_0 \sin\beta\}^2}{2g} \text{ Ans.}$$

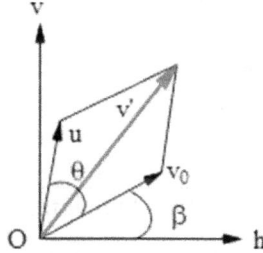

(iii) The vertical component of the velocity of the shell relative to the trolley car is

$$v'_{\text{vertical}} = u \sin(\theta + \beta)$$

The maximum height attained by the shell relative to the trolley car is given as

$$H = \frac{v'^2_{\text{vertical}}}{2g'} = \frac{u^2 \sin^2(\theta + \beta)}{2g \cos\beta} \text{ Ans.}$$

Problem 15 A body is projected up with a speed v_0 along the line of greatest slope of an inclined plane of angle of inclination β. If the body collides elastically perpendicular to the inclined plane, find the (a) angle of projection relative to the inclined plane, (b) time after which the body passes through its point of projection, (c) velocity of striking for the first time, (d) range of the body along the inclined plane. (e) The height at which the projectile strikes the inclined plane, (f) maximum height attained by the projectile from the point of projection.

Solution

(a) Referring to the theory of the section 3.5, we have

$$v_x = v_0 \cos \theta (1 - 2 \tan \theta \tan \beta) \qquad (3.79)$$

$$v_y = -v_0 \sin \theta \qquad (3.80)$$

Using last two equations, we have

$$\tan \phi = \frac{v_y}{v_x} = \frac{v_0 \sin \theta}{v_0 \cos \theta (1 - 2 \tan \theta \tan \beta)}$$

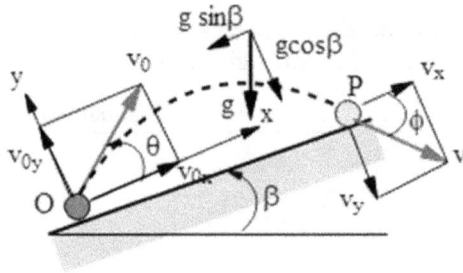

The body is projected up on to the inclined plane

$$\Rightarrow \tan \phi = \frac{\tan \theta}{1 - 2 \tan \theta \tan \beta} \qquad (3.81)$$

As the body strikes the inclined plane perpendicularly, putting $\phi = \frac{\pi}{2}$ radian in the last equation, we have,

$$\tan \frac{\pi}{2} = \frac{\tan \theta}{1 - 2 \tan \theta \tan \beta}$$

$$\Rightarrow \tan \theta = \frac{\cot \beta}{2} \quad \text{Ans.}$$

(b) By using trigonometrical calculation, we have

$$\sin \theta = \frac{\cos \beta}{\sqrt{1 + 3 \sin^2 \beta}} \quad \text{and} \quad \cos \theta = \frac{2 \sin \beta}{\sqrt{1 + 3 \sin^2 \beta}}$$

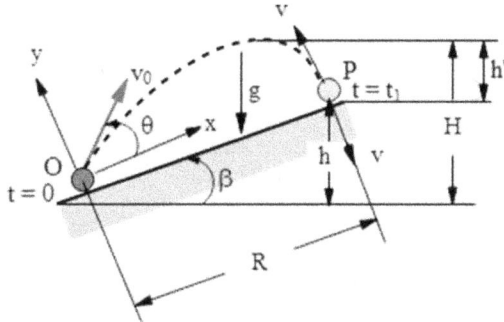

Substituting $\sin \theta$ in the expression of time of flight (please refer the theory of section 3.5),

$$t = \frac{2v_0 \sin \theta}{g \cos \beta}$$

we have,

$$t = \frac{2v_0}{g\sqrt{1 + 3 \sin^2 \beta}}$$

Since the body collides elastically it will rebound perpendicular to the inclined plane with the same speed v. In consequence, it retraces its path elapsing equal time t for a backward journey to the point of projection. Hence, the required time of motion is

$$t' = 2t = \frac{4v_0}{g\sqrt{1 + 3 \sin^2 \beta}} \quad \text{Ans.}$$

(c) The velocity of striking is given by equation (3.80) as

$$v_y = -v_0 \sin \theta = v, \quad \text{say}$$

Putting $\sin \theta = \frac{\cos \beta}{\sqrt{1 + 3 \sin^2 \beta}}$, we have

$$v_y = v = \frac{v_0 \cos \beta}{\sqrt{1 + 3 \sin^2 \beta}} \quad \text{Ans.}$$

(d) The range of the projectile along the inclined plane is

$$R = \frac{1}{2}(g \sin \beta)t^2$$

Putting the obtained value of time, we have

$$R = \frac{1}{2}(g \sin \beta)\left(\frac{2v_0}{g\sqrt{1 + 3\sin^2 \beta}}\right)^2$$

$$R = \frac{2v_0^2 \sin \beta}{g(1 + 3\sin^2 \beta)}$$

(e) The height at which the projectile strikes the inclined plane is

$$h = R \sin \theta = \frac{2v_0^2 \sin \beta}{g(1 + 3\sin^2 \beta)} \sin \theta$$

Putting $\sin \theta = \frac{\cos \beta}{\sqrt{1 + 3\sin^2 \beta}}$, we have

$$h = \left(\frac{2v_0^2 \sin \beta}{g(1 + 3\sin^2 \beta)}\right)\left(\frac{\cos \beta}{\sqrt{1 + 3\sin^2 \beta}}\right)$$

$$\Rightarrow h = \frac{2v_0^2 \sin \beta \cos \beta}{g(1 + 3\sin^2 \beta)^{3/2}} \quad \text{Ans.}$$

(f) The maximum height attained by the projectile is

$$H = \frac{v_0^2 \sin^2 \beta}{2g \cos \beta} + h$$

where

$$h = \frac{2v_0^2 \sin \beta \cos \beta}{g(1 + 3\sin^2 \beta)^{3/2}}$$

$$\Rightarrow H = \frac{v_0^2 \sin^2 \beta}{2g \cos \beta} + \frac{2v_0^2 \sin \beta \cos \beta}{g(1 + 3\sin^2 \beta)^{3/2}}$$

$$\Rightarrow H = \frac{v_0^2 \sin^2 \beta}{2g \cos \beta}\left(1 + \frac{4\cos^2 \beta}{(1 + 3\sin^2 \beta)^{3/2} \sin \beta}\right) \quad \text{Ans.}$$

Problem 16 A ball is projected up from O with a speed v_0 along the line of greatest slope of an inclined plane of angle of inclination β. The ball collides elastically with the inclined plane at P and moves vertically up to point Q; then it comes down and again collides elastically with the inclined plane at P and O to retrace its path. Find the (a) angle of projection relative to the inclined plane, (b) velocity of striking for the first time, (c) time after which the ball collides first time from O to P, (d) time taken from P to Q range of the body along the inclined plane, (e) time after which the ball collides with the point of projection O again, (f) the height PQ, and (g) range of the projectile along the inclined plane.

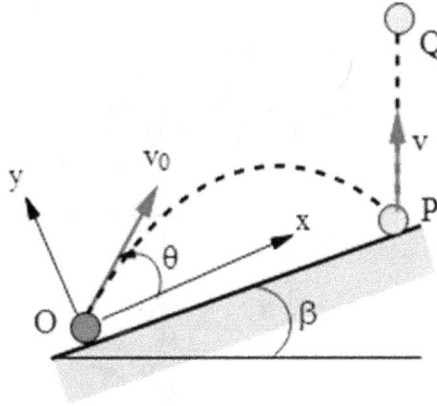

Solution

(a) Referring to the theory of section 3.5, we have

$$v_x = v_0 \cos \theta (1 - 2 \tan \theta \tan \beta) \tag{3.82}$$

$$v_y = -v_0 \sin \theta \tag{3.83}$$

Using last two equations, we have

$$\tan \phi = \frac{v_y}{v_x} = \frac{v_0 \sin \theta}{v_0 \cos \theta (1 - 2 \tan \theta \tan \beta)}$$

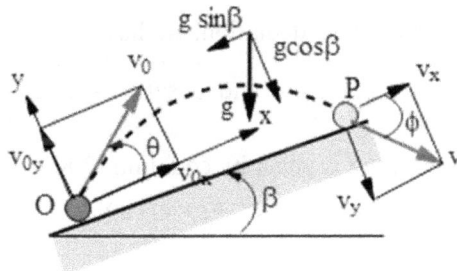

The body is projected up on to the inclined plane

$$\Rightarrow \tan\phi = \frac{\tan\theta}{1 - 2\tan\theta\tan\beta} \qquad (3.84)$$

As the body strikes the inclined plane at P and moves vertically up, referring to the following figure, we have

$$\phi = \frac{\pi}{2} - \beta \qquad (3.85)$$

Using the last equations, we have,

$$\cot\beta = \frac{\tan\theta}{1 - 2\tan\theta\tan\beta} \qquad (3.86)$$

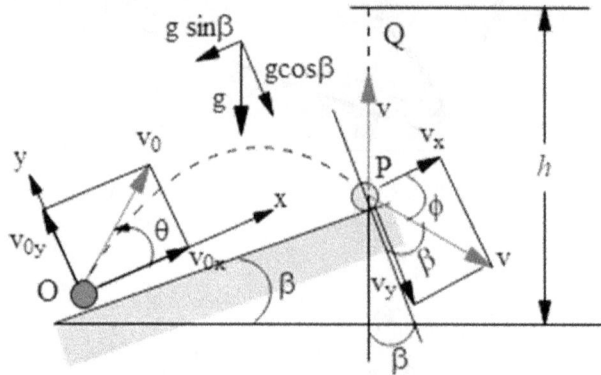

$$\Rightarrow \cot\beta - 2\tan\theta = \tan\theta$$

$$\Rightarrow 3\tan\theta = \cot\beta$$

$$\Rightarrow \theta = \tan^{-1}\left(\frac{\cot\beta}{3}\right) \text{ Ans.}$$

(b) By using trigonometrical calculation, we have

$$\sin\theta = \frac{\cos\beta}{\sqrt{1 + 8\sin^2\beta}} \text{ and } \cos\theta = \frac{3\sin\beta}{\sqrt{1 + 8\sin^2\beta}}$$

The velocity of striking is given by equation (3.83) as

$$v_y = -v_0\sin\theta = v, \text{ say}$$

Putting $\sin\theta = \frac{\cos\beta}{\sqrt{1 + 8\sin^2\beta}}$, we have

$$v_y = v = \frac{v_0\cos\beta}{\sqrt{1 + 8\sin^2\beta}} \qquad (3.87)$$

Putting $\tan \theta = \cot \beta / 3$ and $\cos \theta = \dfrac{3 \sin \beta}{\sqrt{1 + 8 \sin^2 \beta}}$ in equation (3.82), we have

$$v_x = v_0 \cos \theta (1 - 2 \tan \theta \tan \beta)$$

$$= v_0 \dfrac{3 \sin \beta}{\sqrt{1 + 8 \sin^2 \beta}} \{ (1 - 2(\cot \beta / 3) \tan \beta)$$

$$v_x = v_0 \dfrac{\sin \beta}{\sqrt{1 + 8 \sin^2 \beta}} \qquad\qquad (3.88)$$

Using the last two equations, we have

$$v = \sqrt{v_x^2 + v_y^2} = \dfrac{v_0}{\sqrt{1 + 8 \sin^2 \beta}} \quad \text{Ans.}$$

(c) Substituting $\sin \theta$ in the expression of time of flight (please refer the theory of the section 3.5), the time taken from O to P is

$$t = \dfrac{2 v_0 \sin \theta}{g \cos \beta}$$

we have,

$$t = \dfrac{2 v_0}{g \sqrt{1 + 8 \sin^2 \beta}} \quad \text{Ans.}$$

(d) Just after the elastic collision, the ball moves vertically up from P to the highest point Q elapsing a time

$$t' = \dfrac{v}{g}$$

Putting $v = \dfrac{v_0}{\sqrt{1 + 8 \sin^2 \beta}}$, we have

$$t' = \dfrac{v}{g} = \dfrac{v_0}{g \sqrt{1 + 8 \sin^2 \beta}} \quad \text{Ans.}$$

(e) Then, the total time for one-way journey (from O to Q) of the ball is

$$T' = t + t' = t = \dfrac{2 v_0}{g \sqrt{1 + 8 \sin^2 \beta}} + \dfrac{v_0}{g \sqrt{1 + 8 \sin^2 \beta}} = \dfrac{3 v_0}{g \sqrt{1 + 8 \sin^2 \beta}}$$

As it will take the same time to retrace the path, the total time for to-and-fro journey of the ball is

$$T = 2 T' = 2(t + t')$$

$$\Rightarrow T = \frac{6v_0}{g\sqrt{1 + 8\sin^2\beta}} \quad \text{Ans.}$$

(f) The height PQ is given as

$$h' = \frac{v^2}{2g} = \left(\frac{v_0}{\sqrt{1 + 8\sin^2\beta}}\right)^2 / 2g = \frac{v_0^2}{2(1 + 8\sin^2\beta)g}$$

(g) By energy conservation or kinematics, the maximum height is given as

$$h = \frac{v_0^2}{2g}$$

Then, the range along the inclined plane is given as

$$R = (h - h') / \sin\beta$$

$$= \left\{\frac{v_0^2}{2g} - \frac{v_0^2}{2(1 + 8\sin^2\beta)g}\right\} / \sin\beta$$

$$= \frac{4v_0^2 \sin\beta}{(1 + 8\sin^2\beta)g} \quad \text{Ans.}$$

Example 10 A shell is projected from a gun with a muzzle velocity v. The gun is fitted with a trolley car at an angle θ as shown in the figure. If the trolley car is made to move with constant velocity v towards right, find the (a) horizontal range of the shell relative to ground, (b) the horizontal range R of the shell relative to the trolley car, and (c) difference in range of the shell in the ground frame and the trolley car.

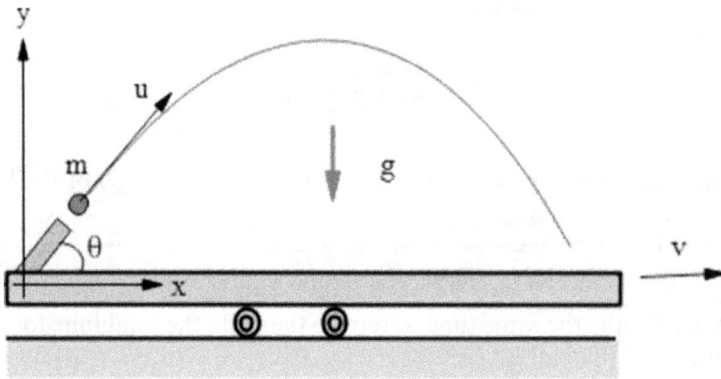

Solution

(a) The velocity of projection of the shell is
$$\vec{v}_s = \vec{v}_{sc} + \vec{v}_c$$

The velocity of the shell relative to the trolley car is
$$\vec{v}_{sc} = u \cos \theta \hat{i} + u \sin \theta \hat{j}$$

Adding it with $\vec{v}_c = v\hat{i}$, the velocity of the shell relative to the trolley car is
$$\vec{u} = \vec{v}_s = \vec{v}_{sc} + \vec{v}_c = (u \cos \theta + v)\hat{i} + u \sin \theta \hat{j}$$

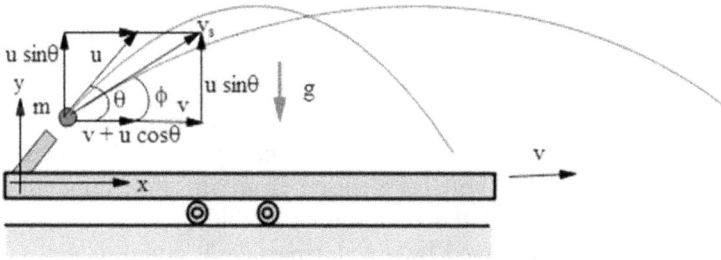

For horizontal range R of the shell its displacement is $\vec{s} = R\hat{i}$. Substituting $\vec{a} = -g\hat{j}$, $\vec{s} = R\hat{i}$ and $\vec{u} = \vec{v}_s = (u \cos \theta + v)\hat{i} + u \sin \theta \hat{j}$ in
$$\vec{s} = \vec{u}t + \frac{1}{2}\vec{a}t^2,$$

we have
$$R\hat{i} = (u \cos \theta + v)t\,\hat{i} + \left(ut \sin \theta - \frac{1}{2}gt\right)\hat{j}$$

Comparing the coefficients of $\hat{i}$ and $\hat{j}$, we obtain
$$R = (u \cos \theta + v)t \tag{3.89}$$

$$ut \sin \theta - \frac{1}{2}gt^2 = 0 \tag{3.90}$$

From equation (3.90), we find
$$t = \frac{2u \sin \theta}{g} \tag{3.91}$$

Finally, substituting the value of time in equation (3.89), we have
$$R = \frac{2u \sin \theta(u \cos \theta + v)}{g} \quad \text{Ans.}$$

(b) The horizontal range R of the shell relative to the trolley car is

$$R' = (u \cos \theta)t = \frac{2u^2 \sin \theta \cos \theta}{g} \quad \text{Ans.}$$

(c) So, the difference in the two ranges is

$$\Delta R = R - R' = \frac{2u \sin \theta (u \cos \theta + v)}{g} - \frac{2u^2 \sin \theta \cos \theta}{g}$$

$$\Rightarrow \Delta R = \frac{2uv \sin \theta}{g}$$

N.B.: In both reference frame the time of flight is the same.

Problem 17 A disc rotates about the horizontal axis passing through the pivot O with a constant angular velocity ω, as shown in the figure. If we pour water on the rotating disc, water particles will be released from the disc tangentially with a constant speed $v = R\omega$. Find the minimum possible angular speed so that the maximum possible height attained by a water particle is equal to H.

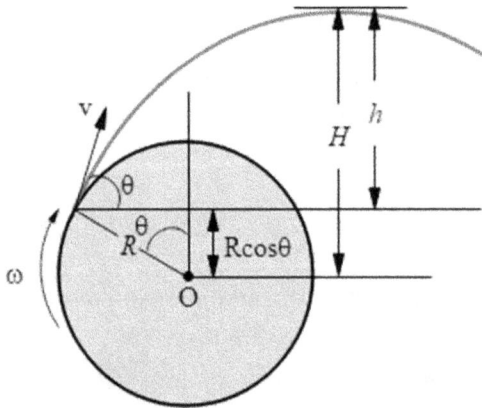

Solution
Let a water particle leave the disc at angular position θ with a velocity $v = R\omega$. The maximum height attained by the water particle is

$$H = R \cos \theta + \frac{(v \sin \theta)^2}{2g} \tag{3.92}$$

For maximum possible height, its derivative relative to θ will be zero.

$$\frac{dH}{d\theta} = -R \sin\theta + \frac{v^2}{2g}2 \sin\theta \cos\theta = 0$$

$$\Rightarrow \cos\theta = \frac{gR}{v^2} \tag{3.93}$$

Putting the obtained value $\cos\theta$ from equation (3.93) in equation (3.92),

$$H = v^2 + \frac{v^2}{2g}\left\{1 - \left(\frac{gR}{v^2}\right)^2\right\}$$

$$\Rightarrow H = \frac{v^2}{2g} - \frac{gR^2}{2v^2}$$

$$\Rightarrow H = \frac{v^4 - g^2R^2}{2gv^2}$$

$$\Rightarrow H = \frac{v^4 - g^2R^2}{2gv^2}$$

$$\Rightarrow v^4 - 2gHv^2 - g^2R^2 = 0$$

$$\Rightarrow v^4 - 2gHv^2 - g^2R^2 = 0$$

$$\Rightarrow v^2 = \frac{2gH \pm \sqrt{(2gH)^2 - 4(-g^2R^2)}}{2}$$

$$\Rightarrow v^2 = g(H + \sqrt{H^2 + R^2})$$

$$\Rightarrow v = \sqrt{g(H + \sqrt{H^2 + R^2})}$$

So, the minimum possible angular speed is

$$\omega = v/R = \sqrt{g(H + \sqrt{H^2 + R^2})}/R \text{ Ans.}$$

Example 11 A man is standing on the ground in rain steadily falling vertically with a speed of 2 m s^{-1}. If he starts moving with 1 m s^{-1} due east, what will be the velocity of the rain? In which direction should he hold the umbrella to protect himself from the rain in the best possible way?
Solution
Practically the man must hold the umbrella in the direction of motion of the rainfall relative to him $(\vec{v}_{rm})$.

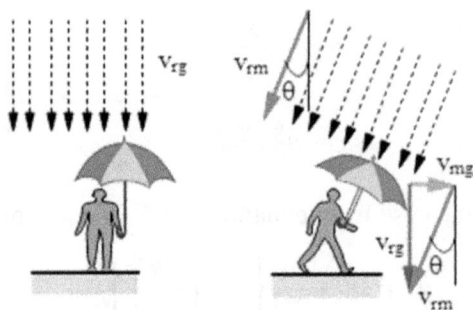

The rain falls vertically down relative to ground, but falls
backward at an angle $\theta = \tan^{-1}(v_{mg}/v_{rg})$ with vertical

Therefore, we need to find $\vec{v}_{rm}$ which can be given as:

$$\vec{v}_{rm} = \vec{v}_r - \vec{v}_m$$

Substituting $\vec{v}_r = -2\hat{j}$ and $\vec{v}_m = 1\,\hat{i}$, we have,

$$\vec{v}_{rm} = -(\hat{i} + 2\hat{j})\,ms^{-1}$$

This yields, $\phi = \tan^{-1}\left(\frac{v_m}{v_r}\right) = \tan^{-1}\left(\frac{1}{2}\right)$

That means the umbrella should be held at an angle $\phi = \tan^{-1}\left(\frac{1}{2}\right)$ to the vertical
in a forward direction as shown in the above figure.

The umbrella should be held in the direction of the rainfall relative to the man
(running in the rain). In other words, he should hold the umbrella in the direction of
$\vec{v}_m = \vec{v}_r - \vec{v}_m$. When the man is stationary, $\vec{v}_{r_m} = \vec{v}_{rg}$; this means that velocity of rain
relative to man and ground are equal as a special case.

Example 12 To a man moving with 1 m s^{-1} due east, the rain appears to fall
vertically with a speed of 2 m s^{-1}. If the man stops walking, in which direction will
the rain appear to fall relative to the man?

Solution

When the man stops walking, he can see the velocity of rain relative to ground
which is given as

$$\vec{v}_r = \vec{v}_m + \vec{v}_{rm}$$

Substituting $\vec{v}_{rm} = -2\hat{j}$ and $\vec{v}_m = 2\,\hat{i}$, we have,

$$\vec{v}_{rg} = -(2\hat{i} - 2\hat{j})\,\text{m s}^{-1}$$

The rain falls at an angle

$$\phi = \tan^{-1}\left(\frac{v_m}{v_r}\right) = \tan^{-1}\left(\frac{2}{2}\right) = 45°$$

with the vertical from backward as shown in the figure.

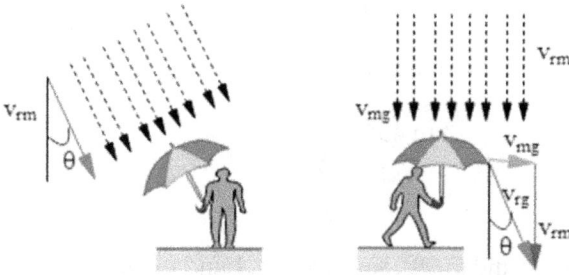

The rain falls vertically down relative to man, but falls forward at an angle $\theta = \tan^{-1}(v_{mg}/v_{rm})$ relative to ground

Example 13 *(Man in rain)*

To a man moving with a velocity with a speed of 2 m s^{-1} to the east, the rain appears to fall vertically at an angle of 37° with the vertical. When the man increases his speed by 1 m s^{-1}, the rain appears to fall at an angle of 45°. Find the actual velocity of rain.

Solution

Method 2

Let the actual velocity of rain be

$$\vec{v}_{rg} = (x\hat{i} - y\hat{j})\,\text{ms}^{-1} \tag{3.94}$$

The velocity of rain relative to man in the first case is

$$\vec{v}_r = \vec{v}_m + \vec{v}_{rm} \tag{3.95}$$

Using the last two equations, we have

$$\vec{v}_r = \vec{v}_{rg} = (x\hat{i} - y\hat{j}) + 2\hat{i}$$

$$\Rightarrow \vec{v}_r = (x\hat{i} - y\hat{j}) + 2x\hat{i} = (x + 2)\,\hat{i} - y\hat{j} \tag{3.96}$$

If the rain appears to fall at an angle of 37° with the vertical, using equation (3.95),

$$\tan 37° = \left(\frac{x + 2}{y}\right) = \frac{3}{4}$$

$$\Rightarrow 3y - 4x = 2 \tag{3.97}$$

Substituting $\vec{v'}_m = (2 + 1)\hat{i} = 3\hat{i}$ in equation (3.95), we have

$$\vec{v}_r = (x\hat{i} - y\hat{j}) + 3\hat{i}$$

$$\Rightarrow \vec{v}_r = \vec{v}_{rg} = (x\hat{i} - y\hat{j}) + 3\hat{i} = (x+3)\,\hat{i} - y\hat{j} \tag{3.98}$$

If the rain appears to fall at an angle of 45° with vertical, using equation (3.95),

$$\tan 45° = \left(\frac{x+3}{y}\right) = 1$$

$$\Rightarrow y - x = 3 \tag{3.99}$$

Solving equations (3.97) and (3.99), we have $x = 1$ and $y = 4$; so, the velocity of rain is

$$\vec{v}_{rg} = x\hat{i} - y\hat{j} = (\hat{i} - 4\hat{j}) \text{ m s}^{-1} \text{ Ans.}$$

Method 2

Let the actual velocity of rain be given as

$$\vec{v}_r = \vec{v}_m + \vec{v}_{rm} = \vec{v}'_m + \vec{v}'_{rm}$$

This forms two vector triangles.

$$\vec{v}_m = \vec{v}_{rm} + \vec{v}_m = \vec{v}'_{rm} + \vec{v}'_m$$

Putting $\vec{v}_m = 2\hat{i}$, $\vec{v}'_m = 3\hat{i}$ and the angles of inclination $\theta = 37°$ and $\beta = 45°$, we can solve for the velocity of rain relative to ground v_r which is represented by the side AD.

Let $BE = x$ and $DE = y$; then, $EC = x + 1$. By using the properties of a triangle, in triangle BDE,

$$\tan \theta = \tan 37° = \left(\frac{x}{y}\right) = \frac{3}{4}$$

$$\Rightarrow y = 4x/3 \tag{3.100}$$

In triangle BDE,

$$\tan \beta = \tan 45° = \left(\frac{BE + BC}{ED}\right) = \left(\frac{x+1}{y}\right) = 1$$

$$\Rightarrow y - x = 1 \tag{3.101}$$

Solving equations (3.100) and (3.101), we have $x = 1$ and $y = 4$; so, the velocity of rain is

$$\vec{v}_{rg} = AE\hat{i} - ED\hat{j} = (\hat{i} - 4\hat{j})\,\text{ms}^{-1}\ \text{Ans.}$$

Example 14 (*Minimum time of crossing a river*)

If v_{bw} is the velocity of a boat relative to water and v_w is the speed of water flow, assuming d is the width of river, (a) derive an expression for time of crossing of the boat, (b) find the minimum time of crossing the river, and (c) find the drift corresponding to the minimum time of crossing the river. Assume $\theta = $ angle of heading of the boat.

Solution

(a) The velocity of the boat is

$$\vec{v}_b = \vec{v}_{bw} + \vec{v}_w$$
$$= -v_{bw}\cos\theta\hat{i} + v_{bw}\sin\theta\hat{j} + v_w\hat{i}$$
$$= (v_w - v_{bw}\cos\theta)\hat{i} + v_{bw}\sin\theta\hat{j}$$

The y-component of $\vec{v}_b$ is responsible for crossing the river.

Hence, $t = \dfrac{d}{v_{bg}} = \dfrac{d}{v_{bw}\sin\theta}$ Ans.

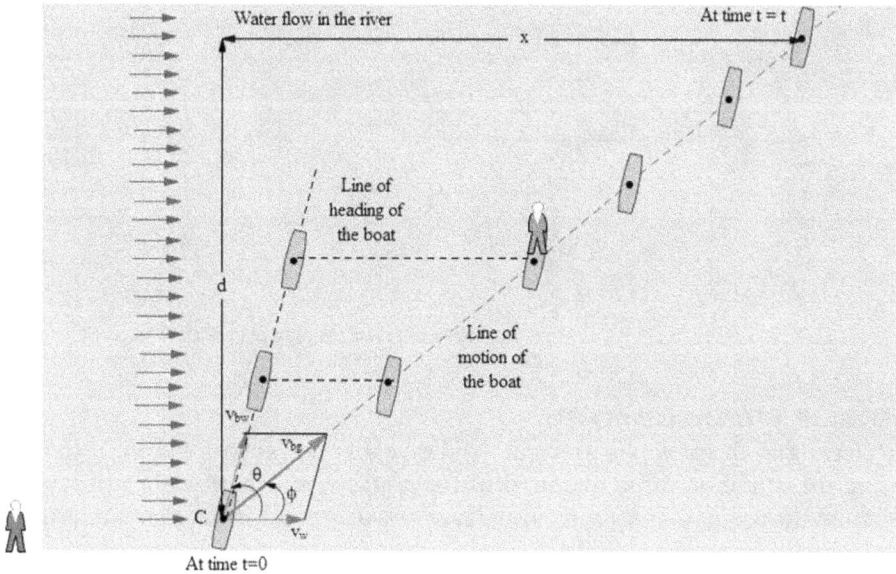

(b) In the last expression, if $\sin\theta = 1$, the time of crossing will be minimum. So, the minimum time of crossing is given as

$$t_{min} = \frac{d}{v_{bw}} \quad \text{Ans.}$$

(c) As $\sin\theta = 1$, $\cos\theta = 0$; so, the boat must head perpendicular to the water flow velocity and move with a velocity

$$\vec{v_b} = v_{bw}\hat{j}$$

Then the drift of the boat will be

$$x = v_w t = v_w\left(\frac{d}{v_{bw}}\right) = \left(\frac{v_w}{v_{bw}}\right)d \quad \text{Ans.}$$

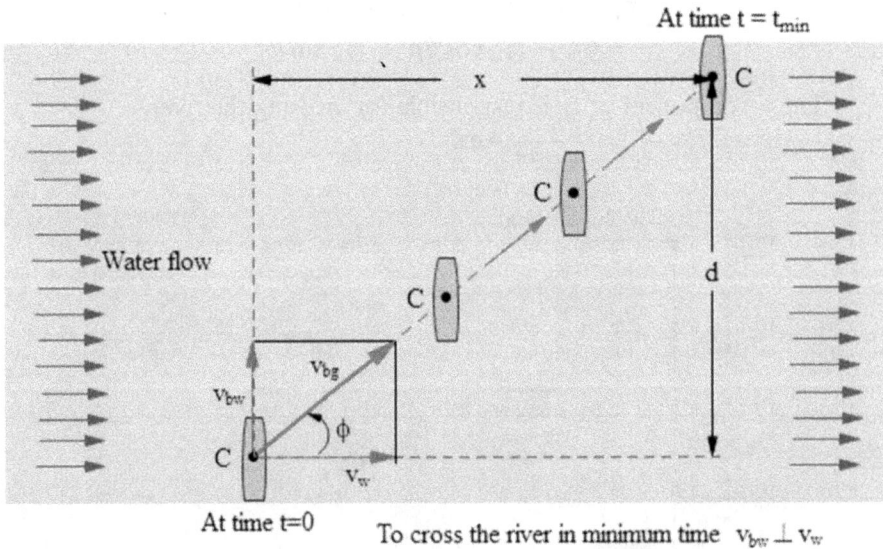

At time t = t_{min}

Water flow

At time t=0   To cross the river in minimum time $v_{bw} \perp v_w$

Problem 18 *(Minimum drift)*

Referring to the previous problem, (i) derive an expression for drift of the boat, (ii) find the condition for minimum drift for (a) $v_{bw} < v_w$, (b) $v_{bw} > v_w$, (iii) find the minimum drift x_{min} for $v_{bw} > v_w$. (iv) Trace the diagram for the minimum drift.

Solution

(i) The x-component of v_b is responsible for the drift x-displacement of the boat.

$$x = v_{bx}t$$
$$= (v_w - v_{bw}\cos\theta)\frac{d}{v_{bw}\sin\theta}$$
$$= \left(\frac{v_w}{v_{bw}}\cos ec\theta - \cot\theta\right)d \quad \text{Ans.}$$

(ii) (a) For x to be minimum,

$$\frac{dx}{d\theta} = 0$$

or, $\dfrac{v_w}{v_{bw}}(-\cos ec\theta\cot\theta) + \cos ec^2\theta = 0$

Then, $\cos\theta = \dfrac{v_{bw}}{v_w}$

So, the angle of heading of the boat is

$$\theta = \cos^{-1}\left(\frac{v_{bw}}{v_w}\right); \quad v_{bw} < v_w \quad \text{Ans.}$$

(b) If $v_{bw} > v_w$, x can be equal to zero then, $\theta = \cos^{-1}\left(-\dfrac{v_w}{v_{bw}}\right)$ Ans.

(iii) As obtained earlier, for minimum drift, $\cos\theta = \left(-\dfrac{v_{bw}}{v_w}\right); \quad v_{bw} < v_w$. So, we have

$$\sin\theta = \frac{\sqrt{v_w^2 - v_{bw}^2}}{v_w}$$

Putting these values in the expression

$$x = \left(\frac{v_w}{v_{bw}}\cos ec\theta - \cot\theta\right)d$$

and simplifying the factors, we have

$$x_{\min} = \frac{\sqrt{v_w^2 - v_{bw}^2}}{v_{bw}}d \quad \text{Ans.}$$

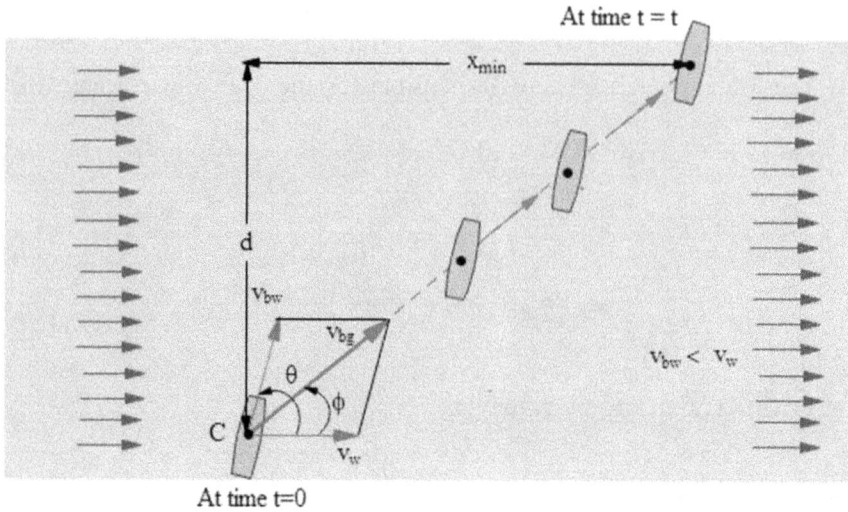

At time t=0

For $v_{bw} > v_w$, we can have zero drift; so, we can attain the right opposite point of the shore if the angle of heading will be given as $\cos\theta = -v_w/v_{bw}$.

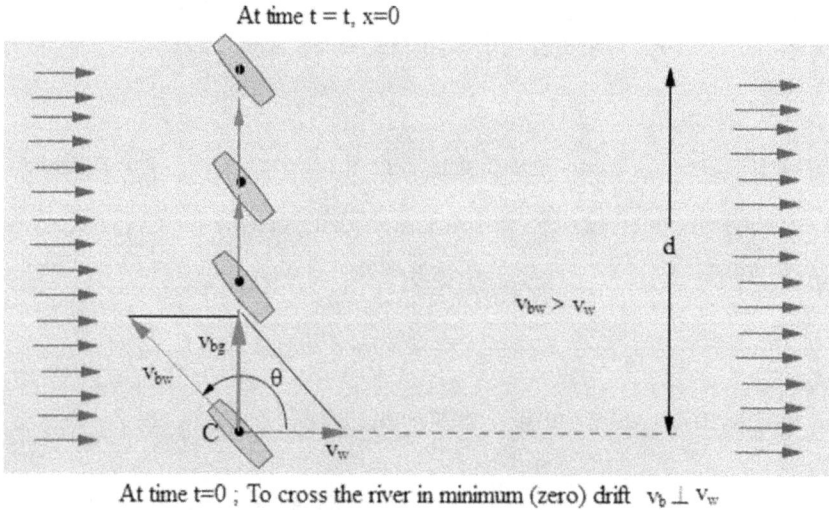

At time t=0 ; To cross the river in minimum (zero) drift $v_b \perp v_w$

(iv) In the above diagrams of minimum drift we have shown that when velocity of the boat is greater than the velocity of water, we can get the right opposite point. Otherwise, we can get the nearest point of the right opposite point. Ans.

Problem 19 (*Time for minimum drift and drift for the minimum time*)

Referring to the previous problem, for $v_{bw} < v_w$, $v_{bw} = v_w$ and $v_{bw} > v_w$, find the time of crossing for the minimum drift of the boat.

Solution

As obtained earlier, for minimum drift, $\cos\theta = \left(-\dfrac{v_{bw}}{v_w}\right)$; $v_{bw} < v_w$

So, we have

$$\sin\theta = \frac{\sqrt{v_w^2 - v_{bw}^2}}{v_w}$$

Then, the time taken to cross the river is

$$t = \frac{d}{v_{bg}} = \frac{d}{v_{bw}\sin\theta},$$

Using the last two equations, we have

$$t = \frac{d}{v_{bw}\left(\dfrac{\sqrt{v_w^2 - v_{bw}^2}}{v_w}\right)} = \frac{v_w d}{v_{bw}\sqrt{v_w^2 - v_{bw}^2}} \quad \text{Ans.}$$

For $v_{bw} > v_w$, we can have zero drift; so, the boat can attain the right opposite point of the shore if the angle of heading will be given as

$$\cos\theta = \left(-\frac{v_w}{v_{bw}}\right).$$

So, we have

$$\sin\theta = \frac{\sqrt{v_{bw}^2 - v_w^2}}{v_{bw}}$$

Then, the time taken to cross the river is

$$t = \frac{d}{v_{bg}} = \frac{d}{v_{bw}\sin\theta}$$

Using last two equations, we have

$$t = \frac{d}{v_{bw}\left(\dfrac{\sqrt{v_{bw}^2 - v_w^2}}{v_{bw}}\right)} = \frac{d}{\sqrt{v_w^2 - v_{bw}^2}} \quad \text{Ans.}$$

At time t = t, x=0

At time t=0 ; To cross the river in minimum (zero) drift $v_b \perp v_w$

Problem 20 A boat rows with a velocity v_{bw} in still water. If the water flows with a velocity $v_w(<v_{bw})$ the boat crosses the river in minimum time T_1. If the boat crosses the river perpendicularly in time T_2. The boat rows in the downstream by a distance equal of the width of the river and comes back to his initial position by swimming upstream along the shore a time T_3. Find the (a) relation between T_1, T_2 and T_3, (b) the average speed of the boat in each case in terms of v_b, v_{bw} and v_w.

Solution

(a) Let $d=$ width of the river

As derived earlier, the minimum time of crossing the river is

$$T_1 = \frac{d}{v_{bw}} \tag{3.102}$$

For the minimum drift, the time of crossing the river is

$$T_2 = \frac{d}{v_b} = \frac{d}{\sqrt{v_{bw}^2 - v_w^2}} \tag{3.103}$$

When the boat moves in downstream, its velocity is $v_b = v_{bw} + v_w$ and when he moves in upstream his velocity is $v'_b = v_{bw} - v_w$.

So, the total time for the to and from journey is

$$T_3 = t_{\text{down}} + t_{\text{up}} \tag{3.104}$$

Putting $t_{\text{down}} = \frac{d}{v_b} = \frac{d}{v_{bw} + v_w}$ and $t_{\text{up}} = \frac{d}{v'_b} = \frac{d}{v_{bw} - v_w}$ in equation (3.104) and simplifying the factors, we have

$$T_3 = \frac{2dv_{bw}}{v_{bw}^2 - v_w^2}$$

(3.105)

Comparing the above expressions of T_1, T_2 and T_3, we have

$$T_2^2 = 2T_1 T_3 \quad \text{Ans.}$$

(b) The average speed of the boat in the first case is

$$v_{b_1} = \sqrt{v_{bw}^2 + v_w^2}$$

The average speed of the boat in the second case is

$$v_{b_2} = \sqrt{v_{bw}^2 - v_w^2}$$

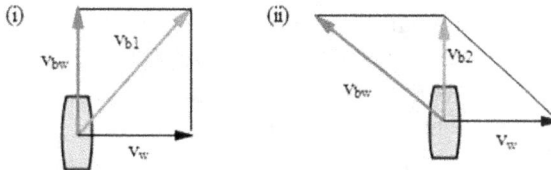

The average speed of the boat in the third case is

$$v_{b3} = \frac{d + d}{t_{down} + t_{up}} = \frac{d + d}{\dfrac{d}{v_{bw} + v_w} + \dfrac{d}{v_{bw} - v_w}}$$

$$\Rightarrow v_{b3} = \frac{(v_{bw}^2 - v_w^2)}{v_{bw}}.$$

Problem 21 A warplane moves horizontally with a constant velocity u at a height h. A guided missile is fired vertically with a speed v when the warplane passes above it. Assuming that the missile always aims at the warplane with the constant speed v, find the (a) time and position after which the missile strikes the warplane, (b) the final distance of separation between the warplane and missile when $u = v$. (c) Find $r = f(\theta)$ in part (a).

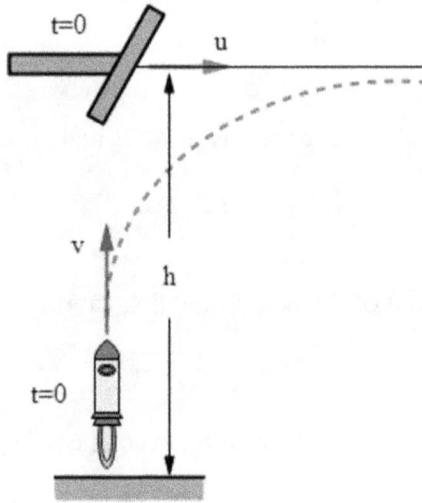

Solution

(a) Let r = distance of separation between the warplane and missile after a time t. Since, the missile always aims at the warplane with a constant speed v, it must move in a curve. At time t if $\vec{r}$ makes an angle θ with the horizontal, the missile approaches the warplane with

$$(v_{rel})_{||} = (v - u \cos \theta)$$

Hence, during a time dt.

$$-dr = (v - u \cos \theta)dt$$

Since, at $t = 0$, $r = l$ and at $t = \tau$, $r = x$

$$-\int_{l}^{0} dr = \int_{0}^{\tau} (v - u \cos \theta)dt$$

$$\Rightarrow v\tau - u \int_{0}^{\tau} \cos \theta = l \tag{3.106}$$

Since, the missile hit the warplane, their x-displacements are same.

$$\Delta x_M = \Delta x_R$$

$$\Rightarrow \int_{0}^{\tau} v_M\, dt = \int_{0}^{\tau} v_R\, dt$$

$$\Rightarrow \int_{0}^{\tau} v \cos \theta\, dt = \int_{0}^{\tau} u\, dt$$

$$\Rightarrow v \int \cos \theta\, dt = u\tau \tag{3.107}$$

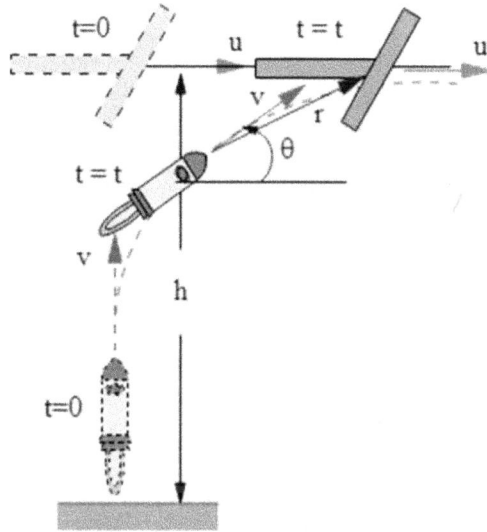

Eliminating $\int_0^T \cos\theta\, dt$ from equation (3.107) by substituting $\int_0^T \cos\theta\, dt = \frac{u}{v}\tau$ from equation (3.106), we have

$$\tau = \frac{vl}{v^2 - u^2} \quad \text{Ans.}$$

At the time of meeting, the rocket moves through a horizontal distance

$$x = v\tau = v\frac{vl}{v^2 - u^2} = \frac{v^2 l}{v^2 - u^2} \quad \text{Ans.}$$

(b) After a long time, the missile will move horizontally. Let the distance of separation between the missile and warplane be x after a long time.

Since, the missile falls through a distance the warplane, their x-displacements are not the same.

$$\Delta x_R - \Delta x_M = x, \quad \text{say.}$$

$$\Rightarrow \int_0^\tau v_R\, dt - \int_0^\tau v_M\, dt = x$$

$$\Rightarrow \int_0^\tau u\, dt - \int_0^\tau v\cos\theta\, dt = x$$

$$\Rightarrow x + v\int \cos\theta\, dt = u\tau \tag{3.108}$$

As the missile approaches the warplane with

$$(v_{rel})_{\parallel} = (v - u \cos \theta)$$

Hence, during a time dt.

$$-dr = (v - u \cos \theta)dt$$

Since, at $t = 0$, $r = h$ and at $t = \tau$, $r = x$

$$-\int_h^x dr = \int_0^\tau (v - u \cos \theta)dt$$

or

$$v\tau - u \int_0^\tau \cos \theta dt = h - x \tag{3.109}$$

Eliminating $\int_0^\tau \cos \theta \, dt$ from equations (3.108) and (3.109) and substituting $u = v$, after simplifying the factors, we have

$$x = h/2 \quad \text{Ans.}$$

(c) As the missile approaches the rocket, at a distance of separation r between the missile and warplane, during a time dt,

$$-dr = (v - u \cos \theta)dt \tag{3.110}$$

Resolving the v perpendicular to r, the angular velocity of the warplane relative to the missile is

$$-d\theta/dt = (u \sin \theta)/r$$

$$\Rightarrow -rd\theta = udt \sin \theta \tag{3.111}$$

Eliminating 'dt' in the last two equations, we have

$$\frac{-dr}{rd\theta} = \frac{(v - u \cos \theta)dt}{udt \sin \theta} = \frac{(v - u \cos \theta)}{u \sin \theta}$$

$$\Rightarrow \frac{-dr}{r} = \left(\frac{v}{u} \cos ec\theta - \cot \theta \right) d\theta$$

Integrating both sides, we have

$$\int_h^r \frac{-dr}{r} = \int_{\pi/2}^\theta \left(\frac{v}{u} \cos ec\theta - \cot \theta \right) d\theta$$

After evaluating the integration, we have

$$r = \frac{h(\cos ec\theta - \cot \theta)^{v/w}}{\sin \theta} \quad \text{Ans.}$$

Problem 22 A ball is projected with a velocity v_0 from (relative to) a large plank that moves with a constant upward acceleration a at an angle θ with the horizontal. The velocity of the plank at the given instant is u (up). Find the (a) time of flight of the ball, (b) maximum height attained by the ball. (c) The maximum height of the ball relative to the ground, (d) horizontal range of the ball relative to the plank. (e) The maximum height of the ball relative to ground.

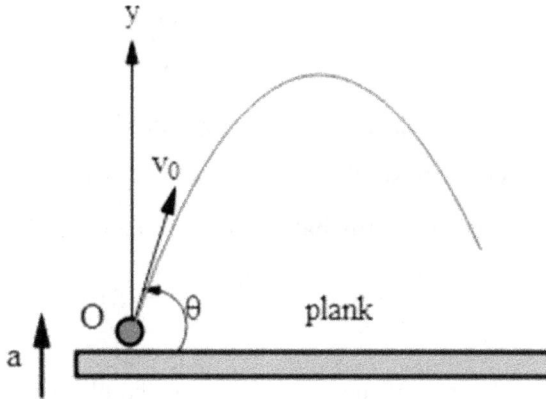

Solution

(a) Relative to the plank the acceleration of the ball is
$$a_r = g + a \ (\text{down})$$
The time of flight is given as
$$t = \frac{(v_{bp})_y}{(a_{bp})_y} = \frac{v_0 \sin \theta}{g + a} \quad \text{Ans.}$$

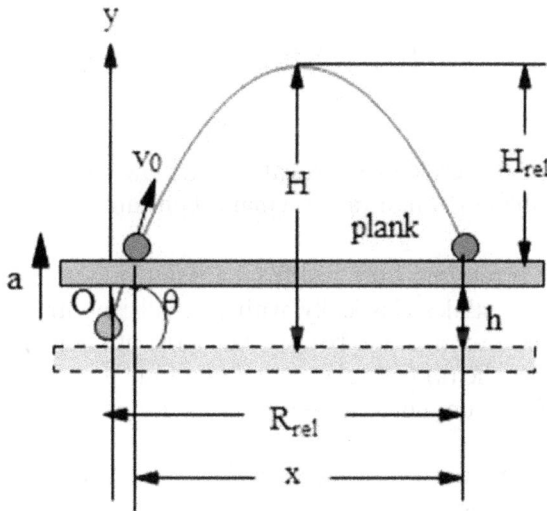

The time of flight does not depend upon the reference frame.

(b) The velocity of the ball relative to the plank is

$$\vec{v}_{bp} = v_0 \cos\theta\hat{i} + v_0 \sin\theta\hat{j}$$

The maximum height of the ball relative to the plank is

$$h_{rel} = \frac{(v_{bp})_y^2}{2(a_{bp})_y} = \frac{v_0^2 \sin^2\theta}{2(g+a)}$$

(c) The velocity of the ball relative to the ground is

$$\vec{u} = v_0 \cos\theta\hat{i} + v_0 \sin\theta\hat{j} + u\hat{j} = v_0 \cos\theta\hat{i} + (v_0 \sin\theta + u)\hat{j}$$

The maximum height of the ball relative to the ground is

$$h = \frac{(v_{bg})_y^2}{2(a_{bp})_y} = \frac{(v_0 \sin\theta + u)^2}{2g} \quad \text{Ans.}$$

(d) The horizontal range of the ball relative to the plank is

$$R = (v_0 \cos\theta\hat{i})T = v_0 \cos\theta\left(\frac{v_0 \sin\theta}{g+a}\right)$$

$$\Rightarrow R = \frac{v_0^2 \sin\theta\cos\theta}{g+a} = \frac{v_0^2 \sin 2\theta}{2(g+a)} \quad \text{Ans.}$$

(e) The horizontal distance covered by the ball relative to the ground is

$$R = (v_0 \cos\theta\hat{i})T = v_0 \cos\theta\left(\frac{v_0 \sin\theta}{g+a}\right)$$

$$\Rightarrow R = \frac{v_0^2 \sin\theta\cos\theta}{g+a} = \frac{v_0^2 \sin 2\theta}{2(g+a)} \quad \text{Ans.}$$

N.B: When the plank accelerates vertically up or down, we can see that the time of flight changes but the horizontal range remains constant.

Problem 23 A ball is released from rest from a point A located at a height h above the ground level. It collides elastically with a small smooth inclined plank after falling through a distance h. If the ball moves horizontally just after collision, find the value of h so that the (a) ball takes maximum time to reach the ground and (b) the ball covers a maximum horizontal distance or range.
Solution
The speed of the ball just before and after the collision will be equal due to elastic collision which is given as

$$v = \sqrt{2gh} \quad (3.112)$$

The time taken from A to B is

$$t = v/g = \sqrt{2gh}/g = \sqrt{2h/g} \tag{3.113}$$

The time taken from B to C is

$$t' = \sqrt{2g(H-h)/g} \tag{3.114}$$

The horizontal distance is

$$x = vt' \tag{3.115}$$

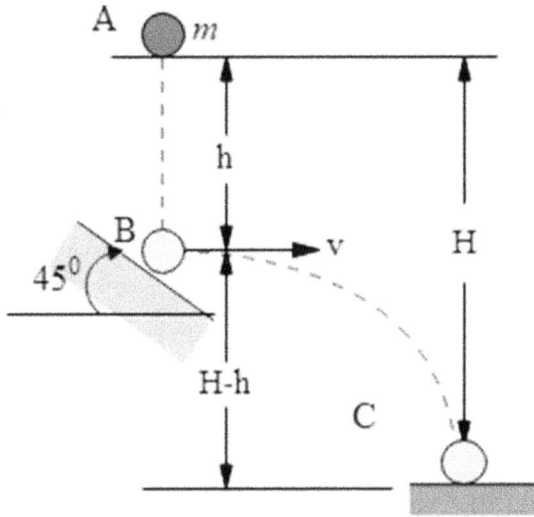

Using the last two equations,

$$x = vt' = (\sqrt{2gh})(\sqrt{2g(H-h)/g})$$

$$\Rightarrow x = 2\sqrt{h(H-h)} \tag{3.116}$$

For maximum distance

$$\frac{dx^2}{dh} = \frac{d}{dh}(4h(H-h)) = 0$$

$$h = H/2 \text{ Ans.}$$

Putting $h = H/2$ in equation (3.116), we have

$$x_{max} = 2\sqrt{(H/2)(H-H/2)} = H \text{ Ans.}$$

Using equations (3.113) and (3.114), the total time of flight is

$$T = t + t' = \sqrt{2h/g} + \sqrt{2(H-h)/g} \tag{3.117}$$

3-72

For maximum time

$$\frac{dT}{dh} = \frac{dT}{dh}\left(\sqrt{2h/g} + \sqrt{2(H-h)/g}\right) = 0$$

After evaluation,

$$h = H/2 \text{ Ans.}$$

Putting $h = H/2$ in equation (3.116), the maximum time is

$$T_{max} = 2\sqrt{H/g} \text{ Ans.}$$

Problem 24 A ball is released from rest from a point A located at a height h above the ground level. It collides with a small smooth inclined plane after falling through a distance h. If e = coefficient of restitution of collision such that the ball moves horizontally just after collision, find the (a) value of h so that the (b) ball takes maximum time to reach the ground and (c) ball covers a maximum horizontal distance or range.

   Solution

(a) The speed of the ball just before and after the collision will be equal due to elastic collision, which is given as

$$v = \sqrt{2gh} \tag{3.118}$$

The time taken from A to B is

$$t = v/g = \sqrt{2gh}/g = \sqrt{2h/g} \tag{3.119}$$

The time taken from B to C is

$$t' = \sqrt{2g(H-h)/g} \tag{3.120}$$

The horizontal distance is

$$x = vt' \tag{3.121}$$

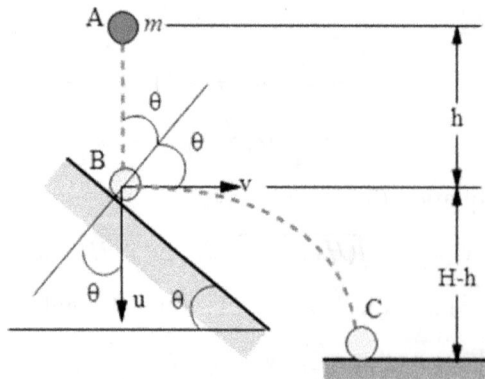

Referring to the theory of collision of chapter 1 in the book *Problems and Solutions in Many-Particle Systems*, we can write the following equations.

Along the inclined plane the velocities of the ball just before and after the collision are equal because the inclined plane is smooth.

$$v \cos \theta = u \sin \theta \tag{3.122}$$

Normal to the inclined plane the velocities of the ball just before and after the collision are related as

$$v \sin \theta = eu \cos \theta \tag{3.123}$$

Using the last two equations, we have

$$\tan^2 \theta = e$$

$$\Rightarrow \tan \theta = \sqrt{e} \tag{3.124}$$

Using the last two equations,

$$v = eu \cot \theta = eu/\sqrt{e} = \sqrt{e}\,u \tag{3.125}$$

Using equations (3.118) and (3.125), we have

$$v = \sqrt{2ghe} \tag{3.126}$$

Using equations (3.120), (3.121) and (3.126), we have

$$x = vt' = (\sqrt{2geh})(\sqrt{2g(H - h)/g})$$

$$\Rightarrow x = 2\sqrt{eh(H - h)} \tag{3.127}$$

For maximum distance

$$\frac{dx^2}{dh} = \frac{d}{dh}(4eh(H - h)) = 0$$

$$h = H/2 \text{ Ans.}$$

Putting $h = H/2$ in equation (3.121), we have

$$x_{\text{max}} = 2\sqrt{e(H/2)(H - H/2)} = H\sqrt{e} \text{ Ans.}$$

(b) Using equations (3.119) and (3.120), the total time of flight is

$$T = t + t' = \sqrt{2h/g} + \sqrt{2(H - h)/g} \tag{3.128}$$

For maximum time

$$\frac{dT}{dh} = \frac{dT}{dh}\left(\sqrt{2h/g} + \sqrt{2(H-h)/g}\right) = 0$$

After evaluation,

$$h = H/2 \text{ Ans.}$$

Putting $h = H/2$ in equation (3.121), the maximum time is

$$T_{max} = 2\sqrt{H/g} \text{ Ans.}$$

Problem 25 An umbrella rotates with an angular velocity ω about its vertical axis. If a water drop escapes from the umbrella at a point A horizontally, find the distance from O where the drop will hit the ground.

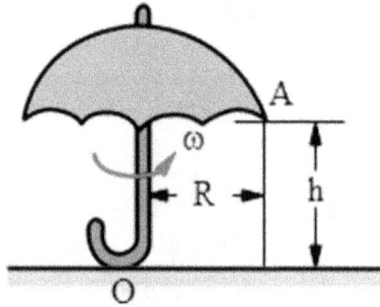

Solution
The speed of the water drop at A is

$$v = R\omega = R\sqrt{g/R} = \sqrt{gR} \tag{3.129}$$

The time taken from A to B is

$$t = \sqrt{2h/g} \tag{3.130}$$

The horizontal distance BC is

$$x = vt = \sqrt{gR}\sqrt{2h/g} = \sqrt{2hR} \tag{3.131}$$

The distance OC is

$$OC = d = \sqrt{AB^2 + BC^2} = \sqrt{h^2 + x^2} \tag{3.132}$$

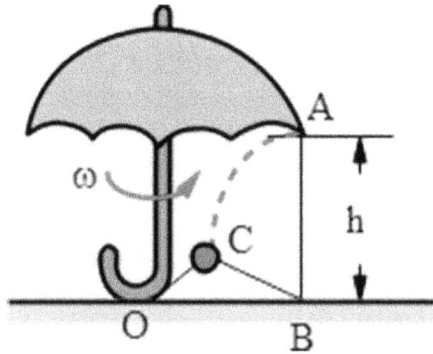

Using the last two equations, we have

$$d = \sqrt{h^2 + x^2} = \sqrt{h^2 + 2hR} = \sqrt{h(h + 2R)} \quad \text{Ans.}$$

Problem 26 A particle slides with a speed of 3 m s^{-1} at P. When it reaches Q, it acquires a speed of 4 m s^{-1} after describing an angle of $\theta = 60°$ about the center O of the circular track, as shown in the figure. Find the (a) magnitude of change in position (b) magnitude of change in velocity, and (c) magnitude of change in acceleration (d) change in total acceleration vector of the particle between P and Q. Assume the radius $R = 1$ m.

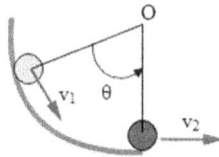

Solution

(a) As the particle moves from P to Q, the magnitude of change in position vector is

$$\begin{aligned}
|\Delta \vec{r}| &= |\vec{r}_2 - \vec{r}_1| = \sqrt{r_1^2 + r_2^2 - 2r_1 r_2 \cos \theta} \\
&= \sqrt{R^2 + R^2 - 2RR\cos \theta} \\
&= 2R\sin \tfrac{\theta}{2} = 2(1)\sin 30° = 1 \text{ m} \quad \text{Ans.}
\end{aligned}$$

(b) As its speed increases from $v_1 = 3$ m s^{-1} to $v_2 = 4$ m s^{-1} and the angle described by the velocity vector is equal to $\theta = 60°$, the change in velocity is given as

$$\Delta v = \sqrt{v_1^2 + v_2^2 - 2v_1 v_2 \cos \theta}$$

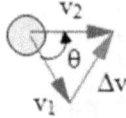

Substituting $v_1 = 3$ m s^{-1}, $v_2 = 4$ m s^{-1} and $\theta = 60°$ in the above expression, we have

$$\Delta v = \sqrt{3^2 + 4^2 - 2.3.4 \cos 60°}$$

This yields $\Delta v = \sqrt{13}$ m s^{-1} and the change in velocity is directed as shown in the figure.

(c) As its speed increases from $v_1 = 3$ m s^{-1} to $v_2 = 4$ m s^{-1} and the centripetal acceleration changes from a_1 at P to a_2 at Q. The angle described by this acceleration vector is equal to $\theta = 60°$. So, the change in centripetal acceleration is given as

$$\Delta \vec{a} = \sqrt{a_1^2 + a_2^2 - 2a_1 a_2 \cos \theta}$$

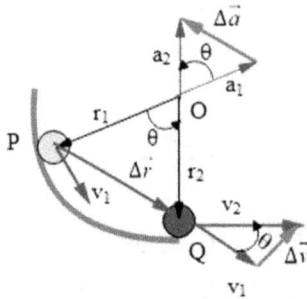

Substituting $a_1 = \frac{v_1^2}{R} = \frac{3^2}{1} = 9$ m s^{-2}, $a_2 = \frac{v_2^2}{R} = \frac{4^2}{1} = 16$ m s^{-2} and $\theta = 60°$ in the above expression, we have

$$|\Delta \vec{a}| = \sqrt{9^2 + 16^2 - 2(9)(16)\cos 60°}$$
$$= \sqrt{9^2 + 16^2 - (16)(9)} = 13.892 \text{ m s}^{-2} \text{ Ans.}$$

(d) At the position P, the particle has tangential acceleration $\vec{a}'$ a radial acceleration $\vec{a}_1$. So, the total acceleration at P is the resultant of these two accelerations, given as

$$\vec{a}_{1net} = \vec{a}' + \vec{a}_1 = a'\cos\theta\hat{i} - a'\sin\theta\hat{j} + a_1\sin\theta\hat{i} + a_1\cos\theta\hat{j}$$
$$= (a'\cos\theta + a_1\sin\theta)\hat{i} + (a_1\cos\theta - a'\sin\theta)\hat{j}$$

Putting the values of $a' = g\sin\theta = (10)(0.866) = 8.7$ m s^{-2} (nearly) and $a_1 = 9$ m s^{-2}, we have

$$\vec{a}_{1net} = (a'\cos\theta + a_1\sin\theta)\hat{i} + (a_1\cos\theta - a'\sin\theta)\hat{j}$$
$$= \{(8.7)(0.5) + (9)(0.866)\}\hat{i} + \{(9)(0.5) - 8.7)(0.866)\}\hat{j}$$
$$= \left(12.144\hat{i} - 3.03\hat{j}\right) \text{ m s}^{-2}$$

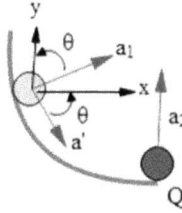

At the position Q, the particle has zero tangential acceleration; so, the net acceleration at Q is equal to its radial acceleration radial acceleration $\vec{a}_2 = 16\hat{j}$ m s^{-2}. Now, the difference in total acceleration at P and Q is given as

$$\vec{a}_2 - \vec{a}_{1net} = 16\hat{j} - \left(12.144\hat{i} - 3.03\hat{j}\right) \text{ m s}^{-2}$$
$$= \left(-12.144\hat{i} + 19.03\hat{j}\right) \text{ m s}^{-2} \text{ Ans.}$$

Chapter 4

Newton's laws of motion of a particle

4.1 Introduction

In kinematics you learned *how* a particle moves and described varieties of motion such as relative motion, circular motion, projectile motion etc, without discussing the cause of its motion. Kinematics just focuses on the calculation of position, velocity and acceleration etc. It does not talk about the cause of motion. This chapter deals with the 'cause of motion' of particles (point mass). So, here you will learn *why* a particle moves and how to find the unknown forces acting on a particle and acceleration caused by the forces acting on it. In this chapter of dynamics we will study the concept of *force* and *mass*. This chapter will cover the three fundamental laws of motion, which were formulated by Sir Isaac Newton based on experimental observations. This will form the basis of understanding the other chapters in mechanics.

4.2 Concept of force

4.2.1 Definition

Force is that which can change the velocity of a particle (a point mass or a point object) either the magnitude, or the direction or both. In other words, force can accelerate an object. Acceleration can be either speeding up, or slowing down or changing the line of motion of the object. Then, we can say that force is the cause of acceleration; loosely we call it cause of motion. We can see or experience the effect of force, that is, acceleration, as the body speeds up or slows down or turns. But we cannot see the forces because the *force* is just an *interaction* between the objects.

Physically we can feel the forces as a muscular push or pull. There are four types fundamental forces depending upon their origin such as gravitational, electromagnetic, strong and weak forces.

4.2.2 Types of forces

These four fundamental forces act upon particles without touching each other. They are called field forces. So, we have four different fields, namely gravitational field, electric field, magnetic field, nuclear field (weak and strong).

The strong interaction existing between subatomic particles inside the nucleus (between protons and neutrons) is also called nuclear force. Nuclear force is a short-range force that acts within a nucleus in a distance less than 10^{-15} m. The weak interaction rises in certain radioactive decay processes like β-decay. The gravitational force acts between any two bodies in the universe. The electromagnetic force acts between any two electric charges as the resultant or vector sum of electric and magnetic forces. The other forces arising from the contact of two bodies such as normal contact force, friction, tension and elastic forces; these are the macroscopic effects of electromagnetic force. Nuclear force is the strongest force that binds the nucleons (protons and neutrons) to form a nucleus. Then we have electromagnetic force between the electric charges and currents. The next level weaker force is the weak nuclear force. It does not hold things together by any attraction or push them apart. Instead, it changes the 'flavor' of a quark; for instance, an up quark becomes a down quark by a weak interaction that involves the exchange of W and Z bosons. However, the weakest of all forces is gravitational force that operates from zero to infinite distances.

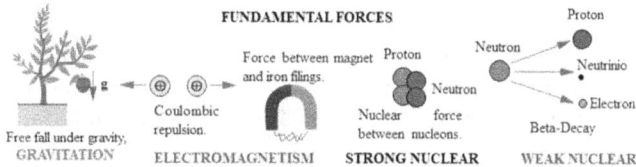

In daily lives we encounter varieties of forces such as tension, friction etc, which are the manifestation of the electromagnetic force. Physically, we experience a force as a *push* or *pull* of an object. One of the properties of a force is to change the shape and size of an object. In other words, a force can deform an object. For instance, when we pull a spring, it elongates. The other property of a force is to change the velocity of a particle or velocity of the center of mass of the object. We will talk about the point mass or particle being acted upon by a force. In mechanics, force can be broadly classified in two categories, the contact forces and the field forces. If we press a spring or kick a football or pull a box then the force comes in the category of contact forces because a physical contact takes place between two objects. Common examples of contact forces are tension in a string, frictional force, normal force etc.

Field forces (gravitational force, electrostatic and magnetic forces) operate between two objects through empty space without any need of physical contact.

4.3 Newton's laws of motion

4.3.1 Linear momentum

Newton expounded three laws of motion of general objects moving with speed v ($\ggg c$) much less than the speed of light. These laws are known as Newton's laws of motion. To describe these laws, let us first introduce the concept of linear momentum of an object which is defined as the product of mass and velocity. It is given as

$$\vec{P} = m\vec{v} \tag{4.1}$$

This physically signifies the quantity of motion contained in the body, where $m =$ quantity of matter that explains the *inertia* of the first law of motion.

4.3.2 First law (law of inertia)

In the absence of external forces, an object at rest remains at rest and an object in motion continues to move with the same state (velocity).

So, you can say that to maintain the same velocity or move with a zero acceleration, no force is needed; neither do you need to push or pull the body. In a reverse way you can say that 'when no (net) force acts on a body, its acceleration is zero'. This tendency of a body to maintain its original state of motion in the absence of a force is called *inertia*, and thus, this law, that is, Newton's first law is called the law of inertia. Inertia of a body is measured in terms of its *mass*, that is, content (stuff) in the body. Due to this stuff (mass), any object resists or tends to resist the change in its state (velocity). More broadly, inertia refers to the 'resisting property of a body to change in its content or state of motion' (or linear momentum). This will form the basis of interpreting the second law of motion. Newton's first law is valid in inertial frame of reference involving all real forces (forces arising from interactions) only. A reference frame that is at rest with ground or moves with uniform velocity is known as the inertial frame of reference. However, in a non-inertial reference frame, Newton's second law is modified to involve the pseudo (non-inertial) forces in addition to all real forces

Example 1 When your school bus driver suddenly presses the bakes, you lean forward, and after the accelerator is pressed suddenly you will automatically lean backward, why?

Explanation

The bottom part of your body which touches the seat was moving with the velocity equal to the velocity of the bus before the brakes were applied. When the driver applies the brakes, the bus suddenly slows down. Hence the lower part of your body slows down immediately; however; due to the inertia of motion, the upper part of your body tends to move with the same velocity but unable do so as it is attached to the lower part of your body. In consequence, to oppose the change in state of motion, that is, due to inertia of motion of the body, the upper part of your body will have to lean forward. Similarly, due to the inertia of rest, you will have to lean back if the driver presses the accelerator suddenly.

4.3.3 Second law (law of accerleration)

Newton developed calculus to find the rate of change or derivative or differentiation to define velocity and acceleration. In his second law he just measured the effect of a force, that is, change in momentum. He experimentally found that the rate of change in linear momentum of an object is directly proportional to the net force impressed upon the object. He quantifies a force by choosing the constant of proportionality as one. Thus, Newton's second law can be stated as:

The net force acting on a body is numerically equal to the rate of change in momentum of a body.

Mathematically, it is given as

$$\vec{F}_{net} = \frac{d\vec{P}}{dt} \tag{4.2}$$

Putting $\vec{P} = m\vec{v}$ from equation (4.2) in equation (4.3),

$$\vec{F}_{net} = \frac{d(m\vec{v})}{dt} \tag{4.3}$$

For a particle moving with a velocity very small compared with that of light, its mass remains constant over time; so taking the mass out of the derivative, we have

$$\vec{F}_{net} = m\frac{d\vec{v}}{dt} \tag{4.4}$$

From kinematics, you know that

$$\frac{d\vec{v}}{dt} = \vec{a} \tag{4.5}$$

Using last two equations, we have

$$\vec{F}_{net} = m\vec{a}. \tag{4.6}$$

4.3.4 Alternative explanation

The last equation tells us that a body accelerates when a net force acts on it; the acceleration of the body is directly proportional to the net force acting on it and inversely proportional to its mass. So, Newton's second law can also be stated as,

The net force acting on a body is numerically equal to the mass of the body and its acceleration.

If the net force acting on the body, that is, $\vec{F}_{net} = 0$, the acceleration of the body is zero; $\vec{a} = 0$; this is the essence of the first law.

4.3.5 Third law (law of action–reaction)

Every action force has equal and opposite reaction force. The forces appear and disappear in pairs, called an action–reaction pair. The action and reaction pair do not act on the same body, The interacting forces $\vec{F_1}$ and $\vec{F_2}$ are the action–reaction pairs in the diagram. They are equal in magnitude, opposite in direction and act along the line joining the interacting particles.

$$\vec{F_1} = -\vec{F_2} \tag{4.7}$$

F_1 = Force acting on the ball 1 by the ball 2, F_2= Force acting on the ball 2 by the ball 1. These two forces are action-raection pair.

If we take particles 1 and 2 as a single system, then the resultant force acting on the system is zero.

4.4 Types of forces

(i) **Field forces:** These forces arise due to the four fundamental interactions. The field forces continue to act on the objects even though they do not touch each other. Some of the field forces are discussed below.

(a) **Gravitational force:** If a particle of mass m is kept in a gravity of gravitational field intensity $\vec{E}$ then the gravitational force on it is given as $\vec{F} = m\vec{E}$. Since the bodies fall with an acceleration $a = g$ in gravity, the gravitational force acting on a body of mass is given by Newton's second law as $\vec{F_g} = m\vec{a} - = m\vec{g}$.

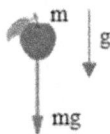

F_g = mg = The gravitational force acting on the apple of mass m moving in gravitational field of strength g.

This $m\vec{g}$ is called weight of the body. The gravitational force between two point masses m_1 and m_2 separated by a distance r is given by, $F = \frac{Gm_1m_2}{r^2}$; where G is universal gravitation constant.

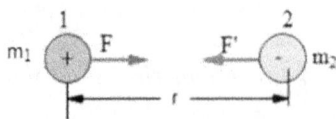

F = Gravitational force acting on the ball 1 by the ball 2, F'=Gravitationl force acting on the ball 2 by the ball 1.

(b) **Electrostatic force**

Two charged particles 1 and 2 attract each other by the forces F and F' such that $F = -F'$; these forces are an action–reaction pair. The electrostatic force between two point charges 1, and 2 separated by a distance r is given by Coulomb's law as,

$$F = \frac{kq_1q_2}{r^2}$$

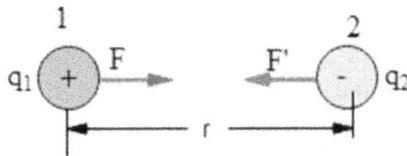

F = Electrostatic force acting on the ball 1 by the ball 2,
F'=Electrostatic force acting on the ball 2 by the ball 1

(c) **Magnetic force**

The charged particle q experiences a magnetic force $\vec{F_m}$, when it moves with velocity $\vec{v}$ in a magnetic field induction $\vec{B}$ is given as $\vec{F_m} = q(\vec{v} \times \vec{B})$.

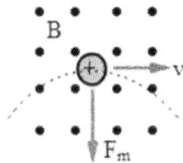

F_m = qvB = The Lorentz magnetic force acting on the point charge q moving in a magnetic field of induction B.

(ii) **Contact forces**

When two bodies are in contact, the force of interaction is called contact force.

(a) **Normal reaction**

When two surfaces are pressed against each other, each surface experiences a 'push' normal to it, called normal reaction. It is an electromagnetic repulsive force of interaction between the atoms of the contacting surfaces. It acts permanently when the surfaces touch permanently or momentarily when the surfaces touch during a short time in collision.

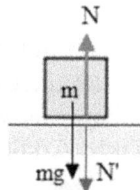

N=Normal contact force acting on the block by the ground,N'=normal contact force acting on the ground by the block.

(b) Friction

This is a contact force acting along the tangent when a body slides or tends to slide. This is called friction, and prevents or opposes the relative sliding between two contacting surfaces. This is also a gross manifestation of electromagnetic force and it will be discussed separately in the next chapter.

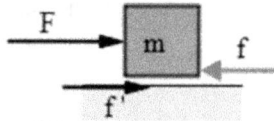

f= Friction (tangential contact force) acting on the block by the ground,
f '=friction (tangential contact force) acting on the ground by the block

(c) Tension

When we pull a string, the adjacent molecules of the string are pulled slightly from their mean positions. As a result, the molecules pull each other, which is manifested as a pulling force, called tension. For a massless smooth string, tension is the same at each point of the string. At any point, the tension is given as the net intermolecular force acting at either (right or left) side of the point. It is not the net force acting on the string. It is a pulling force. The string cannot push, it can only pull, whereas a rigid rod can both pull and push.

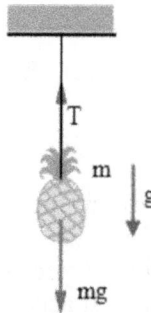

The string pulls the hanging pine apple up by a tension force T of mass m due to its weight in gravity;T=mg.

(d) **Elastic force**

Experimentally, Robert Hooke proved that, when an elastic body (a spring, say) is pulled through an additional length x, it pulls the external agent (which pulls the spring) by a force $F_{sp} = -kx$, where k = stiffness of the spring

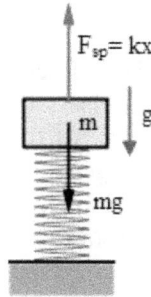

The spring pushes the loaded block of mass m up by a force $F_{sp} = kx$
(x= compression of the spring) due to its weight in gravity; $F_{sp} = mg$.

The negative sign signifies that, an elongated spring will pull and a compressed spring will push. The other contact forces like viscous and hydrostatic forces will be discussed in the relevant chapters.

4.5 Application of Newton's laws of motion

Newton's second law can be applied to an object to find its acceleration and other unknown forces acting on it. This is done by the following steps:

(a) *Drawing a rough sketch*

First of all, you need to draw a rough sketch of a system comprising different objects. Let the system be 'a smooth box pulled by a man by a string'. We intend to find the acceleration of the box.

(b) *Isolation of the bodies*

So, we need to isolate the body (box) from the other contacting surfaces such as the man and the ground.

(c) *Setting a reference frame*

For most of the cases we choose the ground as an inertial reference frame. Then we fix the coordinate axes with the ground. We have two options of taking the direction of x- and y-axes; parallel and perpendicular to the contacting surface (or horizontal and vertical), respectively.

N.B: If you choose a non-inertial frame you have to impose a pseudo-force and find the acceleration of the body relative to the reference frame (section 4.10).

(d) *Free-body diagram*

Draw the object as a block or a point mass. Identify the field (weight etc) and contact forces (applied force, tension, normal reaction etc) acting on the body by the others (but not by the body on the others). In figure (i), the field force acting on the box is downward weight $W(= mg)$. As the man pulls the box by the string, a pulling force, that is, tension T points along the string towards the man. As the box is in contact with the ground, upward normal reaction N acts on the box. In figure (ii) the field force acting on the box is downward weight $W(= mg)$. The inclined plane, pushes the block up with a reaction N normal to the plane. Each force is given as a vector arrow. All forces must act on the point object. In figure (i) the x-axis is horizontal and the y-axis is vertical. In figure (ii) the x-axis is parallel and the y-axis is normal to the plane. So, the general way of taking the axis is along the tangent (parallel) and normal (perpendicular) to the contacting surfaces (ground or inclined plane). In some cases, you can choose the x- and y-axes along the horizontal and vertical, respectively. Newton's third law is helpful to draw the action–reaction pairs on different connecting (interacting) bodies while you are drawing a free-body diagram for each body.

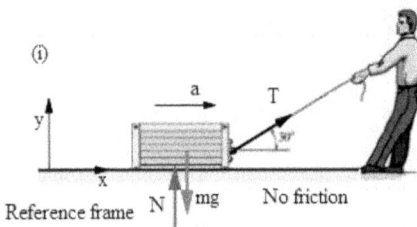

The forces acting on the box are tension T, gravity mg normal reaction N .

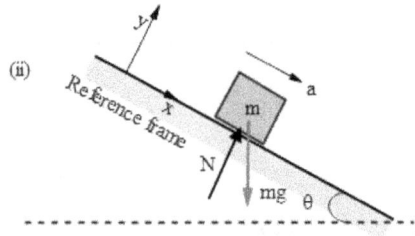

The forces acting on the block are gravity mg normal reaction N.

(e) *Force equation*

Then you have to resolve the forces along the chosen axes by using the component method. If you do not know the exact direction of acceleration, you can assume its directions, and also its magnitudes. Then find the net force along each axis by vector addition, and equate these forces with the product of mass and the assumed accelerations in the corresponding axes by using Newton's second law; so, the equations can be given as in the axes, $F_x = ma_x$, $F_y = ma_y$ etc. In other coordinate systems (such as polar, cylindrical and spherical), you can write the force equations in respective axes. For instance, the force equations along the tangential and normal axes can be given as $F_t = ma_t$ and $F_n = ma_n$. The magnitude of the forces can be calculated by using the relevant force laws; For instance, spring force is given by Hooke's law, buoyant force is given by Archimedes' principle, gravity force is given by Newton's law of universal gravitation etc.

(f) *Constraint equations and kinematical equations*

When there are many bodies connected by springs, we can relate their velocities and accelerations by writing the *constraint* equations, those are described separately in section 4.6. Furthermore, the kinematical equations such as $a = dv/dt$ or $a = vdv/dx$ can be written according to the requirements.

(g) *Solution*

After writing the constraint equations, the total number of force equations, constraint equations and kinematical equations *must* be equal to the total numbers of variables such as accelerations (a_x, a_y, etc) and unknown forces (T, N, etc). Then you can solve these equations for the desired accelerations and forces. Once you find acceleration, integrating acceleration you can find velocity and integrating velocity with time you can find the positions.

(h) *Physical interpretation*

After finding the unknown quantities, you should always examine the obtained values to check if they are physically possible or not. For instance, after following the proper method with correct free-body diagrams, if you get the tension in a string $T = -5$ N, you should not accept this value. This is because, the tension in the string must be greater than or at least equal to zero. Since a taut string has a positive tension, you can conclude that the string is slackened for a negative tension. Physically, a slack string has zero tension, but not a negative tension as it cannot push an object like rods or springs. So, the tension in the string is *zero*, even though you get a non-zero (negative) tension mathematically; so don't believe and accept any mathematical result blindly without a physical verification.

The net force acting on the bodies can be found only when we understand the types of forces acting on the bodies in the process of their interaction. For this, Newton's third law is helpful to draw the action–reaction pairs on different connecting (interacting) bodies. The entire process can be illustrated in the following examples.

Example 2 If a body of mass 2 kg is acted upon by a force of 20 Newton for 10^8 s, what will be its velocity?

Solution

If you just use Newton's second law by using the formula $a = F/m$ for constant acceleration, the velocity after a time t is $v = at = Ft/m$. Putting the values we have, $v = at = Ft/m = 20 \times 10^8/2 = 10 \times 10^8$ m s^{-1}. Since the obtained answer is greater than the speed of light, you cannot accept this answer. Therefore, the body cannot move with a constant acceleration when it attains a speed which is comparable with the speed of light. Thus, the velocity–time graph cannot be linear; rather, it will be hyperbolic obeying the relativistic mechanics.

Example 3 A smooth block of mass m is pressed against the vertical wall by a force F. Find the (a) normal reaction offered by the wall on the block and (b) acceleration of the block.

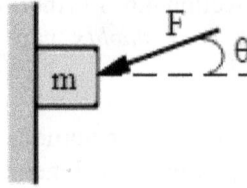

Solution

Free-body diagram

The forces acting on the block are mg (down) and applied horizontal force F. Let the normal reaction be N (horizontal) offered by the vertical wall on the block. As we have assumed a normal reaction, the block will not lose contact with the wall. As the wall does not move, the horizontal acceleration of the block is zero ($a_x = 0$).

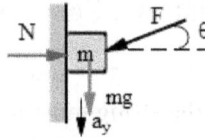

Force equation

(a) The net horizonal force is

$$F_x = ma_x$$

$$\Rightarrow N - F \cos \theta = m(0)$$

$$\Rightarrow N = F \cos \theta \quad \text{Ans.}$$

(b) The net vertical force is

$$F_y = ma_y$$

$$\Rightarrow mg + F \sin \theta = ma_y$$

$$\Rightarrow a_y = g + \frac{F}{m} \cos \theta \quad \text{Ans.}$$

Example 4 A boy is pulling a smooth block of mass m on horizontal ground. Find the (a) acceleration of the block, (b) normal reaction offered by the ground on the block, and (c) range of the force F. Assume that the block does not lose contact with the ground.

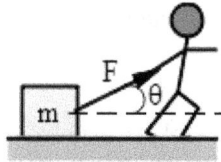

Solution

Free-body diagram

The forces acting on the block are mg (down), tension F (of the string). Let the normal reaction be N (up). As we have assumed a normal reaction, the block will not lose contact with the ground. So, the vertical acceleration of the block is zero $(a_y = 0)$.

Force equation

(a) The net horizonal force is

$$F_x = ma_x$$

$$\Rightarrow F \cos \theta = ma_x$$

$$\Rightarrow a_x = \frac{F}{m} \cos \theta \quad \text{Ans}$$

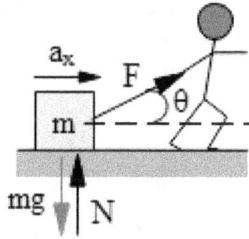

(b) The net vertical force is

$$F_y = ma_y$$

$$\Rightarrow F \sin \theta + N - mg = m(0)$$

$$\Rightarrow N = mg - F \sin \theta \quad \text{Ans.}$$

(c) For the block not to lose contact

$$N = mg - F \sin \theta \geqslant 0$$

$$\Rightarrow F \leqslant mg / \sin \theta \quad \text{Ans.}$$

N.B.: If the applied force is $F \geqslant mg / \sin \theta$, you will get a negative value of N which physically signifies that the block will lose contact $(N = 0)$ with the ground. So, the block also has a vertical acceleration a_y, given by applying the force equation

$$F_y = F \sin \theta - mg = ma_y$$

$$a_y = F \sin \theta / m - g$$

So, the acceleration of the block is given as

$$\vec{a} = a_x \hat{i} + a_y \hat{j} = (F \cos \theta / m)\hat{i} + (F \sin \theta / m - g)\hat{j}$$

(d) String pulls the block and boy with tensions T and T', respectively. Apart from this, the block receives normal contact force N from the ground and gravitational field (hanging bob); force, that is, weight mg. The ground receives a force N' due to the block. If the block moves, we assume a rightward acceleration 'a'.

Example 5 A block of mass m is loaded onto a vertical spring of stiffness k. If it oscillates, find the compression of the spring when the (a) velocity of the block is maximum and (b) acceleration of the block is equal to $g/2$ (i) down (ii) up.

Solution

Free-body diagram

The forces acting on the block are mg (down), spring force F_s (up) as the compressed spring pushes the block up.

If the block has maximum velocity, its acceleration is zero. So, the vertical acceleration of the block is zero ($a_y = 0$).

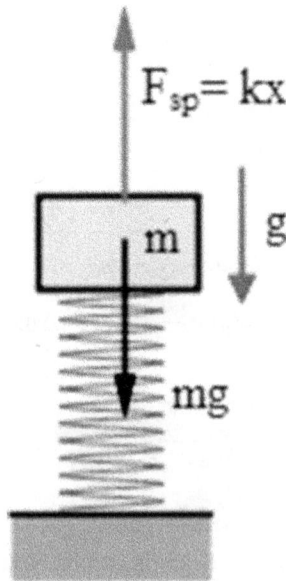

Force equation

(a) The net vertical force acting on the block is

$$F_y = ma_y$$

$$\Rightarrow kx - mg = m(0)$$

$$\Rightarrow x = mg/k \quad \text{Ans.}$$

(b)

    (i) If the block has downward acceleration $g = g/2$, then the vertical force on the block is

$$F_y = ma_y$$

$$\Rightarrow kx - mg = m(-g/2)$$

$$\Rightarrow x = mg/2k \quad \text{Ans.}$$

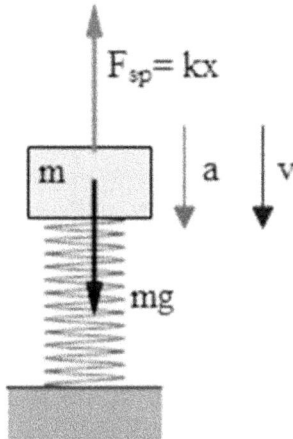

    (ii) If the block has upward acceleration $g = g/2$, then the vertical force on the block is

$$F_y = ma_y$$

$$\Rightarrow kx - mg = m(g/2)$$

$$\Rightarrow x = 3mg/2k \quad \text{Ans.}$$

N.B.: If we slowly load the block so that it remains at rest permanently. So, we can put $a = 0$ in the equation $mg - kx = ma$.

In this condition, $x = mg/k$. However, at the lowest position, the velocity of the block will be instantaneously zero; but it still accelerates up with $a = g$. At the lowest position the compression is maximum; $x_{max} = 2mg/k$. Similarly, at the highest position, again the block will stop instantaneously, but it still accelerates down with $a = g$. At the highest position the compression is minimum; $x_{min} = 0$.

Example 6 An apple is thrown into the air with a velocity v at an angle θ with the horizontal. Find the (a) net acceleration, (b) tangential acceleration and (c) centripetal acceleration of the apple at the given position.

   Solution

(a) The projectile experiences the gravitational pull of earth, disregarding the air drag and the force of other objects. Then, the net force acting on the body is its weight $\vec{W} = m\vec{g}$ which is vertically downward. So, the acceleration of the apple is

$$\vec{a} = \vec{W}/m = m\vec{g}/m = \vec{g}$$

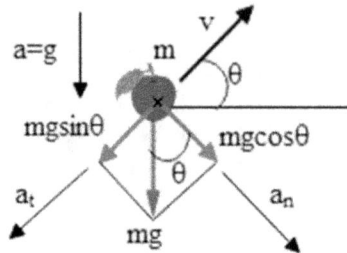

(b) The tangential force is

$$F_t = ma_t$$

$$\Rightarrow -mg \sin \theta = m(-a_t)$$

$$\Rightarrow a_t = g \sin \theta \ \text{Ans.}$$

(c) The normal (centripetal) force is

$$F_n = ma_n$$

$$\Rightarrow -mg \cos \theta = m(-a_n)$$

$$\Rightarrow a_t = g \cos \theta \ \text{Ans.}$$

N.B.: It is a well-known fact that all bodies fall freely with a downward acceleration of $g = 9.8$ m s^{-2}.

Example 7 A block of mass m is kept on an accelerating plank. Find the reaction force offered by the plank on the block for upward and downward acceleration a of the plank.

Solution

(a) *Free-body diagram*

The forces acting on the block are mg (down), normal reaction N (up) as the plank spring pushes the block up. As we have assumed a non-zero N, the block does not lose contact with the plank. Then the acceleration of the block is equal to the acceleration of the plank, that is equal to a (up or down).

(b) *Force equation for upward acceleration*

The net vertical force acting on the block is

$$F_y = ma_y$$

$$\Rightarrow N - mg = m(a)$$

$$\Rightarrow N = m(g + a) \text{ Ans.}$$

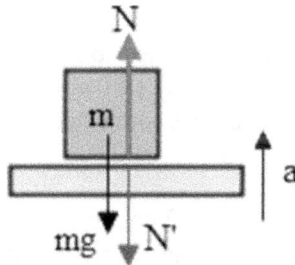

(c) *Force equation for downward acceleration*

The net vertical force acting on the block is

$$F_y = ma_y$$

$$\Rightarrow N - mg = m(-a)$$

$$\Rightarrow N = m(g - a) \text{ Ans.}$$

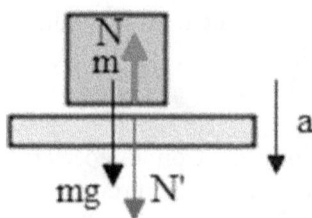

N.B.: For downward acceleration of the plank, as the normal reaction is $N = m(g-a)$, for N to be greater than zero, the plank's acceleration a must be less than g. If the plank's acceleration a is greater than g, then we will get a negative value of N which means that the direction of N is downward. As the block is just placed on the plank (without any glue), the plank cannot pull the block down; it can only push the block up. So, the block will fall freely with an acceleration g and the distance of separation between the block and plank will go on increasing with time. If we glue the block with the plank, the plank can also pull the block down and the block will never lose contact with the plank for its downward acceleration. Please note that upward acceleration is the same as the downward deceleration and downward acceleration is the same as the upward deceleration because the plank moves with a downward acceleration greater than g (acceleration due to gravity). If the block is not glued with the plank, the plank can only push the block up but cannot pull the block down. This signifies that for the ungluing case, normal reaction can be zero for upward deceleration or downward acceleration of the plank.

4.6 Constraint equations

In particle dynamics, objects are constrained to move under the action of tension of strings, reactions of rigid surfaces etc. The kinematical equations which relate the velocities and accelerations of the bodies connected by strings, springs etc, are called constraint equations.

    (i) *String–particle system*

        (a) Let two particles 1 and 2 connected by an inextensible string. Since the string is inextensible, the components of their velocities and accelerations along the string can be equated.

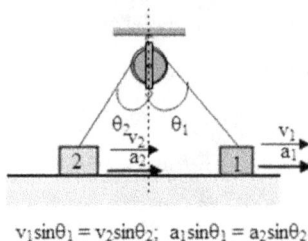

$$v_1\sin\theta_1 = v_2\sin\theta_2; \quad a_1\sin\theta_1 = a_2\sin\theta_2$$

        (b) If the pulley moves with a velocity v and the particles move with velocities v_1 and v_2 along the inextensible string of length l, we can write

$$(x_p - x_1) + (x_p - x_2) = l$$

$$\Rightarrow 2\frac{dx_p}{dt} = \frac{dx_1}{dt} + \frac{dx_2}{dt} \text{ (because } \frac{dl}{dt} = 0 \text{ as the string is inextensible)}$$

$$2v_p = v_1 + v_2$$

$$2\frac{d^2x_p}{dt^2} = \frac{d^2x_1}{dt^2} + \frac{d^2x_2}{dt^2}$$

$$\Rightarrow 2a_p = a_1 + a_2$$

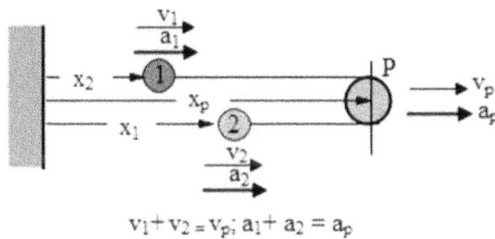

$$v_1 + v_2 = v_p; \ a_1 + a_2 = a_p$$

(ii) *Spring–particle system*

If two particles 1 and 2 are connected by a spring. for any elongation x, the length of the spring of natural length l_0 is,

$$l_0 + x = x_2 - x_1$$

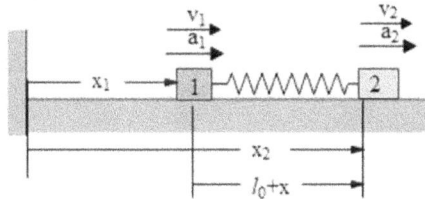

$$d^2x/dt^2 = a_2 - a_1$$

$$\Rightarrow \frac{dx}{dt} = v_2 - v_1$$

$$\Rightarrow \frac{d^2x}{dt^2} = a_2 - a_1$$

(iii) *Wedge–block system*

(a) Let us number the wedge as 1 and the block as 2. If the block is constrained to move along the slant of the wedge, the slope of the wedge is,

$$\tan \theta = \frac{h - y_2}{x_2 - x_1}$$

Differentiating both sides two times relative to time t,

$$\tan \theta = \frac{-\dfrac{d^2 y_2}{dt^2}}{\dfrac{d^2 x_2}{dt^2} - \dfrac{d^2 x_1}{dt^2}}$$

As we have taken a_y in the increasing order of $y = y_2$, we have

$$\frac{d^2 y_2}{dt^2} = +a_y$$

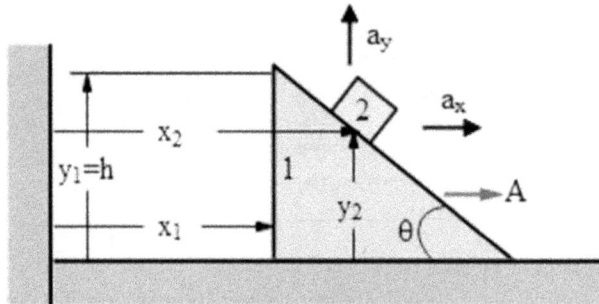

$$\tan\theta = -a_y/(a_x - A)$$

$$\Rightarrow \tan \theta = \frac{-a_y}{a_x - A}$$

Example 8 In this pulley–particle system, in figures (i) and (ii) the accelerations of the pulley and particle 2 are given; find the acceleration of particle 1. In figure (iii) the accelerations of particles 1 and 2 are given; find the acceleration of the pulley.

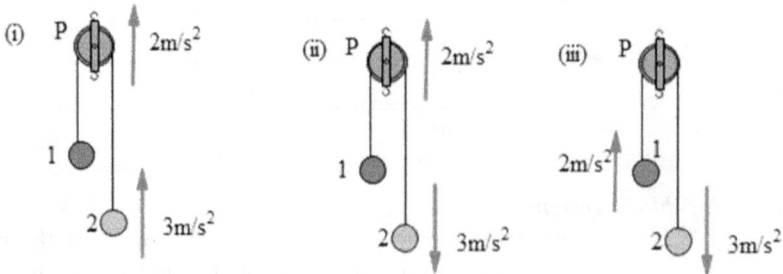

Solution

As derived earlier,

$$2a_P = a_1 + a_2$$

For figure (i),

$$2(+2) = a_1 + (+3)$$

$$\Rightarrow a_1 = 1 \text{ m s}^{-2}$$

Since a_1 is positive, it is directed up. Ans.

For figure (ii),

$$2(+2) = a_1 + (-3)$$

$$\Rightarrow a_1 = 7 \text{ m s}^{-2}$$

Since a_1 is positive, it is directed up. Ans.

For figure (iii),

$$2a_p = (+2) + (-3)$$

$$\Rightarrow a_p = -1/2 \text{ m s}^{-2}$$

Since a_p is negative, it is directed down. Ans.

Example 9 In the pulley–particle system, the pulley P is fixed and the particles 1 and 2 are constrained to move vertically and horizontally. At the given position, find $\frac{v_1}{v_2}$.

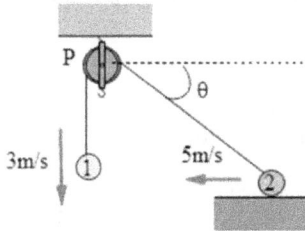

Solution

Equating the components of the velocities v_1 and v_2 along the string, we have

$$v_1 = v_2 \cos \theta$$

$$\Rightarrow \cos \theta = \frac{a_1}{a_2} = \frac{3}{5}$$

$$\Rightarrow \theta = \cos^{-1}\frac{3}{5} = 53° \text{ Ans.}$$

Example 10 In a wedge–block system, the horizontal velocities of wedge and block are $+4$ and $+12$ m s^{-1}, respectively. Find the vertical velocity v_y of the block.

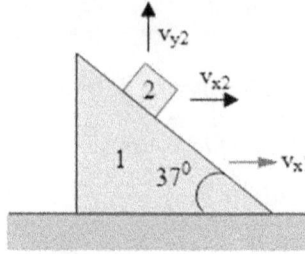

Solution

According to the constraint equation for the wedge–block system,

$$(v_{2x} - v_{1x})\tan\theta = -v_{2y}$$

$$\Rightarrow (12 - 4)\tan 37° = -v$$

$$\Rightarrow v = -6 \text{ m s}^{-1} \text{ down}$$

Example 11 In a spring–particle system the accelerations of particles 1 and 2 are $+2$ m s^{-2} and $+4$ m s^{-2}, respectively. Find the value of $\frac{d^2 l}{dt^2}$.

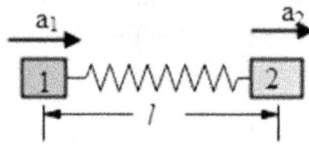

Solution

According to the constrained equation of the spring:

$$\frac{d^2 x_2}{dt^2} - \frac{d^2 x_1}{dt^2} = \frac{d^2 l}{dt^2}$$

$$\Rightarrow \frac{d^2 l}{dt^2} = a_2 - a_1$$

$$\Rightarrow \frac{d^2 l}{dt^2} = (+2) - (+4)$$

$$\Rightarrow \frac{d^2 l}{dt^2} = -2 \text{ m s}^{-2} \text{ Ans.}$$

Example 12 Three bodies are interconnected by two strings 1 and 2. If the
  (a) both strings are inextensible,
  (b) only string 1 is extensible,
  (c) only string 2 only is extensible, find the tension in the strings and acceleration of the bodies just after the application of the force F on the block m.

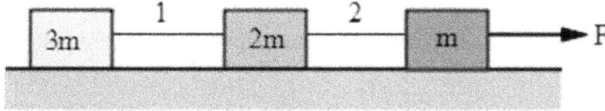

Solution

  (a) *Free-body diagram*
        If both strings are inextensible, tension in both strings will be non-zero just after application of the force and the bodies have the same acceleration. The forces acting, on $3m$, $T' \rightarrow$, on $2\,m$, $\leftarrow T'$ and $T \rightarrow$, on m, $T \leftarrow$ and $F \rightarrow$.

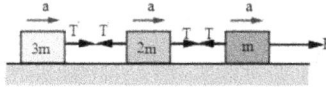

Force equation: $\Sigma F_x = ma$

$$3m: \quad T' = (3m)a \tag{4.8}$$

$$2m: \quad T - T' = (2\,m)a \tag{4.9}$$

$$m: \quad F - T = ma \tag{4.10}$$

Solving the last three equations,

$$T' = \frac{F}{2}, \quad T = \frac{5F}{6}, \quad a = \frac{F}{6m} \text{ Ans.}$$

  (b) If string 1 is extensible, it cannot develop a tension just after the force is applied. So, the tension in string 1 is zero and string 2 will be non-zero. Then, the forces acting on $3\,m$ is zero; on $2m$ is $T \rightarrow$ and on m are $T \leftarrow$ and $F \rightarrow$.

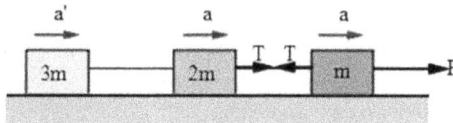

Force equation: $\Sigma F_x = ma$

$$3m: \quad 0 = (3m)a' \tag{4.11}$$

$$2m: \ T = (2m)a \tag{4.12}$$

$$m: \ F - T = ma \tag{4.13}$$

Solving the last three equations,

$$T' = 0, \ T = \frac{2F}{3}, \ a = \frac{F}{3m} \ \text{Ans.}$$

(c) If only string 2 is extensible, it cannot develop a tension just after the force is applied. So, the tension in string 2 is zero. Then, the driving force on $2\,m$ is zero; so, $T^{*} = 0$. Now, the force acting on $3\,m$ is zero; on $2m$ it is zero and on m it is $F\rightarrow$. Then by applying Newton's second law, we can directly write $T = T^{*} = 0$; acceleartions of the blocks $3\,m$ and $2\,m$ is zero. Body m has acceleration F/m to the right. Ans.

Example 13 A system containing three blocks of mass m, $3m$ and $2m$ are connected by two light inextensible strings. The block m is pulled by an upward force $F = 10\,mg$. Find the ratio of tensions in strings 1 and 2.

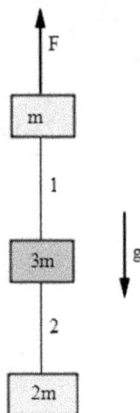

Solution

Free-body diagram

The forces acting on m are, $F\uparrow T_1\downarrow mg\downarrow$. The forces acting on $3\,m$ are, $T_1\uparrow T_2\downarrow 3mg\downarrow$. The forces acting on $2\,m$ are, $T_2\uparrow 2mg\downarrow$.

Force equation on m:

$$\Sigma F_y = ma$$

$$F - T_1 - mg = ma$$

$$10mg - T_1 - mg = ma \tag{4.14}$$

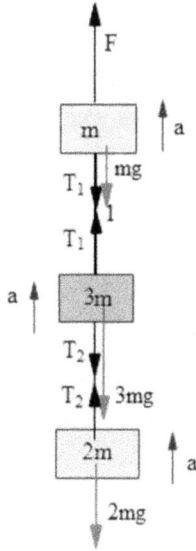

Force equation on $3m$:

$$\Sigma F_y = 3ma$$

$$T_1 - T_2 - 3mg = 3ma \tag{4.15}$$

Force equation on $2m$:

$$\Sigma F_y = 2ma$$

$$T_2 - 2mg = 2ma \tag{4.16}$$

Solving the last three equations, we have $a = 2g/3$, $T_1 = 25mg/3 \cdot T_2 = \frac{10}{3}mg$

So, the ratio of tensions is $\frac{T_1}{T_2} = \frac{5}{2}$ Ans.

4.7 Spring–particle system

4.7.1 Introduction

Springs play an important role in science and technology. For instance, the springs used as shock absorbers give us mechanical comfort. In mechanics, we come across the system of bodies connected with springs. Hence, we need to know the response of a spring to an applied force.

Let us consider a light spring whose one end is fixed with the wall. When we pull the other end of the spring gradually to the right or left, the spring will pull you in the opposite direction with a gradually increasing force F_s.

(i)

If you pull the spring with a force F to right, the spring will pull you with F_s (=F) to left.

(ii)

If you do not pull or push the spring, it will neither push nor pull you.

(iii)

If you push the spring with a force F to left, the spring will push you with a force F_s (= F) to right.

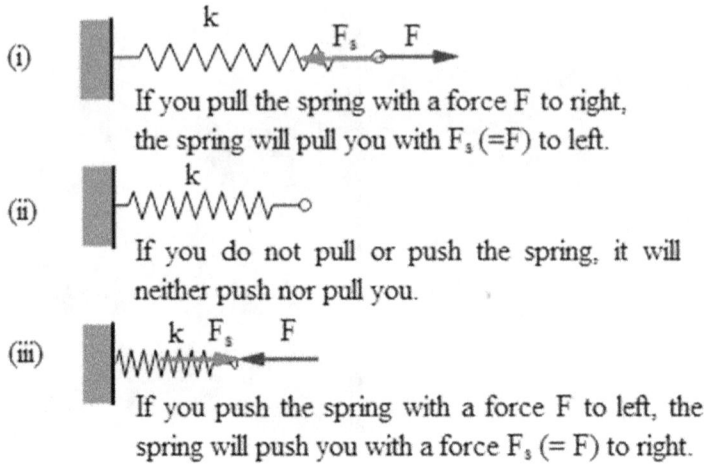

4.7.2 Hooke's law

From the above simple experiment, we can easily understand that,

(i) the force offered by the spring, that is, 'spring force' F_s points (acts) opposite to the displacement of the free end of the spring,

(ii) the amount of spring force increases linearly with the deformation (compression or elongation) of the spring.

When we plot the variation of F, versus x, we obtain a straight line up to a certain (limited) value of x, which is known as elastic limit.

Mathematically, all the above facts can be given as

$$F_s = -kx$$

where, $F_s =$ spring force and $x =$ displacement of one end of the spring (relative to the other end) along the spring. A negative sign signifies that $\vec{F_s}$ and $\vec{x}$ are directed opposite to each other. The amount of the displacement x is called deformation (both compression and elongation). However, F_s varies linearly with x up to certain deformation which is known as 'elastic limit' of the spring. Beyond that point, the spring cannot regain its natural length if the applied force is withdrawn.

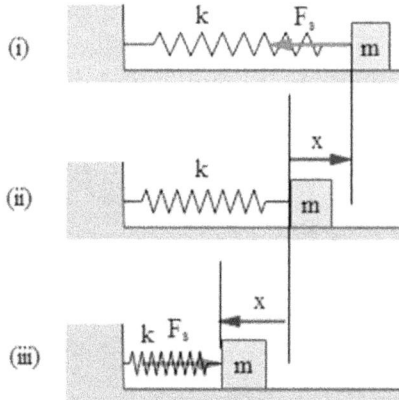

(i)

(ii)

(iii)

Since the spring force F_s and displacement x of the end of the spring are opposite, $F_s = -kx$

4.7.3 Stiffness k

Since, the spring force varies linearly with the deformation of the spring, the ratio of spring force and the deformation of the spring is a constant quantity within the elastic limit. This is what we call 'stiffness' of the spring. If a spring has greater stiffness, it is more difficult to deform (compress or elongate) it. As $k = \frac{|F|}{|x|}$, putting $x = 1$, we have $k = |F|$. This tells us that *the stiffness of the spring is numerically equal to the force required to deform it by a unit length.*

Example 14 A smooth block of mass $m = 2$ kg is acted upon by a horizontal force $F = 10$ N such that the spring of stiffness $k = 1000\frac{N}{m}$ is elongated. (a) Find the elongation of the spring when the block does not accelerate. (b) If the force F is removed suddenly at the given equilibrium position of the block, find the acceleration of the block at that instant.

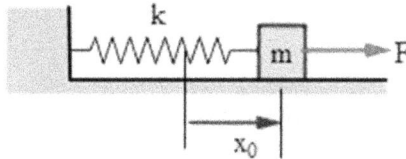

Solution

(a) The horizontal forces acting on the block are applied force F and spring force kx. The net force is $F_{net} = F - kx = ma$. Since the block is stationary $a = 0$,

$$x = \frac{F}{k} = \frac{10}{1000} = 0.01 \text{ m Ans.}$$

(b) If we remove the applied force, the net force acting on the block is the spring force.

$$\text{So, } F_{net} = kx = ma$$

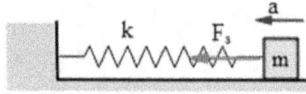

$$\Rightarrow a = \frac{kx}{m} = \frac{F}{m}$$
$$= \frac{10}{2} = 5 \text{ m s}^{-2} \text{ to left Ans.}$$

4.8 String–particle system

When we pull a string, each molecule of the string is pulled away from its neighboring molecule along the string. This generates an intermolecular force of attraction between all neighboring pairs of molecules. Its gross effect is called tension in the string. So, tension in a string arises from the intermolecular forces. When tension develops in a string by pulling it by an external force, each element of the string experiences an infinitesimal stretch. If you take a point P, say, in the string, it is pulled from both sides of the string with equal and opposite forces, which is known as tension, denoted as T.

So, tension is the intermolecular (internal) force but not the net force acting on the string.

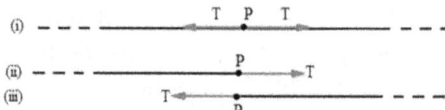

Tension at point P of the light string is a pair of equal and opposite forces each of magnitude, T, say, shown in figure(i). The free body diagram of left and right portion of the string are given with a right-tension T and left-tension T at the point P in figure(ii) and (iii) respecticvely.

Example 15 The tensions at the points A and B of a smooth horizontal string are F_1 and F_2, respectively. Find the value of $\frac{F_1}{F_2}$ if the string is (a) light, (b) stationary (or moves with constant velocity).

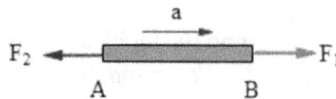

Solution

Free-body diagram
The forces acting along the string are $F_1 \leftarrow$ and $F_2 \rightarrow$:
Force equation

$$\Sigma F_x = ma_x$$

$$F_1 - F_2 = ma$$

(a) If $m = 0$, $\frac{F_1}{F_2} = 1$ Ans.

(b) If $a = 0$, $\frac{F_1}{F_2} = 1$ Ans.

N.B.:

1. A smooth string kept on a horizontal floor has the same tension when its (a) mass is zero, (b) acceleration is zero.
2. Since equal tension is felt at each point of the string, when a spring balance S is pulled in both sides by two forces F and $-F$, the reading of the spring balance will show the spring force $F_s = F$. However, the net force acting on the spring balance S is zero.

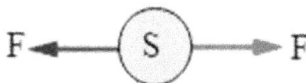

$$F \longleftarrow \boxed{S} \longrightarrow F$$

This means that a spring balance records the tension in the string but not the net force acting on it.

4.9 Wedge–block system

Method 1

In the wedge–block system, first of all draw free-body diagram (FBD). Let the external force acting on the wedge of mass M be F_1, on the block F of mass m along the slant be F_2, normal to the slant be F_3. Furthermore, the magnitude of the reaction forces acting on the wedge and block be N.

Force equation: Resolving the forces in the horizontal on the wedge,

$$F_x = F_1 - N \sin \theta = MA \tag{4.17}$$

Resolving the forces in the horizontal and vertical on the block,

$$F_x = F_2 \cos \theta + N \sin \theta + F_3 \sin \theta = ma_1 \tag{4.18}$$

$$F_y = mg - (N + F_3)\cos \theta = ma_2 \tag{4.19}$$

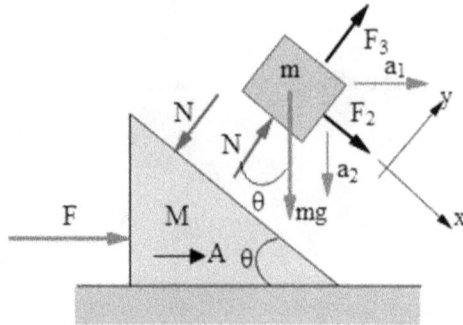

Constraint equation: Referring to section 4.6, we can write

$$h - y_2 = (x_2 - x_1)\tan \theta$$

$$\frac{-d^2y_2}{dt^2} = \left(\frac{d^2x_2}{dt^2} - \frac{d^2x_1}{dt^2}\right)\tan \theta$$

$$\Rightarrow -(-a_2) = (a_1 - A)\tan \theta$$

$$\Rightarrow a_2 = (a_1 - A)\tan \theta$$

Now, we have four equations and four unknown quantities a_1, a_2, A and N.

Method 2

Force equation

In this method we resolve the forces acting on m parallel and perpendicular to the inclined plane. Then we can write the force equation as follows:

$$\text{On } M: \quad F_1 - N \sin \theta = MA \tag{4.20}$$

$$\text{On } m: \quad F_2 + mg \sin \theta = ma_t \tag{4.21}$$

$$N - mg \cos \theta = ma_n \tag{4.22}$$

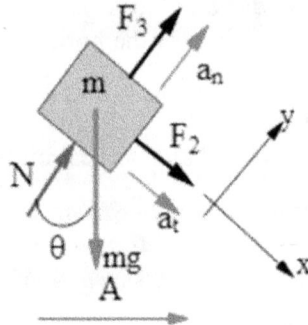

Constraint equation:

Since the block is in contact with the wedge, their accelerations along the normal are equal.

$$a_n = A \sin \theta \qquad (4.23)$$

Now, we have four equations (three force equations on M and m, and one constraint equation for the wedge–particle system), and four unknown quantities a_1, a_n, A and N. We can solve four equations to obtain four unknown quantities.

Example 16 A smooth block of mass m is placed on a prismatic wedge which moves with constant acceleration A. Find (a) normal reaction between wedge and block; (b) acceleration of the block relative to ground; and (c) acceleration of the block relative to wedge.

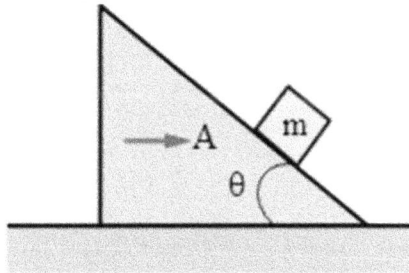

Solution

We have followed method 2 of section 4.9.

(a) *Free-body diagram*

The forces acting on the block are mg and N.

Force equation on the block m:

$$\Sigma F = ma$$
$$mg \sin \theta = ma_n$$

or

$$a_t = g \sin \theta \qquad (4.24)$$
$$\Sigma F_n = ma_n$$
$$-mg \cos \theta + N = ma_n$$
$$N = m(a_n + g \cos \theta) \qquad (4.25)$$

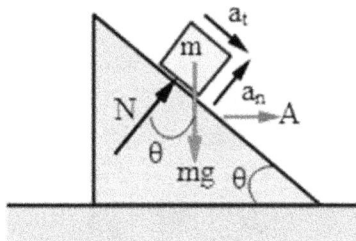

Constraint equation:

$$a_n = A \sin \theta \qquad (4.26)$$

Using equations (4.25) and (4.26),

$$N = m(A \sin \theta + g \cos \theta) \ \text{Ans.}$$

(b) The acceleration of the block is $\vec{a} = \vec{a_t} + \vec{a_n}$

$$\text{or } a = \sqrt{a_t^2 + a_n^2} \qquad (4.27)$$

Substituting a_t and a_n from equations (4.24) and (4.25), in equation (4.27),

$$a = \sqrt{g^2 \sin^2 \theta + A^2 \sin^2 \theta}$$
$$= \left(\sqrt{g^2 + A^2} \right) \sin \theta \ \text{Ans.}$$

(c) The acceleration of the block relative to the wedge along the slant is
$$\vec{a}_{mM} = \vec{a}_{mM_t} + \vec{a}_{mM_n}$$
$$= \vec{a}_{mM_t}(\because \ \vec{a}_{mM_n} = 0)$$
$$= \vec{a}_{m_1} - \vec{a}_{M_t},$$

Putting $a_t = g \sin \theta \searrow$ and $a_{M_t} = A \cos \theta \searrow$; we have

$$a_{mM} = (g \sin \theta - A \cos \theta) \ \text{Ans.}$$

N.B.: The following points regarding the wedge–particle system are worth noting:
1. When $A = g \tan \theta(\rightarrow)$, $N = \frac{mg}{\cos \theta}$, $a_{mM} = 0$, $a_{m_y} = 0$.
2. When $A = -g \cos \theta(\leftarrow)$, $N = 0$, $a_m = g \downarrow$.
3. When $A = 0$, $N = mg \cos \theta$, $a_n = 0$.
4. When $A > g \tan \theta$, $a_{mM} = (A \cos \theta - g \sin \theta)\nwarrow$; the block accelerates up relative to the wedge.
5. When $A < g \tan \theta$, $a_{mM} = (g \sin \theta - A \cos \theta)\searrow$; the block accelerates down relative to the wedge.

Example 17 A block of mass m is placed on a smooth wedge of mass M and angle of inclination θ. If the wedge–block system is released from rest, find the acceleration of the wedge.
  Solution
  Method 1
  Free-body diagram
  Refer to the figure of method 1 of section 4.9.
  The forces acting on the block are mg $\downarrow$, $N \nearrow$ and the forces acting on the wedge are $Mg \downarrow$, $N \swarrow$ and $N' \uparrow$.

Resolving the forces along horizontal and vertical.
For m:

$$N \sin \theta = ma_x \tag{4.28}$$

$$mg - N \cos \theta = ma_y \tag{4.29}$$

For M:

$$-N \sin \theta = MA \tag{4.30}$$

Constraint equation:

$$a_y = (a_x - A)\tan \theta \tag{4.31}$$

Substituting a_x, a_y, N from equations (4.28), (4.29) and (4.30) in equation (4.31) and simplifying the factors, we have

$$A = -\frac{mg \sin \theta \cos \theta}{M + m \sin^2 \theta}$$

Method 2
Refer to the figure of method 2 of section 4.9.
  Force equations:
  For m:

$$-mg \cos \theta + N = ma_n \tag{4.32}$$

For M:

$$-N \sin \theta = MA \tag{4.33}$$

Kinematics

$$a_n = A \sin \theta \tag{4.34}$$

Substituting a_n, from equation (4.34) in equation (4.32) and eliminating N by substituting

$$N = -\frac{MA}{\sin \theta}$$

From equation (4.33), we have

$$A = -\frac{mg \sin \theta \cos \theta}{M + m \sin^2 \theta} \quad \text{Ans.}$$

N.B: The -ve sign of A signifies that the wedge will accelerate to the left.

4.10 Non-inertial frame and pseudo-force

Let us imagine a man sitting in a moving car of velocity $\vec{v_0}$ and acceleration $\vec{a_0}$.

  The momentum of a body acted upon by a net real force $\vec{F}$ moving with a velocity $\vec{v}$, relative to the man is

$$\vec{P} = m\left(\vec{v} - \vec{v_0}\right)$$

The rate of change of momentum of the body relative to the man is

$$\frac{d\vec{P}}{dt} = m\frac{d\vec{v}}{dt} - \frac{md\vec{v}_0}{dt}$$

$$\Rightarrow \frac{d\vec{P}}{dt} = \vec{F} - m\vec{a}_0 \quad \left(\because \frac{d\vec{v}_0}{dt} = \vec{a}_0 \text{ and } m\frac{d\vec{v}}{dt} = \vec{F} \right)$$

The above expression tells us that the rate of change of momentum of a body relative to an accelerating (non-inertial) frame is not equal to the net real force $\vec{F}$ acting on the body. Rather it is equal to the net real force $\vec{F}$ plus the factor ' $- m\vec{a}_0$'.

Since ' $- m\vec{a}_0$' is added with the real force $\vec{F}$ –, it is called a force. As it does not arise from the interaction of the body with the surrounding bodies, it is not a real force, we can call it an imaginary or fictitious or pseudo-force. This arises due to the relative acceleration between the body and the non-inertial frame. As it is not a *real* force, it has no action–reaction pair.

When we solve the problem by pseudo-force method observing the bodies from an accelerating frame, we add real forces with the pseudo-force to obtain the net force,

$$\vec{F}_{net} = \vec{F}_{real} + \vec{F}_{pse}$$

and then equate it with 'ma' where a = acceleration of the body relative to the accelerating frame.

$$\vec{F}_{net}\left(= \vec{F}_{real} + \vec{F}_{pse} \right) = m\vec{a}, \quad \text{where } \vec{F}_{ps} = -m\vec{a}_0.$$

Example 18 A pendulum bob hangs from the roof of a trolley car which moves with a horizontal acceleration a. If the string makes a constant angle θ with the vertical, find:
   (a) horizontal acceleration a,
   (b) tension in the string.

Solution
Method 1 (Ground frame method)
   (a) *Free-body diagram*
         The forces acting on the bob are $T \nearrow$ and $mg\downarrow$.

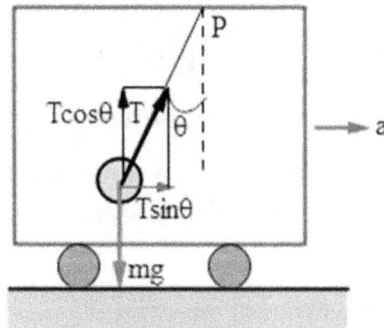

Force equation:

$$\Sigma F_x = ma_x$$

$$T \sin \theta = ma \tag{4.35}$$

$$\Sigma F_y = ma_y$$

$$T \cos \theta - mg = 0 \tag{4.36}$$

Solving the last two equations,

$$a = g \tan \theta \quad \text{Ans.}$$

(b) From equation (4.36),

$$T = mg \sec\theta \quad \text{Ans.}$$

Method 2 (Pseudo-force method)

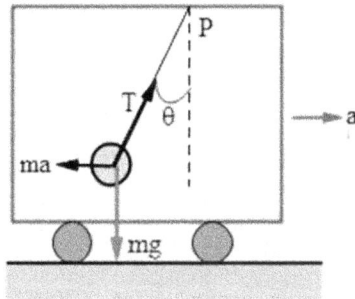

Free-body diagram

The forces acting on the bob are $T \nearrow$ and $mg\downarrow$ and pseudo-force $ma\leftarrow$.

Force equation:

$$\Sigma F_x = ma_x$$

$$T \sin \theta - ma = m(a_x)_{\text{rel}} = 0 \tag{4.37}$$

$$\Sigma F_y = ma_y$$

$$T \cos \theta - mg = m(a_y)_{\text{rel}} = 0 \tag{4.38}$$

Solving the last two equations,

$$a = g \tan \theta \quad \text{Ans.}$$

N.B.: If the trolley car suddenly starts moving with constant velocity, you can find that the acceleration of the (a) bob (b) tension in the string at the given instant will be equal to $g \sin \theta$ and $mg \cos \theta$.

1. In the first case $a_x = a$, $a_y = 0 \Rightarrow F_y = T \cos \theta - mg = 0$.
2. In the second case the acceleration of the bob along the string is $a_r = 0$ because $v_{rel} = 0 \Rightarrow \Sigma F_r = T - mg \cos \theta = 0$

Example 19 An apple of mass m is glued with a light string, as shown in the figure below. If the apple hangs in gravity, find the tension in portions 1 and 2 of the string.

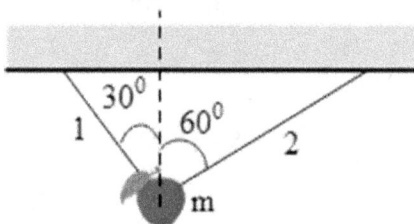

Solution
As the apple is glued with the string, the tensions in portions 1 and 2 are different.
Free-body diagram
The forces acting on the apple are $T_1 \nearrow$, $T_2 \searrow$ and $mg \downarrow$.

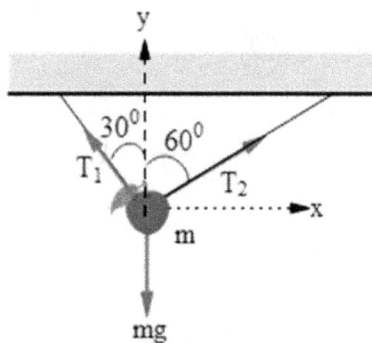

Force equation:

$$\Sigma F_x = ma_x$$

$$\Rightarrow T_1 \sin 30° - T_2 \sin 60° = m(0)$$

$$\Rightarrow T_1 = \sqrt{3}\, T_2 \qquad\qquad (4.39)$$

$$\Sigma F_y = ma_y$$

$$\Rightarrow T_1 \cos 30° + T_2 \cos 60° - mg = 0$$

$$\Rightarrow \sqrt{3}\, T_1 + T_2 = mg \qquad\qquad (4.40)$$

Solving equations (4.39) and (4.40),

$$T_1 = \frac{\sqrt{3}\,mg}{4} \quad \text{and} \quad T_2 = \frac{mg}{4} \quad \text{Ans.}$$

Problem 1 (Tension in the string)
Two forces F_1 and F_2 act on blocks of masses m_1 and m_2, respectively, as shown in the figure. If the horizontal surface is smooth find the tension in the light inextensible string.

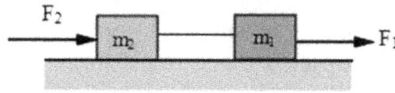

Solution:
Free-body diagram:
Assuming tension T in the string, the horizontal forces acting on the bodies m_1 are $F_1 \rightarrow$, $T \rightarrow$ and on m_2 are $T \leftarrow$, $F_2 \rightarrow$. Let the acceleration of the bodies be $a_1 \rightarrow$ and $a_2 \rightarrow$, respectively.
Force equation:

$$\text{on } m_1: \ \Sigma F_x = ma_x$$

$$\text{or,} \ F_1 + T = m_1 a_1 \tag{4.41}$$

$$\text{on } m_2: \ \Sigma F_y = ma_y$$

$$\text{or,} \ F_2 - T = m_2 a_2 \tag{4.42}$$

Constraint equation:
As the string is inextensible, assuming a non-zero tension in the string,

$$a_1 = a_2 = a(\text{say}) \tag{4.43}$$

Putting a_1 from equation (4.41) and a_2 from equation (4.42) in equation (4.43),

$$\frac{F_1 + T}{m_1} = \frac{F_2 - T}{m_2}$$

or,

$$T = \frac{m_2 F_1 - m_1 F_2}{m_1 + m_2} \quad \text{Ans.}$$

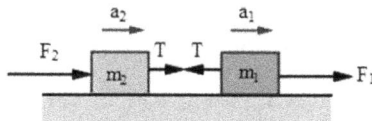

N.B.:

1. If $\vec{F_1}$ is reversed, find the tension is given as

$$\frac{m_2 F_1 + m_1 F_2}{m_1 + m_2}$$

2. If $\frac{\vec{F_1}}{m_1} = \frac{\vec{F_2}}{m_2}$, tension is zero. This condition for tension to induce is that the bodies must tend to move (accelerate) away from each other in the absence of the string. For a connecting light rod, if $a_{\text{rel}} \neq 0$ in the absence of the rod, a non-zero compressive or tensile reaction will be developed. when the horizontal forces are applied along the string.

3. If $\frac{F_1}{m_1}$ is greater that $\frac{F_2}{m_2}$, the string will slacken; $T = 0$ and the bodies will move with different accelerations of $\frac{F_1}{m_1}$ and $\frac{F_2}{m_2}$ to the right, respectively.

Problem 2 (Contact forces)

A block of mass m is stationary on an inclined plank or wedge of an inclined plane (of angle of inclination θ, say) which is moving with an arbitrary acceleration $\vec{a}$. Find the contact or reaction force offered by the inclined plane on the body.

Solution

As the direction of acceleration is arbitrary, we should not assume it is horizontal. So, we need to follow vector approach to apply Newton's second law.

Method 1

Free-body diagram:

The forces acting on the block are weight $m\vec{g}$ and contact force $\vec{F_c}$.

Force equation:

Applying Newton's second law, the net force is

$$\Sigma \vec{F} = \vec{F_c} + m\vec{g} = m\vec{a}$$

So, the contact force is given as

$$\vec{F_c} = m(\vec{a} - \vec{g}) \text{ Ans.}$$

Method 2 (Pseudo-force method)

Free-body diagram:

In this method we have to write Newton's second law sitting on the plank (accelerating frame). So, we have to impose a pseudo-force $\vec{F_{ps}} = -m\vec{a}$. The real forces acting on the block are weight $m\vec{g}$ and contact force $\vec{F_c}$.

Force equation:

Applying Newton's second law, the net force is

$$\Sigma \vec{F} = \vec{F_c} + m\vec{g} + \vec{F_{ps}} = m\vec{a}',$$

where $\vec{F}_{ps} = -m\vec{a}$ and $\vec{a}' =$ acceleration of the block relative to the plank. As the block is at rest relative to the plank, putting $\vec{a}' = 0$, the contact force is given as

$$\Sigma\vec{F} = \vec{F}_c + m\vec{g} + \vec{F}_{ps} = m\vec{a}'$$

$$\Rightarrow \vec{F}_c = m(\vec{a} - \vec{g}) \text{ Ans.}$$

N.B.:

1. If the inclined plane has acceleration $\vec{a}$ and no other forces are applied other than gravity and contact force, the contact or reaction force can be given as

$$\vec{F}_c = -m(\vec{g} - \vec{a}).$$

2. If the block is stationary, $\vec{a} = 0$; then, the contact force is

$$\vec{F}_c = -m\vec{g}$$

Resolving $\vec{F}_c$ along the tangent and normal to the slant, we have $(F_c)_t = f = mg \sin \theta \searrow$ and $(F_c)_n = N = mg \cos \theta \nearrow$.

Problem 3 Two identical blocks each of mass m are on a smooth horizontal floor. If an external force of F is acting on the right block, find the reaction forces acting on the blocks offered by the horizontal surface.

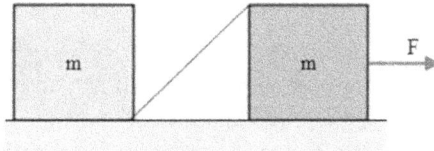

Solution

Free-body diagram:

The forces acting on the blocks are tension T, mg and the normal reactions N_1 and N_2.

Force equation:

Applying Newton's second law, on the left block

$$F_x = T \cos \theta = ma \qquad (4.44)$$

$$F_y = -mg + N_1 + T \sin \theta = ma_y = 0 \qquad (4.45)$$

Applying Newton's second law, on the right block

$$F_x = F - T \cos \theta = ma \qquad (4.46)$$

$$F_y = -mg + N_2 - T \sin \theta = ma_y = 0 \qquad (4.47)$$

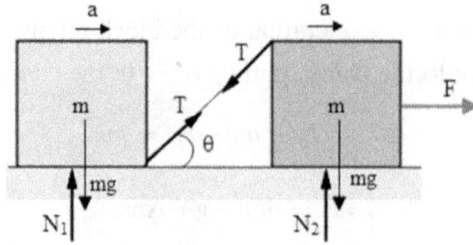

Solving equations (4.44) and (4.46), $a = F/(2m)$, $N_1 = mg - \frac{F}{2} \tan \theta$ and $N_2 = mg + \frac{F}{2} \tan \theta$. Ans.

Problem 4 Two blocks of masses $m = 1$ kg and $M = 2$ kg are placed on a plank which moves up with an upward acceleration $a = 2$ m s^{-2}. If an external force of $F = 10$ N is acting on the 1 kg block, find the reaction forces in the contacting surfaces.

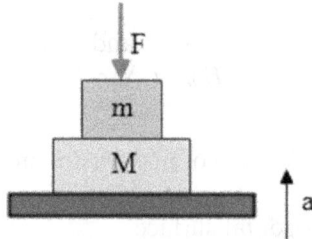

Solution

Free-body diagram

The forces acting on the 1 kg block are weight 10 $N \downarrow$, applied forces $10N \downarrow$ normal reaction $N \uparrow$, and the force acting on the 2 kg block are normal reaction $N \downarrow$, weight 20 N.

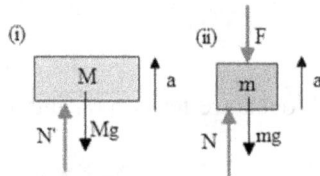

Force equation:

For 1 kg: $\Sigma F_y = ma_y$

$$N - 10 - 10 = (1)(2) \tag{4.48}$$

For 2 kg: the net vertical force acting on m is

$$N' - N - 20 = (2)(2) \tag{4.49}$$

Solving equations (4.48) and (4.49),

$$N = 22 \, N \quad \text{and} \quad N' = 46 \, N \quad \text{Ans.}$$

Problem 5 A body of mass m is placed on an elevator. When the elevator accelerates up and down with equal amount of acceleration, the body experiences normal reactions N_1 and N_2, respectively. Find the acceleration of the elevator.

Solution:

Free-body diagram

The body is acted upon by $N\uparrow$ and $mg\downarrow$.

Force equation: Referring to FBD, we have following force equations.

For m: If the elevator moves with constant upward acceleration a,

$$\Sigma F_y = N_1 - mg = ma \tag{4.50}$$

When the elevator accelerates down,

$$\Sigma F_y = mg - N_2 = ma \tag{4.51}$$

Equation (4.50) plus (4.51) yields

$$N_1 - N_2 = 2ma$$

This gives $a = \frac{N_1 - N_2}{2m}$ Ans.

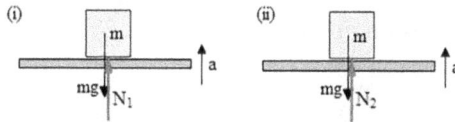

Problem 6 A block of mass m is resting inside a smooth horizontal tube. If the tube accelerates vertically downward obeying the relation $a = \alpha t$, where α is a positive constant, plot the variation of normal reaction received by the block as the function of time.

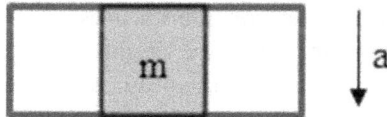

Solution

Arbitrarily considering the direction of N, the equation of motion of the block can be given as

$$N - mg = m(-a)$$

Then $N = m(g - a)$, where $a = \alpha t$

When $t = 0$, $a = 0$; then $N = +mg$

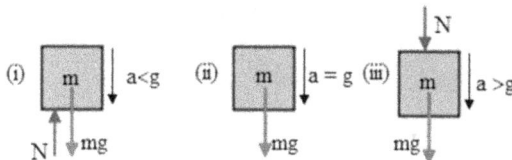

Positive sign of N signifies an upward direction to nullify the gravitational force mg.

When $a = g\downarrow$ at $t = \frac{g}{\alpha}$, we have $N = 0$; the contact force will vanish.

When $a > g$, for $t > \frac{g}{a}$, N becomes negative which signifies a downward N. Assuming upward N is positive in the graph, we have shown the variation of $\vec{N}$ and $|\vec{N}|$ using the equation

$$N = m(g - \alpha t)$$

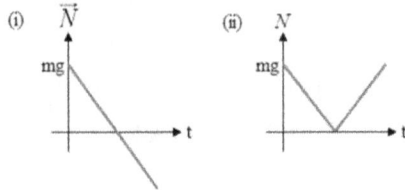

Problem 7 A smooth bead of mass m can slide on an inextensible string. If it hangs in the gravity, find the tension in the string. Put $b = 20$ cm, $H = 9$ cm and $h = 6$ cm.

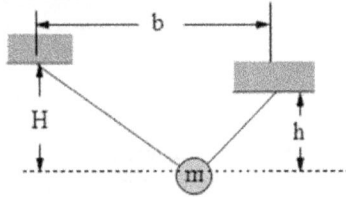

Solution

Since the bead is smooth, the tension at both sides of the string remains the same. So, the bead is in equilibrium under the action of two tension forces each of magnitude T, say and the weight mg of the bead. As the net horizontal force acting on the bead is zero, the strings must make equal angle θ, say, with the horizontal at both sides of the bead.

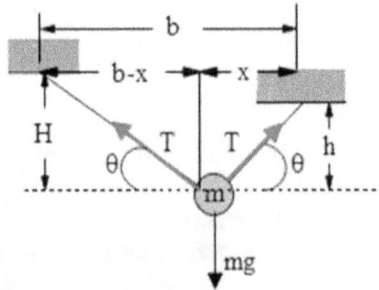

Applying Newton's second law, we have

$$F_y = 2T \sin \theta - mg = ma_y = 0$$

$$\Rightarrow T = \frac{mg}{2 \sin \theta}.$$

From the geometry of the figure,

$$\tan \theta = \frac{h}{x} = \frac{H}{b - x}$$

$$\Rightarrow x = \frac{bh}{H + h} = \frac{(20)(6)}{9 + 6} = 8 \text{ cm}$$

$$\tan\theta = \frac{h}{x} = \frac{6}{8} = \frac{3}{4}$$

$$\Rightarrow \sin\theta = \frac{3}{5}$$

So, the tension in the string is

$$T = \frac{mg}{2 \sin\theta} = \frac{mg}{2(3/5)} = \frac{5\,mg}{6} \text{ Ans.}$$

Problem 8 (Horizontal spring—blocks)

A smooth block of mass m is welded with a spring of stiffness k, on a horizontal plane. If it is given as horizontal velocity v_0, (a) derive the speed of the block as the function of deformation of the spring. Draw the $v-x$ graph, (b) find deformation of the spring, velocity and acceleration of the block as the function of time. (c) Find the amplitude, frequency and time period of the cyclic motion of the block.

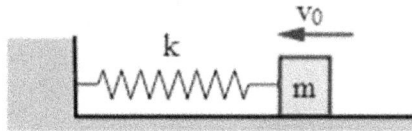

Solution

(a) *Hooke's law*:

As the block is given an initial velocity, it oscillates back and forth. When the block is located at the left side of the initial position, at a distance x, according to Hooke's law the spring pushes the block to right with a force of magnitude.

$$F_s = kx \tag{4.52}$$

Force equation:

$$\Sigma F = ma$$

$$\Rightarrow F_s = ma \qquad (4.53)$$

Kinematics:

Since the assumed direction of $\vec{a}$ opposes the displacement $\vec{x}$.

$$a = -\frac{vdv}{dx} \qquad (4.54)$$

Using the last four equations, we have

$$\frac{vdv}{dx} = -\frac{k}{m}x \text{ Ans.}$$

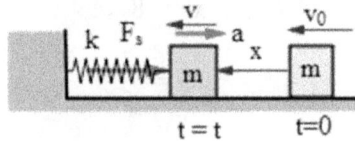

If the speed of the block increases from zero to v, when the deformation of the spring changes from 0 to x, integrating both sides,

$$\int_{v_0}^{v} v\, dv = -\frac{k}{m}\int_{0}^{x} x\, dx$$

$$\Rightarrow \frac{v^2 - v_0^2}{2} = -\frac{kx^2}{2m}$$

$$\Rightarrow v = \sqrt{v_0^2 - \frac{k}{m}x^2} \text{ Ans.}$$

$$\Rightarrow v = \frac{dx}{dt} = \sqrt{v_0^2 - \frac{k}{m}x^2} \text{ Ans.}$$

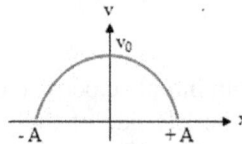

(b) Separating the variables,

$$\frac{dx}{\sqrt{v_0^2 - \frac{k}{m}x^2}} = dt$$

As the spring gets elongated by x during a time t, integrating both sides,

$$\int_{o}^{x} \frac{dx}{\sqrt{v_0^2 - \frac{k}{m}x^2}} = \int_{0}^{t} dt$$

$$\Rightarrow \sqrt{\frac{m}{k}} \sin^{-1} \sqrt{\frac{k}{mv_0^2}} x = t$$

$$\Rightarrow x = \sqrt{\frac{m}{k}} v_0 \sin \sqrt{\frac{k}{m}} t \quad \text{Ans.}$$

Differentiating with time, the velocity is

$$\Rightarrow v = dx/dt = v_0 \sin \sqrt{\frac{k}{m}} t \quad \text{Ans.}$$

Again, differentiating with time, the acceleration is

$$\Rightarrow a = -\sqrt{\frac{k}{m}} v_0 \sin \sqrt{\frac{k}{m}} t \quad \text{Ans.}$$

(c) So, the nature of motion of the block is simple harmonic having the amplitude A from the mean position, given as

$$x = A \sin \omega t,$$

$$A = \pm \sqrt{\frac{m}{k}} v_0 \quad \text{Ans.}$$

The angular frequency of oscillation is

$$\omega = \sqrt{\frac{k}{m}}$$

Then, the frequency of oscillation is

$$f = \frac{\omega}{2\pi} = \frac{1}{2\pi} \sqrt{\frac{k}{m}} \quad \text{Ans.}$$

The time of oscillation is

$$T = \frac{1}{f} = 2\pi \sqrt{\frac{m}{k}} \quad \text{Ans.}$$

N.B.:
1. $\frac{d^2x}{dt^2} + \frac{k}{m}x = 0$ is the differential equation of a horizontal spring–mass system.
2. Due to its inertia of motion, the block compresses and elongates the spring repeatedly. Since the compressed spring pushes the block and the elongated

spring pulls the block, the block always accelerates towards the mean (equilibrium) position. Hence, the block executes oscillatory motion between two points $x = -x_0$ and $x = +x_0$.

Problem 9 (Vertical spring–mass)

A block of mass m is released from rest when a vertical light spring of stiffness k is relaxed. At any deformation x of the spring, find (a) speed of the block as the function of deformation x, (b) compression of the spring when the velocity of the block is maximum, (c) maximum velocity of the block, (d) maximum compression of the spring, (e) compression of the spring as the function of x, (f) amplitude, frequency and time period of oscillation of the block.

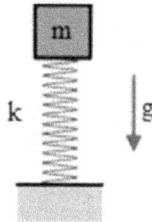

Solution

(a) *Free-body diagram*

At any deformation x of the spring, referring to the free-body diagram, the force acting on the block are spring force $F_x = kx\uparrow$ and weight $mg\downarrow$
Force equations:

$$\Sigma F = ma$$

$$\Rightarrow kx - mg = ma$$

$$a = \frac{k}{m}x - g \tag{4.55}$$

Kinematics:
Since a and x are oppositely directed

$$a = -\frac{vdv}{dx} \tag{4.56}$$

Using equations (4.55) and (4.56),

$$-\frac{vdv}{dx} = \frac{k}{m}x - g$$

$$\Rightarrow vdv = \frac{k}{m}xdx - g\,dx$$

If the block speeds up from 0 to v, integrating both sides, we have

$$\int_0^v v\,dv = \frac{k}{m}\int_0^x x\,dx - g\int_0^x dx$$

$$\Rightarrow v = \sqrt{\frac{k}{m}x^2 - 2gx}\ \ \text{Ans.}$$

(b) The velocity is maximum, when $dv/dx = 0$ or $v\,dv/dx = 0$ or $a = 0$.
By using equation (4.56), we have

$$a = \frac{k}{m}x - g = 0$$

$$\Rightarrow x = \frac{mg}{k}$$

which is called stable equilibrium denoted as x_0. Ans.

(c) Putting this value in the obtained expression of velocity, we have

$$\Rightarrow v_{\max} = \sqrt{2g\left(\frac{mg}{k}\right) - \frac{k}{m}\left(\frac{mg}{k}\right)^2} = g\sqrt{\left(\frac{m}{k}\right)}\ \ \text{Ans.}$$

(d) The compression of the spring is maximum, when velocity is zero.

$$\Rightarrow v = \frac{dx}{dt} = \sqrt{2gx - \frac{k}{m}x^2} = 0$$

$$\Rightarrow x = \frac{2mg}{k}\ \ \text{Ans.}$$

(e) The velocity of the block is

$$v = \frac{dx}{dt} = \sqrt{2gx - \frac{k}{m}x^2}$$

Separating the variables,

$$\frac{dx}{\sqrt{2gx - \frac{k}{m}x^2}} = dt$$

Integrating both sides,

$$\int_0^x \frac{dx}{\sqrt{2gx - \frac{k}{m}x^2}} = \int_0^t dt$$

$$\Rightarrow \int_0^x \frac{dx}{\sqrt{\left(2g - \frac{k}{m}x\right)x}} = \int_0^t dt$$

Let $x = u^2$; then, $2udu = dx$. Putting the values of x and dx in the integration, we have

$$I = \int_o^x \frac{dx}{\sqrt{\left(2g - \frac{k}{m}x\right)x}} = \int_o^u \frac{2udu}{\sqrt{\left(2g - \frac{k}{m}u^2\right)u^2}} = t$$

$$\Rightarrow 2\int_o^u \frac{du}{\sqrt{\left(2g - \frac{k}{m}u^2\right)}} = t$$

$$\Rightarrow 2\sqrt{\frac{m}{k}} \sin^{-1} \sqrt{\frac{k}{2mg}} u = t$$

$$\Rightarrow x = \frac{2mg}{k} \sin^2 \sqrt{\frac{k}{m}} \frac{t}{2}$$

$$x = \frac{mg}{k}\left(1 - \cos \sqrt{\frac{k}{m}} t\right) \quad \text{Ans.}$$

(f) The nature of motion of the block is simple harmonic having the displacement from the mean position, given as

$$y = \left|x - \frac{mg}{k}\right| = \frac{mg}{k}\left(1 - \cos \sqrt{\frac{k}{m}}\right)$$

The amplitude of oscillation is given as

$$A = \pm\frac{mg}{k} \quad \text{Ans.}$$

The angular frequency of oscillation is

$$\omega = \sqrt{\frac{k}{m}}$$

The frequency of oscillation is

$$f = \frac{\omega}{2\pi} = \frac{1}{2\pi}\sqrt{\frac{k}{m}} \quad \text{Ans.}$$

The time of oscillation is

$$T = \frac{1}{f} = 2\pi\sqrt{\frac{m}{k}} \quad \text{Ans.}$$

(i)

Top extreme position

Stable Equilibrium or mean position

Bottom extreme position

(ii) F,a,v

$a = g - kx/m$

$v = 2(kx^2/2m - gx)^{1/2}$

$F = ma = mg - kx$

(iii)

$x = A(1 - \cos\omega t)$

2mg/k

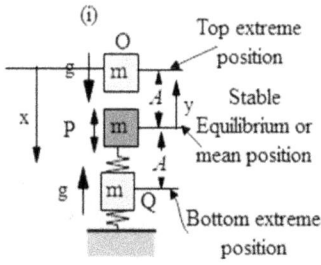

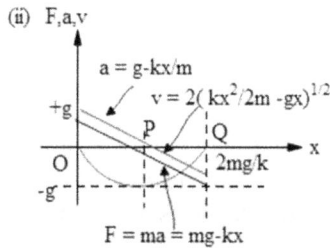

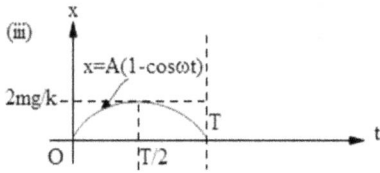

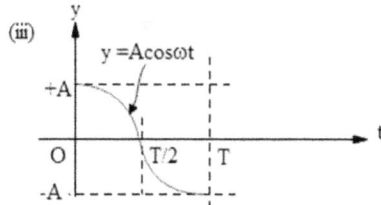

Magnitude of displacement x verses time

(iii)

$y = A\cos\omega t$

The displacement (magnitude and direction) y verses time

N.B.: The following points must be remembered regarding the above example:
1. Velocity is maximum when acceleration is zero at $x = \frac{mg}{k} = x_0$ (mean position).
2. Acceleration and deformation are maximum when velocity is zero (at extreme position).
3. Above the mean position, $kx < mg$ and acceleration is downward.
4. Below the mean position, $kx > mg$ and acceleration is upward.

Problem 10 Two blocks of masses m_1 and m_2 are equilibrium. The block m_2 hangs from a fixed smooth pulley by an inextensible string that is fitted with a light spring of stiffness k, as shown in the figure. Neglecting friction and mass of the string, find the acceleration of the bodies just after the string S is cut.

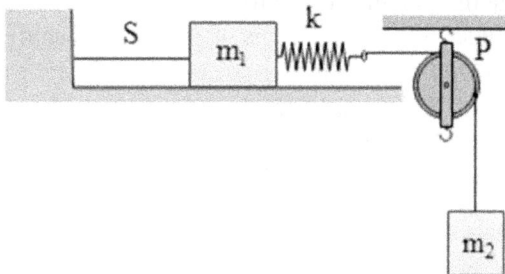

Solution

Free-body diagram

Let the spring forces be $F = kx$, just after cutting the spring. Hence, at that instant the forces acting on m_1 are $T = kx \rightarrow$, $m_1 g \downarrow$ and $N \uparrow$; on m_2 the forces are $m_2 g \downarrow$ and $T \uparrow$.

Force equation (Newton's second law)

Initially, all the particles are stationary. Hence $a_1 = a_2 = 0$. Applying Newton's second law

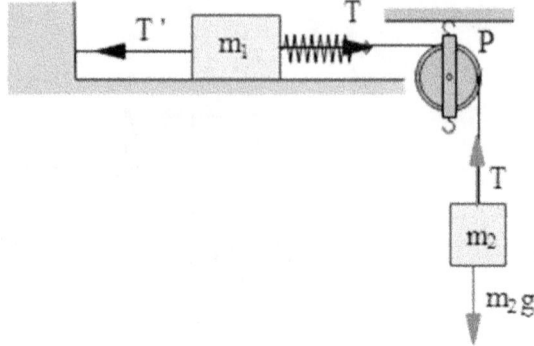

$$\text{For } m_1: \quad \Sigma F = T' - kx = 0 \tag{4.57}$$

$$\text{For } m_2: \quad \Sigma F = T - m_2 g = 0 \tag{4.58}$$

$$\text{For spring:} \quad \Sigma F = T - kx = 0 \tag{4.59}$$

Solving equations (4.57), (4.58) and (4.59), we have

$$T' = T = kx = m_2 g$$

Just after the string S is cut, tension T' vanishes immediately, but the spring force does not change suddenly. As the spring force remains the same just after cutting the string, the net force acting on m_1 is equal to kx, whereas the net force acting on m_2 is equal to $T - m_2 g$. Hence, the accelerations of m_1 and m_2 just after cutting the string are given by the following equations:

$$\Sigma F = m_1 a_1 = kx = m_2 g \tag{4.60}$$

$$\Sigma F = m_2 a_2 = T - m_2 g = 0 \tag{4.61}$$

Solving last two equations, we have

$$a_1 = \left(\frac{m_2}{m_1}\right) g \text{ and } a_2 = 0 \text{ Ans.}$$

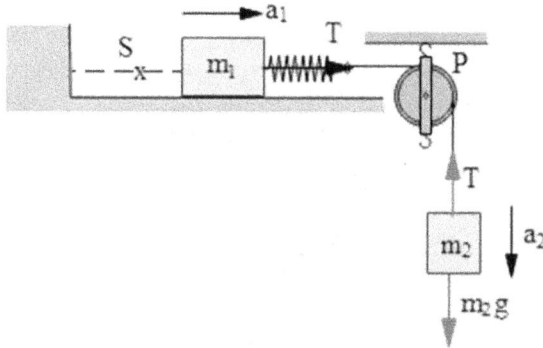

Problem 11 Two blocks of masses m and $3m$ are connected by two vertical springs of stiffness k and k' $(=2k)$, respectively. The blocks hang from a fixed smooth pulley P by an inextensible string that passes over the pulley. The blocks are in equilibrium. Neglecting friction and mass of the string, find the (a) deformation of the spring assuming that the initial elongation of the spring k is $mg/2k$. Find the (b) acceleration of the bodies just after the string S is cut.

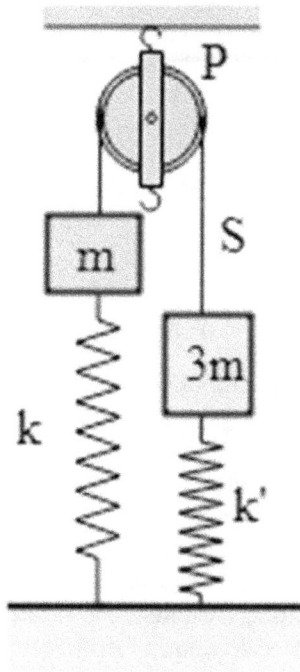

Solution

(a) *Free-body diagram*

Initially let us assume that the spring k is elongated by a length x pulling the body m by a force $kx\downarrow$; the weight of m is $mg\downarrow$. Let the tension in the string be T. Then, on the body m the tension is $T\uparrow$. On the body $3m$ the forces are $3mg\downarrow$ and $T\uparrow$. Let the spring k' push the body $3m$ up with a compression x'. Then an upward spring force $k'x'\uparrow$ acts on the body $3m$.

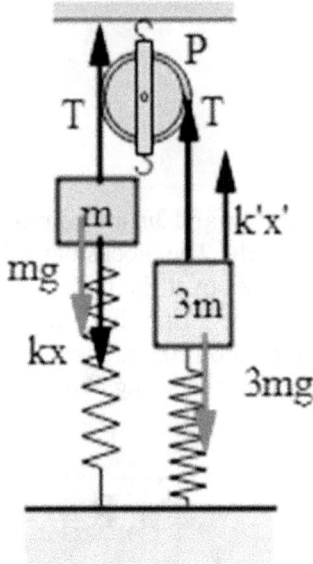

Force equation (Newton's 2nd law)

Initially, all the particles are stationary. Hence $a_1 = a_2 = 0$. Applying Newton's second law

$$\text{For } m: \quad F = T - mg - kx = 0 \tag{4.62}$$

$$\text{For } 3m: \quad F = T - 3mg + k'x' = 0 \tag{4.63}$$

Solving equations (4.62) and (4.63), we have

$$k'x' + kx = 2mg$$

Putting $k' = 2k$ and $x = mg/2k$, we have $x' = 3mg/4k$. Ans.

(b) Just after the cutting the string S, tension T of the string vanishes immediately, but the spring forces do not change suddenly. As the spring forces remains the same just after cutting the string, the net force acting on m and $3m$ are equal to kx, whereas the net force acting on m_2 is equal to $T - m_2 g$. Hence, the acceleration of m_1 and m_2 just after cutting the string is given by the following equations:

Force equation (Newton's second law)

Just after cutting the string, the tension of the string will be zero. Although the bodies are stationary ($v = 0$), the forces will be unbalanced (because the tension vanishes). Then, the net force acting on m is

$$F_1 = -mg - kx = -mg - mg/2 = -3mg/2 \text{ (downward)}$$

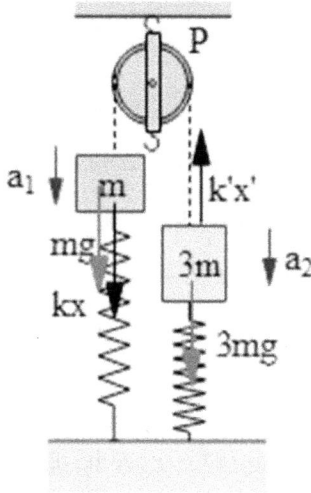

So, the acceleration of the body m is

$$a_1 = F_1/m = -3g/2 (\downarrow) \text{ Ans.}$$

Similarly, the net force acting on $3m$ is

$$F_2 = -3mg + k'x' = -3mg + (2k)(3mg/4k) = -3mg/2 \text{ (downward)}$$

So, the acceleration of the body $3m$ is

$$a_2 = F_2/m = -(3mg/2)/(3m) = -g/2 (\downarrow) \text{ Ans.}$$

Problem 12 Three smooth blocks of masses $3m$, $2m$ and m are connected by two strings 1 and 2. Blocks m and $3m$ are pushed by the forces $3F$ and $4F$, respectively. If (a) both strings are rigid (b) string 2 is rigid and string 1 is flexible, find the tensions in the strings 1 and 2. Assume that both strings are inextensible.

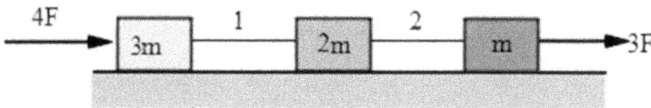

Solution:
(a) *Free-body diagram*
The forces acting on $3m$ are, $4F \rightarrow T' \rightarrow$, on $2m$, $\leftarrow T'$ and $T \rightarrow$, on m, $T \leftarrow$ and $3F \rightarrow$.

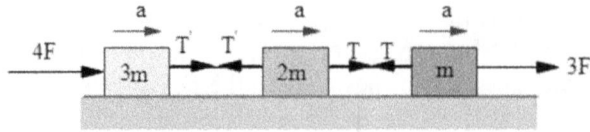

Let us assume that all the bodies move with the same acceleration a, say.

Force equation: $\Sigma F_x = ma$

$$3m: \quad 4F + T' = (3m)a \tag{4.64}$$

$$2m: \quad T - T' = (2m)a \tag{4.65}$$

$$m: \quad 3F - T = ma \tag{4.66}$$

Adding the last three equations,

$$a = \frac{7F}{6m} \tag{4.67}$$

Using equations (4.64) and (4.67), we have

$$T' = -\frac{F}{2} \quad \text{Ans.}$$

Using equations (4.66) and (4.67), we have

$$T = +\frac{11F}{6} \quad \text{Ans.}$$

(b) If string 1 is rigid, it can produce a negative tension of $F/2$, that is, pushing force. Both $2m$ and $3m$, are pushed by string 1 with a force of magnitude $F/2$. If string 1 is flexible, it cannot produce a negative tension, that is, pushing force; so, the tension T' in string 1 will vanish and the block $3m$ will have different accelerations than the other two blocks. So, we have to draw a new free-body diagram by putting $T' = 0$ and assigning new accelerations a_1 for $3m$ and a_2 for the other two blocks.

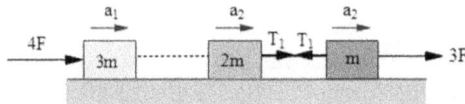

Force equation: $\Sigma F_x = ma$

$$3m: \quad 4F = (3m)a_1 \tag{4.68}$$

$$2m: \quad T_1 = (2m)a_2 \tag{4.69}$$

$$3m: \quad 3F - T_1 = ma_2 \tag{4.70}$$

Adding the last three equations,

$$a_1 = \frac{4F}{3m}, \quad T' = 0; \quad a_2 = \frac{F}{m} \quad \text{and} \quad T = 2F \quad \text{Ans.}$$

N.B.: Flexible string can only pull; rigid string can push and pull; the massless or light string will carry same tension; elastic string cannot develop or change any tension immediately after changing the external forces.

Problem 13

(a) A string of mass m, length l is rotating about its end C with a constant angular velocity ω on a smooth horizontal surface. Find the tension in the string at a distance x from the point C.

(b) If a block of mass M is attached at the end of the string, find the variation of tension with radial distance x.

Solution

(a) *Free-body diagram*

Let T' be the tension at a distance x from the point C. Let us consider an elementary segment of mass dm and length dr at a distance of r from the axis C. Since each portion of the rotating string accelerates toward the center of the circular path, we have the following force equation on the elementary segment of mass δm,

Force equation:

$$\Sigma F = T - (T + \delta T) = \delta m \cdot a,$$

where $a = r\omega^2$ and $dm = \frac{m}{l}dr$

$$\Rightarrow dT = -\frac{m\omega^2}{l} r \, dr$$

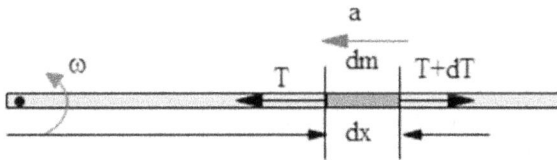

If we move from the free end towards the center, the tension increases from zero to T. So, integrating both sides, we have

$$\int_0^{T'} dT = -\frac{mw^2}{l} \int_l^x r \, dr$$

$$\Rightarrow T = \frac{m\omega^2}{2}(l^2 - x^2) \text{ Ans.}$$

(b) If we attach a block of mass M at the free end of the string, the extra tension is equal to the extra centrifugal force $Ml\omega^2$; so, the tension varies with x as

$$T = \frac{m\omega^2}{2}(l^2 - x^2) + Ml\omega^2 \text{ Ans.}$$

Problem 14 A light, inextensible smooth string is in contact with a fixed pulley through an angle ϕ as shown in the figure below. If the tension in the string is T, find (a) the horizontal and vertical reaction forces acting on the pulley by the ceiling, (b) the force with which the string presses the pulley radially.

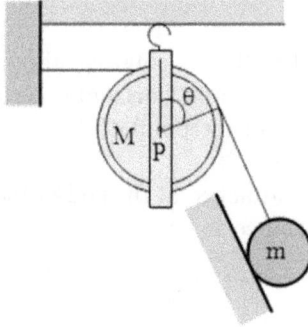

Solution

(a) *Free-body diagram*

Since the string is light, equal tension T is felt at each point of the string. The forces acting on the pulley are $Mg\downarrow$, $F_x(\rightarrow)$, $F_y(\uparrow)$, $T(\leftarrow)$ and $T(\searrow)$. The forces acting on the particle (bead) m are $mg\downarrow$, $T(\searrow)$ and $N'(\nearrow)$.

Force equation on the pulley m:

At the angle of inclination θ, the component of mg parallel to the plane is $mg\sin\theta$ is equal to the tension in the string because particle m is at rest. So, the tension in the string is

$$T = mg\sin\theta \qquad (4.71)$$

Force equation on the pulley M:

$$\Sigma F_x = Ma_x$$

$$F_x + T\cos\theta - T = Ma_x = 0$$

$$\Rightarrow F_x = T(1 - \cos\theta) \qquad (4.72)$$

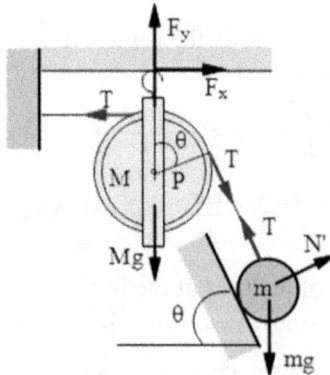

$$\Sigma F_y = M a_y$$

$$F_y - T \sin \theta - Mg = M a_y = 0$$

$$\Rightarrow F_y = T \sin \theta + Mg \tag{4.73}$$

Putting the values of T from equation (4.71) in equations (4.72) and (4.73), we have

$$F_x = mg(1 - \cos \theta)\sin \theta$$

$$F_y = (M + m \sin^2 \theta)g \quad \text{Ans.}$$

(b) We can see that the angle between two tensions acting on the pulley is $180° - \theta$. So, the resultant of these two tensions can be given by the parallelogram law of vectors or components methods. Let their resultant be N which is radially inward, given as

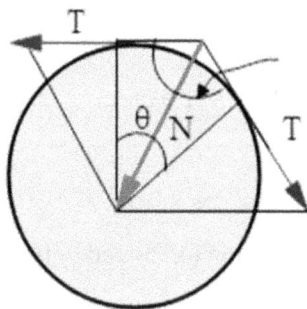

$$N_r = 2T \sin \frac{\theta}{2} \tag{4.74}$$

Putting $T = mg\sin \theta$ from equation (4.71) in equation (4.74), we have

$$N_r = 2mg \sin \theta \sin \frac{\theta}{2} \quad \text{Ans.}$$

Problem 15 Two smooth blocks of masses m_1 and m_2 are connected by a light inextensible string which passes over a smooth pulley. If the hanging mass m_2 is released from rest assuming that m_1 does not lose constant at the given position, find the (a) tension in the string, (b) acceleration of the block m_1, and (c) reaction offered by the ground on the block m_1.

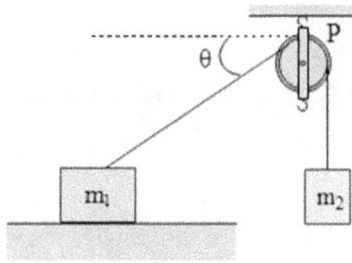

Solution

(a) *Free-body biagram*

The forces on m_1 are $T \nearrow$, $m_1 g \downarrow$ and $N \uparrow$. The forces on m_2 are $T \uparrow$ and $m_2 g \downarrow$

Force equation:

For m_1:

$$F_x = ma$$

$$T \cos \theta = m_1 a_1 \qquad (4.75)$$

For m_2: $F_y = ma_y$

$$m_2 g - T = m_2 a_2 \qquad (4.76)$$

Constraint equation:

$$a_1 \cos \theta = a_2 \qquad (4.77)$$

Substituting a_1 from equation (4.75) and a_2 from equation (4.76) in equation (4.77),

$$\left(\frac{T \cos \theta}{m_1} \right) \cos \theta = \left(\frac{m_2 g - T}{m_2} \right)$$

$$\Rightarrow T = \frac{m_1 m_2 g}{m_1 + m_2 \cos^2 \theta} \quad \text{Ans.}$$

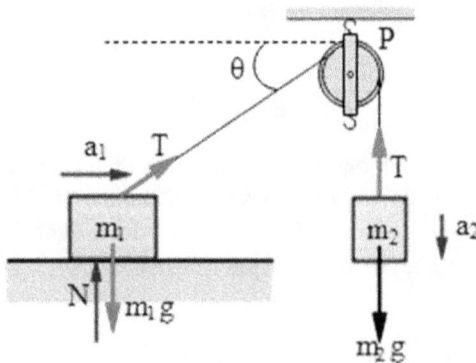

(b) Putting the obtained value of T in equation (4.75),

$$a_1 = T \cos \theta / m_1 = \frac{m_2 g \cos \theta}{m_1 + m_2 \cos^2 \theta} \quad \text{Ans.}$$

(c) The normal reaction offered by the ground on the block is

$$N = m_1 g - T \sin \theta$$

$$\Rightarrow N = m_1 g - \left(\frac{m_1 m_2 g}{m_1 + m_2 \cos^2 \theta} \right) \sin \theta$$

$$\Rightarrow N = m_1 g \left(1 - \frac{m_2 \sin \theta}{m_1 + m_2 \cos^2 \theta} \right) \quad \text{Ans.}$$

Problem 16

Two particles of masses m_1 and m_2 are interconnected by a light inextensible string which passes over a smooth pulley. (i) If the system is released from rest, find the (a) tension in the string, (b) acceleration of the particles, (c) time after which the particles cross each other. (ii) If we remove m_2 and pull the string by a force $m_2 g$, find the acceleration of m_1.

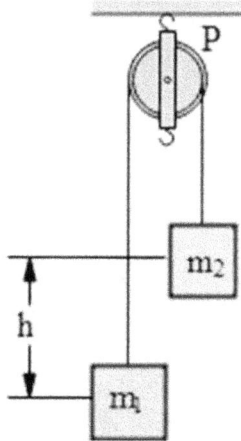

Solution

(i) *Free-body diagram*

The forces acting on m_1 are $T \uparrow$, $m_1 g \downarrow$ and on m_2 are $T \uparrow$ and $m_2 g \downarrow$. Let they move with upward accelerations a_1 and a_2, respectively.

Force equation:
For m_1: $\Sigma F_y = ma_y$

$$T - m_1g = m_1a_1 \tag{4.78}$$

For m_2: $\Sigma F_y = ma_y$

$$T - m_2g = m_2a_2 \tag{4.79}$$

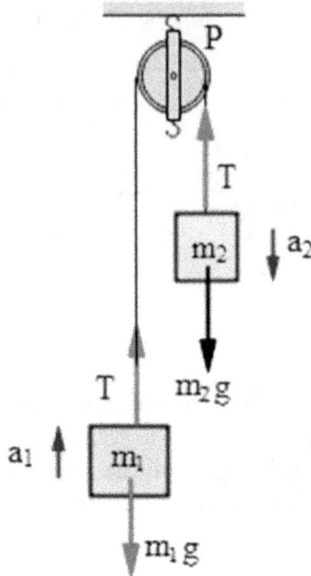

Kinematics (constraint equation):
As derived earlier,

$$a_1 + a_2 = 0 \tag{4.80}$$

(a) Substituting a_1 from equation (4.78) and a_2 from equation (4.79) in equation (4.80),

$$\left(\frac{T_1}{m_1} - g\right) + \left(\frac{T_2}{m_2} - g\right) = 0$$

$$\Rightarrow T = \frac{2m_1m_2g}{m_1 + m_2} \quad \text{Ans.}$$

(b) Substituting T in equation (4.78), we have

$$a_1 = a_2 = \frac{|m_2 - m_1|}{m_1 + m_2}g \quad \text{Ans.}$$

(c) The relative acceleration between the particles is

$$\vec{a}_{12} = \vec{a}_1 - \vec{a}_2 = 2\left(\frac{m_2 - m_1}{m_1 + m_2}\right)g\hat{j}$$

The relative displacement between the particles is

$$s_{12} = -h\hat{j}$$

Then, the time after which they cross each other is

$$t = \sqrt{2s_{12}/a_{12}} = 2\sqrt{\frac{(m_1 + m_2)h}{(m_2 - m_1)g}} \quad \text{Ans.}$$

(ii) If we remove m_2 and pull the string by a force $m_2 g$, the net force acting on m_1 is

$$F = (m_2 - m_1)g$$

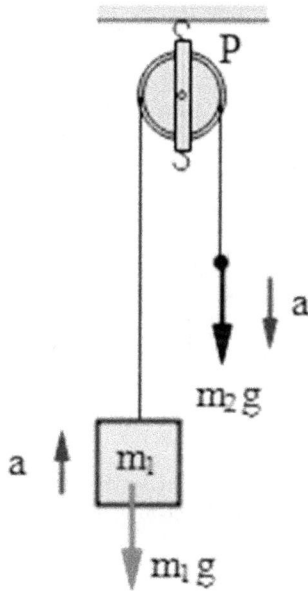

So, the acceleration of m_1 is

$$a = \frac{F}{m_1} = \frac{|m_2 - m_1|}{m_1}g \quad \text{Ans.}$$

Problem 17 In part (i) of the previous example, if the pulley is massless and it is moved with an upward acceleration a_0, find the tension in the string and accelerations of the particles. Put $a_0 = 2 \text{ m s}^{-2}$, $m_1 = 1$ kg, $m_2 = 2$ kg.

Solution

(a) *Force equation:*

The net force acting on m_1 is

$$T - m_1 g = m_1 a_1 \tag{4.81}$$

The net force acting on m_2 is

$$T - m_2 g = m_2 a_2 \tag{4.82}$$

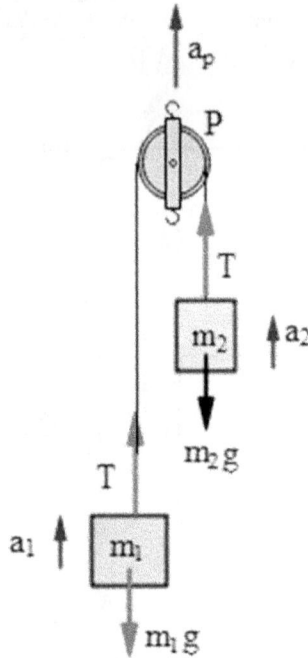

Kinematics:

The acceleration of m_1 is

$$a_1 = a_{1P} + a_p = a + a_0. \tag{4.83}$$

The acceleration of m_1 is

$$a_2 = a_{2p} + a_p = -a + a_0 \tag{4.84}$$

Substituting a_1 from equation (4.83) in equation (4.81),

$$T - m_1 g = m_1(a + a_0) \tag{4.85}$$

Substituting a_2 from equation (4.84) in equation (4.82)

$$T - m_2 g = m_2(-a + a_0) \tag{4.86}$$

Solving equations (4.85) and (4.86),

$$T = \frac{2m_1 m_2}{m_1 + m_2}(g + a_0)$$

$$\Rightarrow T = \frac{2(1)(2)}{1 + 2}(10 + 2) = 16 \text{ N Ans.}$$

(b) The relative acceleration between the bodies is

$$a = \frac{m_2 - m_1}{m_1 + m_2}(g - a_0)$$

$$\Rightarrow a = \frac{2 - 1}{2 + 1}(10 - 2) = 8/3 \text{ m s}^{-2} \text{ Ans.}$$

From equation (4.81),
$$a_1 = T/m_1 - g = 16/1 - 10 = 4 \text{ m s}^{-2} \text{ (up) Ans.}$$
From equation (4.82),

$$a_2 = T/m_2 - g = 16/2 - 10 = -2 \text{ m s}^{-2} \text{ (down) Ans.}$$

N.B.: For a vertically accelerating pulley–particle system, remember the following points.
1. If $m_2 > m_1$, a_{1p} is up and a_{2p} is down.
2. $a_{1p} + a_{2p} = 0$ but $a_1 + a_2 \neq 0$.
3. $a_1 = (a_0 + a)\uparrow$ and $a_2 = (a_0 - a)\uparrow$, if we assume $a_{1p} = a\uparrow$.
4. Since, $2a_0 = a_1 + a_2 \Rightarrow (a_0 - a_1) + (a_0 - a_2) = 0 \Rightarrow a_{1p} + a_{2p} = 0$.

Problem 18 A man of mass m_1 holding one end of a massless, inextensible string that passes over the smooth pulley, is accelerating up while standing on a platform of mass m_2. Find the (a) pressing force offered by the man on the platform, (b) tension in the string if it moves up with an upward acceleration a.

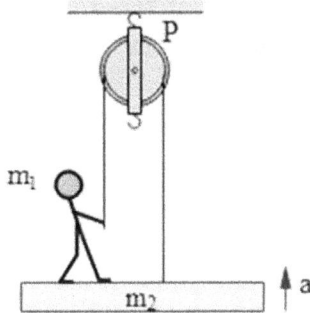

Solution

Free-body diagram

Forces on m_1 are $m_1g\downarrow$. $T\uparrow$ and $N\uparrow$; forces on m_2 are $m_2g\downarrow$, $T\uparrow$ and $N\downarrow$, where N is the normal contact force experienced by both man and platform as an action–reaction pair.

Force equation:

$$\text{For } m_1: \quad \Sigma F = N + T - m_1g = m_1a \tag{4.87}$$

$$\text{For } m_2: \quad \Sigma F = T - N - m_2g = m_2a \tag{4.88}$$

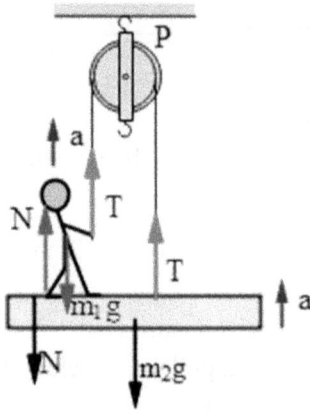

As the man is in contact with the platform, both will move with the same (upward) acceleration in this case: $a_1 = a_2 = a$.

Kinematics:

Eliminating T from equations (4.87) and (4.88), we have

$$N = \frac{(m_1 - m_2)}{2}(g + a) \text{ and } T = \left(\frac{m_1 + m_2}{2}\right)(g + a) \text{ Ans.}$$

N.B.: If $m_1 < m_2$, $N = 0$, then it will be reduced to a general case with

$$|\vec{a_1}| = |\vec{a_2}| = \frac{m_2 - m_1}{m_1 + m_2}g \text{ and } T = \frac{2m_1m_2}{m_1 + m_2}g.$$

Problem 19 Three smooth pulleys of masses m_1, m_2 and m_3 are interconnected by an inextensible light string that passes over two fixed smooth pulleys P_1 and P_2, as shown in the figure below. Find the tension in the string.

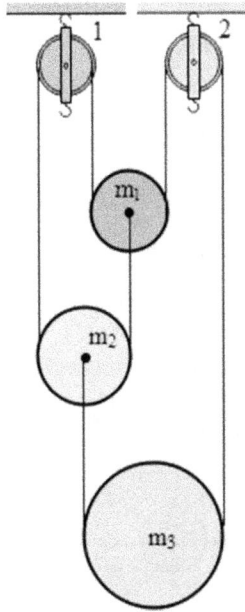

Solution

Free-body diagram

Since the string is single, smooth and massless, tension in the string is the same. The forces acting on m_1 are $m_1 g\downarrow$, $2T\uparrow$ and $T\downarrow$; on m_2 are $m_2 g\downarrow$, $2T\uparrow$ and $T\downarrow$; on m_3 are $m_3 g\downarrow$, $2T\uparrow$. Let us assume that the bodies accelerate downwards with a_1, a_2 and a_3, respectively. Referring to the free-body diagram, the net force acting on bodies 1, 2 and 3 are given as follows.

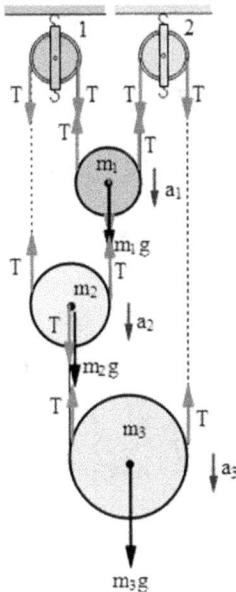

Force equation:
   Then we write the laws of motion;

$$\text{For } m_1: \quad \Sigma F = 2T - T - m_1 g = T - m_1 g = m_1(-a_1) \tag{4.89}$$

$$\text{For } m_2: \quad \Sigma F = 2T - m_2 g - T = T - m_2 g = m_2(-a_2) \tag{4.90}$$

$$\text{For } m_3: \quad \Sigma F = 2T - m_3 g = 2T - m_3 g = m_3(-a_3) \tag{4.91}$$

Kinematics (constraint equation):
   For string: $4y_1 + 3(y_2 - y_1) + 2(y_3 - y_2) = l$

$$\Rightarrow y_1 + y_2 + 2y_3 = l$$

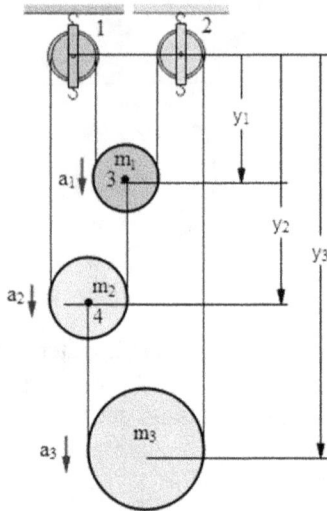

Differentiating twice with respect to time we have

$$\frac{d^2 y_1}{dt^2} + \frac{d^2 y_2}{dt^2} + 2\frac{d^2 y_2}{dt^2} = 0 \left(\text{Since } l = \text{constant}, \frac{dl}{dt} = 0 \right)$$

Since direction of accelerations is assumed in the direction of the position vectors, we can write

$$\frac{d^2 y}{dt^2} = +a,$$

So, we have

$$a_1 + a_2 + 2a_3 = 0 \tag{4.92}$$

Substituting a_1, a_2 and a_3 from equations (4.89), (4.90) and (4.91), respectively, in equation (4.92), we have

$$\left(-\frac{T}{m_1} + g \right) + \left(-\frac{T}{m_2} + g \right) + 2\left(-\frac{2T}{m_3} + g \right) = 0$$

$$\Rightarrow T = \frac{4m_1 m_2 m_3}{4m_1 m_2 + m_2 m_3 + m_1 m_3} g \quad \text{Ans.}$$

N.B.: If either mass is zero, the tension will vanish. Then, try to find the acceleration of the bodies.

Problem 20 A horizontal force F acts on the wedge of mass M and angle of inclination θ. Disregarding friction in all contacting surfaces, find (a) the acceleration of the wedge and discuss some possible cases; (b) the value of acceleration A of the wedge so that the block remains at rest relative to the wedge, (c) the corresponding value of the reaction on the block when it remains stationary relative to the wedge (d) the horizontal force F required to move the wedge so that the block does not slip with the wedge, (e) the minimum force required to move the wedge so that the block loses contact with the wedge, (f) the force required to move the wedge with constant velocity or to prevent its sliding with the ground.

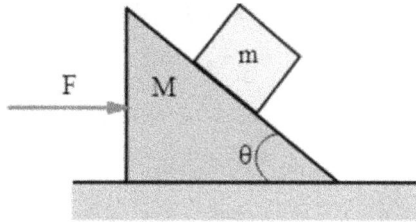

Solution
Let us take the x-axis (horizontal or to the right) and y-axis (vertically up).

(a) *Free-body diagram*:
Let us apply $F = ma$ for the entire wedge–block system $(M + m)$. The net horizontal force acting on $(M + m)$ is F. The forces acting on m are $mg\downarrow$, $N\nearrow$ and pseudo-force $mA\leftarrow$. The accelerations are also specified in the free-body diagrams.

Force equations:

$$\text{For } (M + m): \quad F = ma_x + MA \tag{4.93}$$

$$\text{For } m \text{ (relative to } M): \quad mg \sin \theta - mA \cos \theta = ma \tag{4.94}$$

Kinematics:

$$a_x = (a \cos \theta + A) \tag{4.95}$$

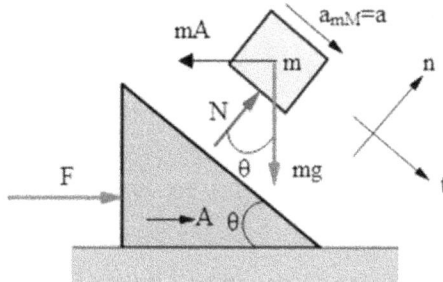

Solving the above three equations, we have

$$A = \frac{F - mg \sin \theta \cos \theta}{M + m \sin^2 \theta} \quad \text{Ans.}$$

1. When $F > mg \sin \theta \cos \theta$, A is directed to the right.
2. When $F = mg \sin \theta \cos \theta$, $A = 0$.
3. When $F < mg \sin \theta \cos \theta$, A is directed to the left. N decreases to zero when the wedge is pulled to the left with gradual increase in its acceleration. Ans.

(b) If the block remains at rest relative to the wedge, equating the relative acceleration to zero by using equation (4.94), we have

$$mg \sin \theta - mA \cos \theta = ma = 0$$

$$\Rightarrow A = g \tan \theta \ \ \text{Ans.}$$

(c) If the block does not slip on the wedge, it does not have a vertical acceleration; so, the net vertical force acting on the block is

$$F_v = -mg + N \cos \theta = ma_v = 0$$

$$\Rightarrow N = mg / \cos \theta = mg \sec\theta \ \ \text{Ans.}$$

In an alternative method, you can put
$$A = g \tan \theta$$
In the equation of normal reaction given as
$$N = m(A \sin \theta + g \cos \theta)$$

to get the same answer.

(d) The force required to drag the wedge so that the block does not slip on the wedge is

$$F = (M + m)A.$$

where $A = g \tan \theta$.

$$\Rightarrow F = (M + m)g \tan \theta \ \ \text{Ans.}$$

(e) As the block just loses contact from the wedge, $N = 0$. Then the only horizontal force acting on M is

$$F = MA$$

where $A = g \cot \theta$ obtained after putting $N = 0$ in the expression $N = m(g\cos \theta - A\sin \theta)$.

$$\Rightarrow F = Mg \cot \theta \ \ \text{Ans.}$$

(f) When the acceleration of the wedge is zero, putting $a = 0$ in

$$A = \frac{F - mg \sin \theta \cos \theta}{M + m \sin^2 \theta},$$

we have $F = mg \sin \theta \cos \theta$ Ans.

Problem 21 Two smooth blocks of masses m and m' connected by a light inextensible strings are moving on a smooth wedge of mass M. A horizontal force F acts on the wedge so that the blocks do not slide relative to the wedge. Find the (a) acceleration of the wedge and (b) value of F.

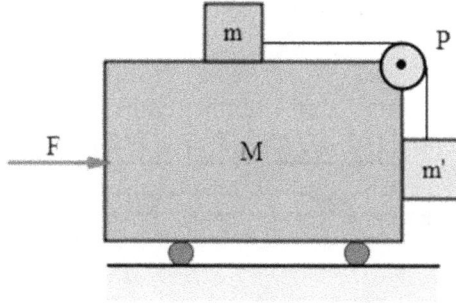

Solution

(a) *Free-body diagram*

The force acting on blocks m and m' parallel to the surfaces in contact are given as $mA \leftarrow$ and $T \rightarrow$ (on m); $mg \downarrow$ and $T\uparrow$ (on m').

Force equation relative to M: Since $a_{mM} = a_0 = 0$

For m: $-mA + T = 0$ (4.96)

For M: $T - m'g = 0$ (4.97)

From equations (4.96) and (4.97),

$$A = \frac{m'}{m}g \text{ Ans.}$$

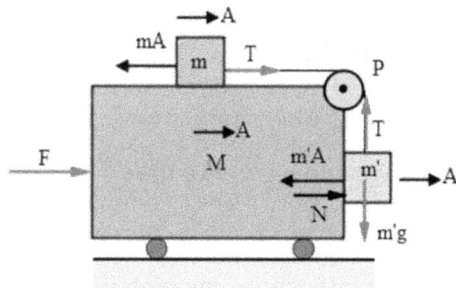

(b) *Force equation on* $M + m + m'$:

$$\Sigma F_x = ma_x$$
$$F = MA + mA + m'A$$
$$= (M + m + m')A$$
$$= (M + m + m')\frac{m'}{m}g \text{ Ans.}$$

Problem 22 A block of mass m is fitted with a light undeformed spring of stiffness k which is rigidly connected with a trolley car. If the trolley car moves with leftward acceleration a_0, find the (a) equation of deformation of the spring as the function of time, (b) minimum and maximum elongation of the spring, (c) minimum and maximum velocity of the block relative to the trolley car.

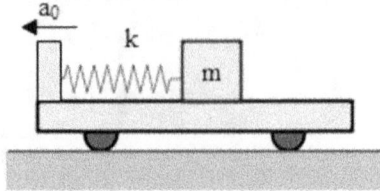

Solution

(a) *Free-body diagram*

Relative to the trolley car the horizontal forces acting on m are $F_p(=kx)\leftarrow$ and pseudo-force $ma_0\rightarrow$.

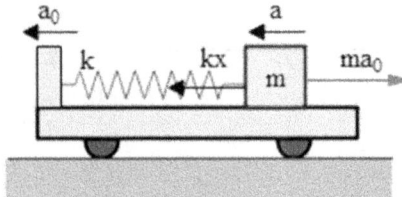

Let the block move with an acceleration a relative to the trolley car.

Force equation: Applying Newton's second law from the trolley car, we have

$$\Sigma F = kx - ma_0 = ma \tag{4.98}$$

Kinematics:

Since x decreases in the direction of a_x we have

$$a = \frac{-d^2x}{dt^2} \tag{4.99}$$

Using equations (4.98) and (4.99), we have

$$\frac{d^2x}{dt^2} + \frac{k}{m}x = a_0$$

Solution of the above differential equation is

$$x = \frac{ma_0}{k}\left(1 - \cos\sqrt{\frac{k}{m}}\,t\right)$$

(b) Then, the minimum and maximum elongation x of the spring are $2ma_0/k$ and zero, respectively.

(c) The velocity of the block relative to the trolley car is

$$\frac{dx}{dt} = \frac{ma_0}{k}\sqrt{\frac{k}{m}}\sin\sqrt{\frac{k}{m}}t = \sqrt{\frac{m}{k}}a_0\sin\sqrt{\frac{k}{m}}t$$

So, the minimum and maximum velocity of the block relative to the trolley car are zero and $(\sqrt{m/k})a_0$, respectively.

Problem 23 Find the acceleration of the particles m_1 and $m_2(>m_1)$ relative to the pulley P using pseudo-force method, if the pulley moves with an upward acceleration A.

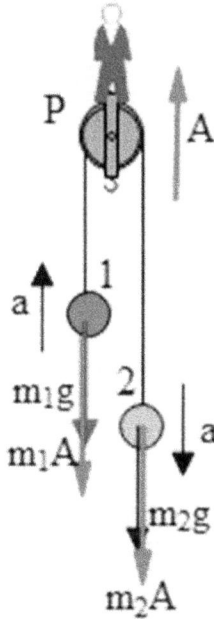

Solution

Free-body diagram

Let us build the equations by using Newton's second law sitting on the accelerating pulley. Hence, we impose pseudo-force $-m_1A\downarrow$, on both m_1 and m_2, in addition to the upward tension and their weights $m_1g\downarrow$ and $m_2g\downarrow$, respectively. If m_1 accelerates up relative to the pulley, m_2 must accelerate down relative to the pulley with acceleration a.

Force equation:

For m_1: $\quad \Sigma T = T - m_1g - m_1A = m_1a \qquad (4.100)$

For m_2: $\quad \Sigma F = m_2g + m_2A - T = m_2a \qquad (4.101)$

Solving equations (4.100) and (4.101), we have

$$a = \frac{m_2 - m_1}{m_1 + m_2}(g + A) \text{ Ans.}$$

Problem 24 A block of mass m is sliding on a smooth accelerating wedge of angle of inclination θ. (a) If the acceleration of the wedge is A, using pseudo-force method find the acceleration of the block relative to the wedge and discuss the possible cases. (b) If acceleration of the wedge is $A = 2g \tan \theta$ to the right, find the time after which the block will escape from the wedge. (c) Find the acceleration of the block in the case of (b).

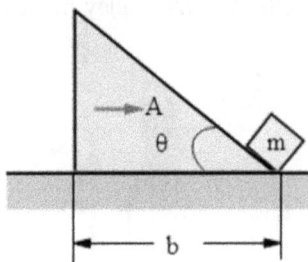

Solution

(a) *Free-body diagram*:

Let us sit on the wedge and try to write the Newton's equations of motion of the block m. For this we apply the pseudo-force $mA\leftarrow$ in opposite direction of $\overrightarrow{A}$, in addition to the forces mg and N. When observed from the trolley car, the forces acting on the block are pseudo-force $mA\leftarrow$, $mg\downarrow$ and $N \nearrow$.

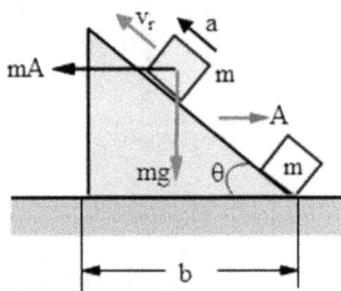

Force equation:

Then resolving the forces along the slant, we have

$$F = mA \cos \theta - mg \sin \theta \text{ Ans.}$$

(i) If $mA \cos \theta > mg \sin \theta$, F points up along the slant. Then, the block will accelerate relative to the wedge with $a = A \cos \theta - g \sin \theta$.

(ii) If $mA\cos\theta = mg \sin \theta$, $F = 0$. Hence the block remains stationary relative to the wedge (or more with constant velocity relative to the wedge).

(iii) If $mA \cos \theta < mg \sin \theta$, F points down along the slant. Hence the block accelerates down the slant with $a = g \sin \theta - A \cos \theta$ Ans.

(b) The acceleration of the block relative to the wedge is

$$a = A \cos \theta - g \sin \theta$$

Putting $A = 2g \tan \theta$, we have

$$a = (2g \tan \theta)\cos \theta - g \sin \theta = g \sin \theta \text{ up the wedge.}$$

Then the time after which the block will reach the top of the wedge is

$$t = \sqrt{\frac{2l}{a}}$$

Putting $a = g \sin \theta$ and $l = b/\cos \theta$, we have

$$t = \sqrt{\frac{2b/\cos \theta}{g \sin \theta}} = \sqrt{\frac{2b}{g \sin \theta \cos \theta}} \quad \text{Ans.}$$

(c) The acceleration of the block is

$$a_m = \left(\sqrt{g^2 + A^2} \right) \sin \theta$$

Putting $A = 2g \tan \theta$, we have

$$a_m = \frac{g \sin \theta}{\cos \theta} \sqrt{1 + 3 \sin^2 \theta} \quad \text{Ans.}$$

IOP Publishing

Problems and Solutions in Particle Mechanics

Pradeep Kumar Sharma

Chapter 5

Friction

5.1 Introduction

Friction is a force that is always around us helping us to do our regular activities such as walking, riding vehicles, sitting, climbing, pulling, writing and much more. Basically, there are two types of friction, namely static and kinetic (or sliding) friction. Static friction prevents the relative sliding and kinetic friction opposes the relative sliding between two surfaces. This chapter talks about the types, nature and laws of friction. Newton's laws and kinematics are applied for mechanical systems such as spring–particle, pulley–particle and wedge–particle systems. In this chapter, you will learn the dynamics of a particle in the presence of friction and other forces for the aforementioned systems. We can understand how friction helps us in moving in a curve. So, you can find the concept of friction used in the next chapter 'Dynamics of circular motion'. First of all, we define *friction* as a tangential contact force prevailing between two surfaces that prevents or opposes their relative sliding (motion).

5.2 Static friction

The friction between two surfaces that prevents their relative sliding is called static friction because the surfaces are static (stationary) relative to each other. For example, a monkey is pushing a box with a horizontal force F, say, and the box does not move; you have to understand that somebody must be pushing the box against the force F applied by the monkey. Since the ground is in contact with the box, only the ground can be responsible for pushing the box back with an equal and opposite force so as to balance the applied force F. As the table remains static relative to the ground, the tangential push offered by the ground on the box is called static friction, denoted as f_s; also, an equal and opposite static friction will act on the ground by the box to obey Newton's third law as an action–reaction pair. Furthermore, the monkey receives a reaction force F from the box in response to the action force F exerted by the monkey on the box. The net horizontal force acting on the monkey is zero if it does not slip; so, $f_s - F = 0$ which means that another static friction of magnitude f_s will act on the monkey to the right.

doi:10.1088/978-0-7503-6442-3ch5

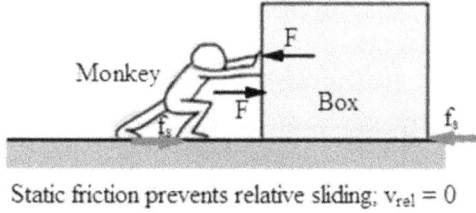

Static friction prevents relative sliding; $v_{rel} = 0$

5.2.1 Static friction is variable

If you gradually increase your push to a critical value F_0, say, till the box begins to slide. In this process the static friction acting on the box gradually increases from zero to a limiting (maximum) value so as to nullify your increasing force. The maximum value of friction till the body begins to slide is called limiting friction $f_{s_{max}}$. In other words, the magnitude of static friction varies between zero to $(f_s)_{max}$.

5.2.2 Direction of static friction

If there were no friction, the box would accelerate towards the right relative to ground by the applied force F. As the box is not sliding in the presence of static friction, we can say that it *tends to slide* (but does not slide) to the right. Then, static friction acts to the left. *Static friction acts opposite to the direction in which the body tends to slide.*

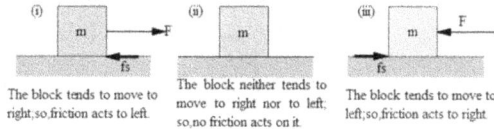

The block tends to move to right; so friction acts to left.

The block neither tends to move to right nor to left; so, no friction acts on it.

The block tends to move to left; so friction acts to right.

5.2.3 Law of static friction

Newton and Leonardo da Vinci conducted experiments on static friction which concluded the following properties.

(a) Static friction varies its magnitude from zero to limiting friction; $0 \leqslant f_s \leqslant f_{s_{max}}$.

(b) Limiting friction does not depend on the area of the contact.

(c) Limiting friction is directly proportional to the normal reaction (pressing force) between the surfaces in contact. Experimentally, we have the relation,

$$f_{s_{max}} = \mu_s N,$$

where μ_s = coefficient of static friction that depends on the nature of surfaces, impurities, temperature etc. μ_s is mostly less than one. However, for some contacting surfaces such as cement–rubber, $\mu_s > 1$.

For the static friction f_s to exist (i) the surfaces must be rough, (ii) the bodies must be pressed against each other, and (iii) the bodies must *tend to move (slide)* relative to each other.

Example 1 Describe how friction helps us in walking.

Solution

To generate friction, we need a pressing force. We are pressed against the ground by Earth's gravity. Once you push the ground by the back foot, it tends to slip backward, and static friction will push your back foot forward. Thus, friction helps us in walking by providing a grip over the surface or ground. Once you start walking, your back foot pushes the ground backward and the front foot pushes the ground forward. In response, the ground pushes your back foot forward by the static friction f'_s, say, and your front foot backward by static friction f_s, say; If you intend to speed up, forward friction is greater than backward friction acting on your feet. The net forward friction pushes your body forward and accelerates you by increasing your speed; the net backward friction can retard your motion. The sidewise or transverse or lateral component of friction can be used in changing the direction of motion. In the process of walking, the center of mass of the body also moves up and down due to the effect of gravity and normal impulsive reaction.

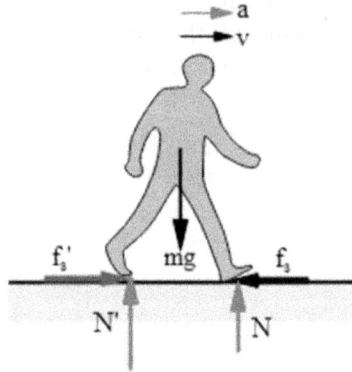

Hence, walking is a process of 'slow collision' in which the static friction and normal reactions are impulsive in nature. Please note that while you are speeding up, the back foot is a driving foot as *static friction* pushes it forward and the front foot retards the motion due to the backward static friction acting on it. This will be opposite for retarding motion.

Example 2 In a tug of war, who wins? Explain the role of static friction.

Solution

Let us assume that two persons 1 and 2 pull a rope. As the rope is light, the same tension T pulls each person back. As persons 1 and 2 pull the rope, their back feet must not slide due to static frictions f_s and f'_s. If person 1 presses the ground harder, f_s will be greater than f'_s. So, person 1 will accelerate toward the left and make the other person move towards left. Thus, person 1 will win the game. If person 2 pushes the ground harder, $f_s' > f_s$. Then, person 2 will win.

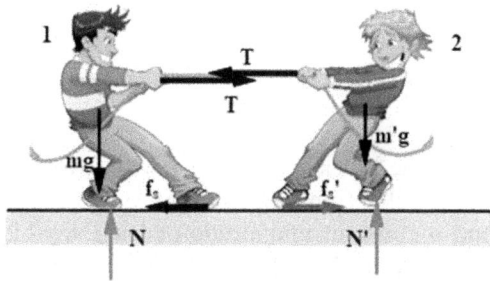

Example 3 A block of mass m placed on a horizontal floor is pulled by a horizontal force $F = 0$ N, 5 N (to right) and 6 N (to left). If $\mu_s = $ coefficient of static friction between the block and the surfaces in contact $= 0.75$, find the friction in each case.

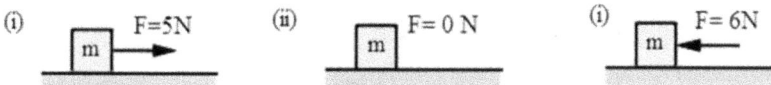

Solution

Referring to the free-body diagram (i), we can write the following equations:

Newton's second law: $F_x = F - f_s = ma_x = 0$

$$\Rightarrow f_s = F \tag{5.1}$$

$$F_y = N - mg = ma_y = 0$$

$$\Rightarrow N = mg \tag{5.2}$$

Law of static friction:

$$f_s \leqslant \mu_s N \tag{5.3}$$

Using the last two equations,

$$f_s \leqslant \mu_s mg \tag{5.4}$$

Using the last equations (5.1) and (5.4),

$$F \leqslant \mu_s mg$$

$$\Rightarrow F \leqslant (0.75)(1)(10)$$

$$\Rightarrow F \leqslant 7.5 \text{ N}$$

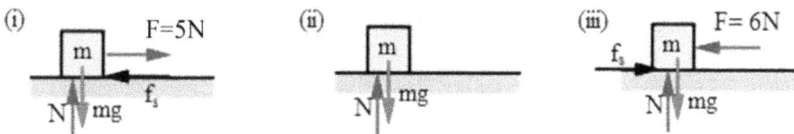

Since the given magnitudes of the applied force $F = 5$ N, 0 N and 6 N are less than its maximum value of 7.5 N, the block will not slide. So, the prevailing static friction is given by equation (5.1) as $f_s = F = 5$ N (to the left), 0 N (no friction) and 5 N (to the right), respectively.

5.3 Kinetic friction

In example 3, if you pull the block horizontally with a force $F = 10$ N, it will be greater than its maximum value (limiting friction) to avoid sliding of the block. So, the block will slide and a constant magnitude of friction which is slightly smaller than the limiting friction that acts on the block so as to resist or oppose its sliding relative to the ground. This constant friction acting on the block during its relative sliding is called sliding or kinetic friction. The kinetic friction acts to the left as the block slides to the right. Kinetic friction is a tangential contact forces between two surfaces so as to oppose their relative motion (or sliding). How do you know that kinetic friction is acting on the block? Let's see.

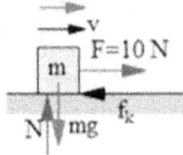

As the applied force F=10 N is greater than its maximum value of 7.5 N for no relative sliding of the block, the kinetic friction comes in to play and the block can slide with an acceleration

After the block slides, you still have to pull or push it with a constant force to move the block with a uniform velocity. This is because the ground pushes the block to the left with a constant force of magnitude which is called kinetic friction denoted as f_k.

5.3.1 Direction of f_k

If you stop pushing a block while it is in motion, you can notice that the block continues to move with decreasing speed and comes to rest after some time. This is because the kinetic friction f_k acts so as to oppose its motion (velocity) relative to the ground. Then, we can say that f_k acting on the block opposes the velocity of the block relative to ground ($\vec{v}_{tg} = \vec{v}$). So, roughly we can say that the kinetic friction f_k opposes the relative motion (velocity) between the contacting surfaces. However, kinetic friction acting on one surface may favor or oppose the velocity of the surface, but it must oppose the relative velocity between the surfaces.

5.3.2 Laws of kinetic friction

It is experimentally verified that (i) kinetic friction increases proportionally with the normal reaction between the surfaces, (ii) it is a force of constant magnitude for a given pressing force and surfaces in contact, and (iii) it does not depend on the 'area of contact' of the surface,

$$f_k = \mu_k N$$

where μ_k = coefficient of kinetic friction. You can practically see that we need to push the block harder just before sliding than during sliding; so, $f_{s_{max}} > f_k$. Then, from the law of static and kinetic friction, $\mu_s > \mu_k$. You can note that $\mu_s \cong 0.75\mu_k$ for most of the practical applications.

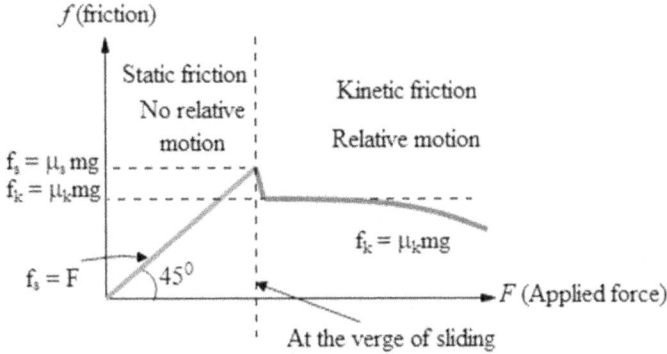

The variation of friction acting on the block with increasing applied horizontal force F

Example 4 A block of mass $m = 1$ kg placed on a horizontal floor is pulled by a horizontal force F of different magnitudes as shown in the figure. If $\mu_s = 0.75$ and $\mu_k = 0.7$, find the (a) friction and reaction force offered by the ground and (b) acceleration of the block.

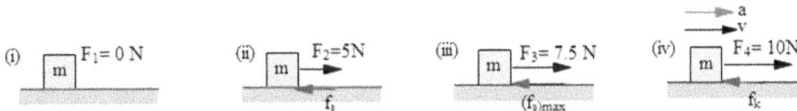

On gradually increasing the force F, the static friction gradually increases to its limiting (maximum) value and the block will be at the verge of sliding; then the block slides as the kinetic friction which is slightly less than the limiting friction will appear

Solution

(a) Referring to the free-body diagram (i), we can write the following equations:

Newton's second law: $F_x = F - f_s = ma_x = 0$

$$\Rightarrow f_s = F \tag{5.5}$$

$$F_y = N - mg = ma_y = 0$$

$$\Rightarrow N = mg \tag{5.6}$$

Law of static friction:

$$f_s \leqslant \mu_s N \tag{5.7}$$

Using the last two equations,

$$f_s \leqslant \mu_s mg \tag{5.8}$$

Using the last equations (5.5) and (5.8),

$$F \leqslant \mu_s mg$$

$$\Rightarrow F \leqslant (0.75)(1)(10)$$

$$\Rightarrow F \leqslant 7.5 \text{ N}$$

Since the given magnitudes of the applied force $F = 0$ N and 5 N are less than the limiting friction of 7.5 N, the block will not slide. For $F = 7.5$ N, the static friction will be maximum. So, there will be no relative sliding; then, the values of existing static friction for cases (i), (ii) and (iii) are given by equation (5.8) as 0 N (no friction), $f_s = F = 5$ N (to the left) and 7.5 N (to the left), respectively. In case (iv), $F = 10$ N which is greater than the limiting friction (7.5 N); so, the block will slide and kinetic friction will prevail which is given as $f_s = \mu_k mg = (0.7)(1)(10) = 7$ N. Then, the reaction offered by the ground on the block is given as the vector sum of the friction and normal reaction. The magnitude of the net reaction R for cases (i), (ii) and (iii) are given as

$$\vec{R} = (-f\hat{i} + 10\hat{j})N,$$

where $f = 0$, 5, 7.5 and 7 respectively. Ans.

(b) For the first three cases till the static friction becomes maximum, the block does not slide; so, the acceleration of the block is zero. In the last case, the net force acting on the block is

$$F_{net} = F - \mu_k mg = 10 - 7 = 3 \text{ N}$$

So, the acceleration of the block is

$$a = F_{net}/m = 3/1 = 3 \text{ m s}^{-2} \text{ to right.} \quad \text{Ans.}$$

5.4 Mechanism of dry friction (micro-interpretation)

When a block of mass m is placed on a horizontal surface, the forces acting on the block are its weight $mg \downarrow$ and normal reaction of surface $N\uparrow$. If a small horizontal force F acts on the block, for the block to remain stationary, in equilibrium, a horizontal component f of the surface reaction R is required. Here, $f = f_s$ is a static-friction force. As F increases, static friction force f_s increases as well until it reaches a maximum value $(f_s)_{max} = \mu_s N$. Further increase in F causes the block to begin sliding as the friction f drops to a little bit smaller kinetic friction force $f_s = \mu_k N$.

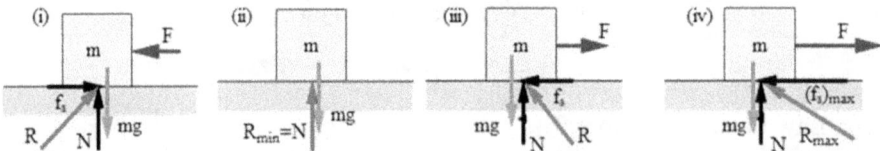

As the applied force F changes, the reaction R and static friction f_s change their magnitudes and directions. The reaction is minimum when F=0 and f_s=0; the reaction is maximum when the static friction becomes maximum.

Maximum static friction or limiting friction force and kinetic friction force are directly proportional to normal reaction force. These frictional forces depend on type and condition of contact surfaces and are independent of contact area.

The coefficients of friction (both static and dynamic or kinetic or sliding) reflect the roughness of the contacting surfaces, which is a geometric property of the surfaces.

A visibly flat surface (to a naked eye) is not really flat under a microscope. We can see lots of microscopic humps and bumps on surfaces. You can call it surface irregularities. The peaks of one surface oppose (block) the motion of the peaks of the other surfaces as they touch and push each other under the action of external pushing or pulling forces. Furthermore, the peaks of one surface touch the peaks of another surface and deform the contact points plastically under the normal pressing force. This is called cold welding (or surface adhesion) which is an intermolecular attraction between the surface atoms at the points of contact.

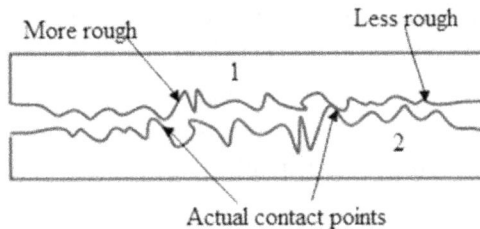

In the microscopic view of irregularities, the actual contact points at the peaks of the irregularities are responsible for producing friction

When you pull or push the surfaces against each other tangentially, you have to fight to rupture (or dislocate or bend) the microscopic peaks and the adhesive bonds between the contacting surface atoms or molecules. These two effects appear as friction on a macroscopic scale.

If the contacting surfaces move relative to each other (relative sliding), the surfaces touch each other nearly along the tops (peaks) of the humps. The tangential (t-) components of the reaction force R are smaller than when there is no relative sliding (relative rest) condition. In other words, the kinetic friction is less than the limiting friction. So, the horizontal force required to maintain the relative motion is generally less than that required to start the block when the surface irregularities are more nearly in mesh; $(f_s)_{max} > f_k$.

5.5 Angle of friction

Let us assume that many external forces act on a block in addition to gravity. Let F = resultant of all external forces excluding gravity and R = net reaction or contact force which is the resultant of tangential force (friction) and normal contact force (normal reaction). Under the action of the forces F, mg and R, the block is either at rest or moving relative to the other surface (the ground here). First of all, resolve the reaction R along the tangential and normal direction. The tangential component of reaction R is the friction; it can be static or kinetic. The normal component of reaction R is called normal reaction N. So, we have two angles of friction.

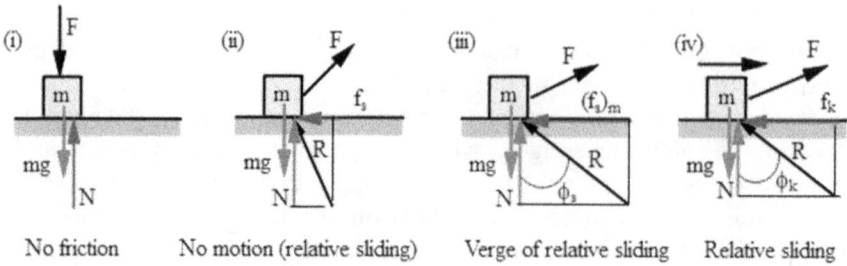

No friction   No motion (relative sliding)   Verge of relative sliding   Relative sliding

5.5.1 Angle of static friction

Let us pull (or push) a block till it begins to slide on the surface in contact. When maximum static friction (limiting friction) $f_{s_{max}}$ comes into play between the block and ground, let the normal contact force (reaction) acting on the block be N.

Then, the angle made by total reaction $\vec{R}$ and $\vec{N}$ is given as

$$\tan \phi_s = \frac{f_{s_{max}}}{N} \left(\text{but not } \frac{f_s}{N} \right),$$

where $\frac{f_{s_{max}}}{N} = \mu_s$

$$\Rightarrow \tan \phi_s = \mu_s$$

$$\Rightarrow \phi_s = \tan^{-1} \mu_s$$

The angle ϕ_s is known as angle of static friction.

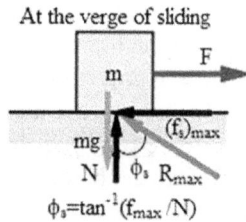

At the verge of sliding

$\phi_s = \tan^{-1}(f_{max}/N)$

When a particle begins to slide on any surface, the angle between the total reaction force and normal reaction offered by the surface (acting on the particle) is called angle of static friction given, as $\phi_s = \tan^{-1} \mu_s$.

5.5.2 Angle of kinetic friction

If the block slides on the surface in contact, kinetic friction f_k comes into play between the block and the ground, let the normal contact force (reaction) acting on the block be N.

Then, the angle made by total reaction $\vec{R}$ and $\vec{N}$ is given as

$$\tan \phi = \frac{f_k}{N},$$

where $\frac{f_k}{N} = \mu_k$

$$\Rightarrow \tan \phi_k = \mu_k$$

$$\Rightarrow \phi_k = \tan^{-1} \mu_k$$

The angle ϕ_k is known as angle of kinetic friction.

When a particle slides on any surface, the angle between the total reaction force and normal reaction offered by the surface (acting on the particle) is called angle of kinetic friction given, as $\phi_k = \tan^{-1} \mu_k$.

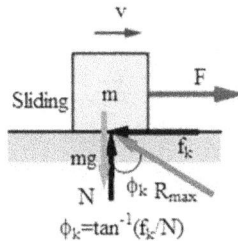

Example 5 A block of mass 2 kg requires a minimum horizontal force of 10 N so that the block will begin to slide. Once it slides, a force of 9.8 N is required to maintain the velocity of the block. Find the value of (a) angle of static friction ϕ_s and (b) angle of kinetic friction ϕ_k.

Solution

(a) As the block is on the verge of sliding, the acceleration of the block is equal to that of the ground, which is equal to zero. So, we can write

Newton's second law: $F = (f_s)_{\max} = 10$ N (5.9)

$$N = mg = 2 \times 10 = 20 \text{ N} \tag{5.10}$$

Law of static friction:

$$\mu_s = \frac{(f_s)_{\max}}{N} \tag{5.11}$$

Using the last three equations, we have

$$\mu_s = \frac{(f_s)_{\max}}{N} = \frac{10}{20} = 0.5 \tag{5.12}$$

Then, the angle of static friction is

$$\phi_s = \tan^{-1} \mu_s \tag{5.13}$$

Using the last two equations,

$$\phi_s = \tan^{-1} \mu_s = \tan^{-1} 0.5 = 26.56° \text{ Ans.}$$

(b) As the block is sliding with constant velocity, the acceleration of the block is zero. So, we can write

Newton's second law: $F = f_k = 9.8 \text{ N}$ (5.14)

$$N = mg = 2 \times 10 = 20 \text{ N} \tag{5.15}$$

Law of kinetic friction:

$$\mu_s = \frac{f_k}{N} \tag{5.16}$$

By using the last three equations,

$$\mu_s = \frac{f_k}{N} = \frac{9.8}{20} = 0.49 \tag{5.17}$$

Then, the angle of kinetic friction is

$$\phi_k = \tan^{-1} \mu_k \tag{5.18}$$

Using the last two equations, we have

$$\phi_k = \tan^{-1} 0.49 = 26.1° \text{ Ans.}$$

Example 6 When a block of mass m is (a) stationary and (b) sliding along the line of greatest slope of an inclined plane of inclination θ, find the net or total reaction (contact) force offered by the ground on the block.
Solution
(a) **Method 1**
Referring to the free-body diagram below, the forces acting on the block are $f_s \nwarrow$, $N \nearrow$ and $mg \downarrow$. As the block is stationary on the inclined plane, the acceleration of the block is zero. Let us resolve all the forces parallel and perpendicular to the inclined plane, to apply Newton's second law.
Force equation:

$$\sum F_x = -f_s + mg \sin \theta = 0 \tag{5.19}$$

$$\sum F_y = N - mg \cos \theta = 0 \tag{5.20}$$

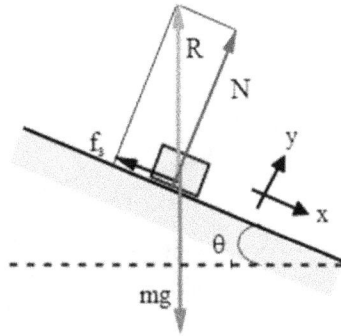

Substituting, $f_s = mg \sin \theta$ from equation (5.19) and $N = mg \cos \theta$ from equation (5.20) we have, the total contact force

$$R = \sqrt{f_s^2 + N^2} = \sqrt{(mg \sin \theta)^2 + (mg \cos \theta)^2} = mg\uparrow \quad \text{Ans.}$$

Method 2

Since, $\sum \vec{F} = \vec{F_c} + m\vec{g} = 0$, we have $\vec{F_c} = -m\vec{g} = mg\uparrow$ Ans.

Now, resolve the total contact force (reaction) $\vec{F_c}$ parallel (tangential) and perpendicular (normal) to the surface to obtain

$$f_s = mg \sin \theta \searrow \text{ and } N = mg \cos \theta \nearrow$$

The normal component of $\vec{F_c}$ is popularly known as normal contact force (or normal reaction) N and its component along the plane is known as tangential contact force which is none other than static friction in the above case.

(b) Referring to the free-body diagram, the forces acting on the block are $f_k \searrow$, $N \nearrow$ and $mg\downarrow$. As the block is sliding on the inclined plane, the acceleration of the block is zero. Let us resolve all the forces parallel and perpendicular to the inclined plane, to apply Newton's second law.

Force equation:

$$\sum F_x = -f_k + mg \sin \theta = 0 \tag{5.21}$$

$$\sum F_y = N - mg \cos \theta = 0 \tag{5.22}$$

Substituting, $f_k = \mu_k mg \cos \theta$ from equation (5.21) and $N = mg \cos \theta$ from equation (5.22) we have, the total contact force

$$R = \sqrt{f_k^2 + N^2} = \sqrt{(\mu_k mg \cos \theta)^2 + (mg \cos \theta)^2}$$
$$= mg \cos \theta \sqrt{\mu_k^2 + 1} = mg \cos \theta \sqrt{\tan^2 \phi_k + 1}$$

Since $\theta_k < \theta$, putting $\cos \theta < \cos\theta_k$, we have $R < mg$ and it will no longer point vertically upward. Ans.

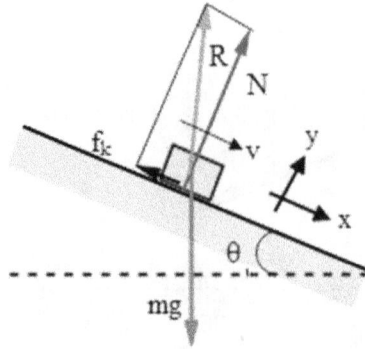

N.B.: When the block is stationary, the net reaction points vertically up and it is equal to weight of the body, but when the block slides, the net reaction will be less than the weight of the block and no longer point vertically up.

Example 7 (Angle of repose)

A block of mass m is placed on an inclined plane μ_s= coefficient of static friction between the block and inclined plane. If the block is at the verge of sliding, find the (a) angle of inclination θ_0 and (b) relation between θ_0 and angle of friction ϕ.

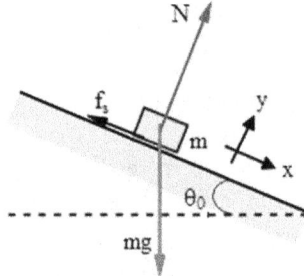

Solution

(a) Referring to the free-boy diagram at the verge of sliding, the forces acting on the block are limiting friction $f_{s_{max}}$ ↗, N ↖ and weight mg↓.

Since, the block does not slide, $a_x = 0$. As the inclined plane is at rest, we have $a_y = 0$.

Newton's second law:

Resolving the force parallel and perpendicular to the inclined plane, we have

$$\sum F_x = f_{s_{max}} - mg \sin \theta_0 = ma_x = 0$$

$$\Rightarrow f_{s_{max}} = mg \sin \theta_0 \tag{5.23}$$

$$\sum F_y = N - mg \cos \theta_0 = ma_y = 0$$

$$\Rightarrow N = mg \cos \theta_0 \tag{5.24}$$

Law of static friction:

$$(f_s)_{max} = \mu N \tag{5.25}$$

Substituting $(f_s)_{max}$ from equation (5.23) and N from equation (5.24) in equation (5.25) we have

$$\mu_s = \tan \theta_0$$

$$\Rightarrow \theta_0 = \tan^{-1} \mu_s \text{ Ans.}$$

(b) Since, $\tan^{-1} \mu_s = \phi$, we have $\theta_0 = \phi$ Ans.

N.B.: The angle of repose is the maximum angle of inclination of the inclined plane made with the horizontal for which the block does not slide; if $\theta > \theta_0$, the block will slide and kinetic friction comes into play. The body will slide down with an acceleration $a = g(\sin \theta - \mu_k \cos \theta)$.

Example 8 A block of mass m projected with a speed v_0 on a horizontal floor. If it comes to rest after a time t and distance x. If the coefficient of kinetic friction is equal to μ_k, find (a) t, (b) x, (c) reaction offered by ground on the block during its motion.

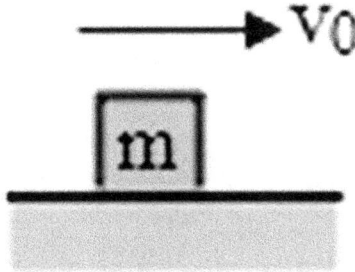

Solution

(a) Referring to the free-body diagram, the net horizontal force acting on the block is

$$f_k = ma \tag{5.26}$$

$$F_y = N - mg = ma_y = 0 \tag{5.27}$$

Law of kinetic friction:

$$f_k = \mu_k N \tag{5.28}$$

By using the last three equations, the deceleration of the block is

$$a = \frac{f_k}{m} = \frac{\mu m_k g}{m} = \mu_k g \text{ (to the left)}$$

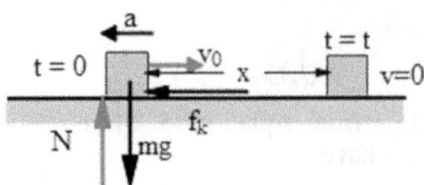

As the block comes to rest after a time t,

$$v = v_0 - \mu_k g t = 0$$

$$\Rightarrow t = \frac{v_0}{\mu_k g} \text{ Ans.}$$

(b) As it comes to rest after covering a distance x,

$$2as = v^2 - u^2$$

$$\Rightarrow 2(-\mu_k g)x = 0 - v_0^2$$

$$\Rightarrow x = \frac{v_0^2}{2g\mu_k} \text{ Ans.}$$

(c) The reaction offered by the ground on the block during its motion is

$$R = \sqrt{f_k^2 + N^2} = \sqrt{(\mu_k mg)^2 + (mg)^2}$$
$$= mg\sqrt{\mu_k^2 + 1} \text{ Ans.}$$

Example 9 A block is projected with a velocity v_1 on a horizontal surface which is moved with a constant velocity v_2. Find the direction of kinetic friction on each body. Can the kinetic friction favor the motion? Explain.

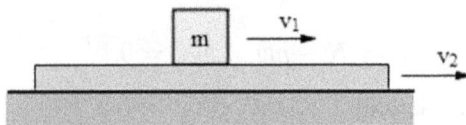

Solution

If $v_1 > v_2$, $\vec{v}_{12}$ is +ve. Then the block moves towards right relative to the horizontal surface. Hence, kinetic friction f_k acts backward on the block 1 to oppose $\vec{v}_{12}$. To obey Newton's third law, the reaction force $f'_k (=f_k)$ acts towards right on the horizontal surface. This means, kinetics friction favors the motion of surface 2 and opposes the motion of surface 1 (if $v_1 > v_2$). However, kinetic friction on each surface opposes their relative motion (relative velocity).

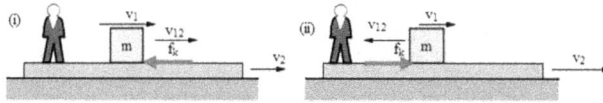

Example 10 When you throw a stone of mass m vertically up, taking the air friction f into account, find the ratio of time of descent and time of ascent.

Solution

When the stone moves up, kinetic friction is down. Then the net force acting on the body is:

$$F_1 = mg + f_k.$$

When the body moves down, kinetic friction is up; then the net force acting on the body is:

$$F_s = mg - f_k.$$

Since the time $t \propto \sqrt{2s/a}$, where s = distance of ascent (and descent), the ratio of times is:

$$\frac{t_2}{t_1} = \frac{a_1}{a_2}$$

$$\Rightarrow \frac{t_2}{t_1} = \frac{F_1}{F_2}$$

$$\Rightarrow \frac{t_2}{t_1} = \frac{mg + f_k}{mg - f_k}$$

$$\Rightarrow t_2 < t_1$$

Then, the time of descent is less than time of ascent. If we ignore friction, time of ascent is equal to time of descent. In the presence of friction, the body moves up with greater downward acceleration ($>g$) and moves down with less downward acceleration ($<g$). Hence, it takes less time to go up and more time to come down.

5.5.3 Rolling friction

Rolling friction or rolling drag or rolling resistance is the force that opposes the motion of a body when it rolls over a surface. It arises from the deformation at the point contact of the rolling object and the surface on which the object rolls.

While rolling, both the surface and the rolling body deform slightly at their point of contact. The object must continuously overcome these tiny deformations (bumps) created in the process of rolling. So, the body loses its kinetic energy as it is a non-elastic process. As the body retards, the horizontal component of the reaction force R acting on the rolling body at the point of contact can be considered as the rolling friction. As it is easier to roll an object when compared to sliding, rolling friction is less than the sliding friction; so, the coefficient of rolling friction less than that of the kinetic(sliding) friction.

Example 11 Three forces each of magnitude mg are acting on the block, the first is horizontal (acting to right), the second is vertically downward and the third is acting at an angle $\emptyset = 37°$. Find the (a) minimum coefficient of static friction so that the block will not slide and (b) friction acting on the block and acceleration of the block. Assume that the coefficient of static and kinetic friction are (i) $\mu_s = 0.7$ and $\mu_k = 0.6$(ii) $\mu_s = 0.8$ and $\mu_k = 0.75$, respectively.

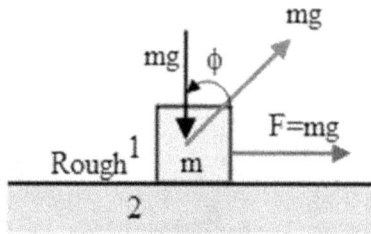

Solution

(a) Let us assume that static friction prevails; so the acceleration of the block is zero as the ground is stationary. The net horizontal force acting on the block is

$$F_h = mg(1 + \sin \phi) - f_s = ma_x = 0$$

$$f_s = mg(1 + \sin \phi) \tag{5.29}$$

The net vertical force acting on the block is

$$F_v = mg(2 - \cos \phi) - N = ma_y = 0$$

$$\Rightarrow N = mg(2 - \cos \phi) \tag{5.30}$$

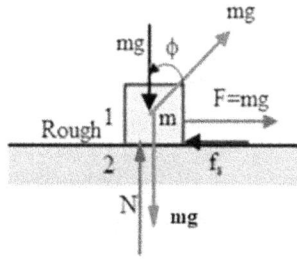

Law of static friction:

$$f_s \leqslant \mu_s N \qquad (5.31)$$

Using the last three equations, we have

$$\Rightarrow mg(1 + \sin \phi) \leqslant \mu_s mg(2 - \cos \phi)$$

$$\Rightarrow (1 + \sin \phi) \leqslant \mu_s(2 - \cos \phi)$$

$$\Rightarrow \mu_s \geqslant \frac{(2 - \cos \phi)}{(1 + \sin \phi)}$$

$$\Rightarrow \mu_s \geqslant \frac{(2 - 4/5)}{(1 + 3/5)} = \frac{3}{4}$$

$$\Rightarrow \mu_s \geqslant \frac{3}{4} \text{ Ans.}$$

(b)

(i) If $\mu_s \geqslant \frac{3}{4}$, the block does not slide and its acceleration is zero. But the given value of coefficient of static friction is $\mu_s = 0.7$ which is less than the required value; so, the block will slide and kinetic friction will prevail which is given as

$$f_k = \mu_k mg(2 - \cos \phi) = (0.6)mg(2 - 4/5) = 0.72 \, mg \text{ Ans.}$$

Then, the acceleration of the block is

$$F_h = mg(1 + \sin \phi) - f_k = ma_x$$

$$\Rightarrow mg(1 + 3/5) - 0.72 \, mg = ma_x$$

$$\Rightarrow a_x = 0.88g \text{ m s}^{-2} \text{ Ans.}$$

(ii) As the given value of coefficient of static friction is $\mu_s = 0.8$ which is greater than the required value of 0.75, the block remains stationary and the static friction will prevail which is given as

$$f_k = \mu_s mg(2 - \cos \phi) = (0.8) \, mg \, (2 - 4/5) = 0.96 \, mg \quad \text{Ans.}$$

Then, the acceleration of the block is zero. Ans.

Example 12 A block of mass m is kept on a plate. A force F acts on the block at an angle θ. If the block is made to move with an acceleration A, find the value/s of the force F so that the block does not slip on the plank. Assume that the coefficient of static and kinetic friction are μ_s and μ_k, respectively; so also assume that the block does not lose contact with the horizontal surface.

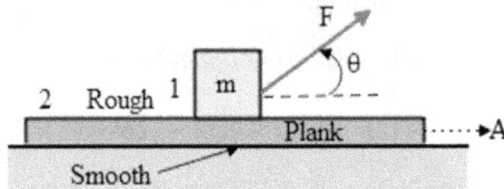

Solution

(a) Let us assume that static friction prevails; so, the acceleration of the block is zero relative to the plank. Then, the acceleration of the block is equal to A. The net horizontal force acting on the block is

$$F_h = F \cos \theta - f_s = ma_x = mA$$

$$\Rightarrow f_s = F \cos \theta - mA \tag{5.32}$$

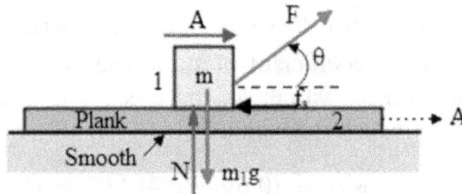

The net vertical force acting on the block is

$$F_v = mg - N - F \sin \theta = ma_y = 0$$

$$N = mg - F \sin \theta \tag{5.33}$$

Law of static friction:

$$f_s \leqslant \mu_s N \tag{5.34}$$

Using the last three equations, we have

$$\Rightarrow F \cos \theta - mA \leqslant \mu_s(mg - F \sin \theta)$$

$$\Rightarrow F(\cos \theta + \mu_s \sin \theta) \leqslant m(\mu_s g + A)$$

$$\Rightarrow F \leqslant \frac{m(\mu_s g + A)}{(\cos \theta + \mu_s \sin \theta)} \quad \text{Ans.}$$

Example 13 An insect of mass m crawls with an upward acceleration a on a light inextensible thread hanging from a rigid support. Find the tension in the string.
Solution
Free-body diagram:
On the insect static friction is $f_s \uparrow$ and gravity $mg \downarrow$, on the string tension is $T \uparrow$ and $f_s \downarrow$.
Force equation on m:

$$\sum F_y = ma$$

$$\Rightarrow f_s - mg = ma$$

$$\Rightarrow f_s = m(g + a) \tag{5.35}$$

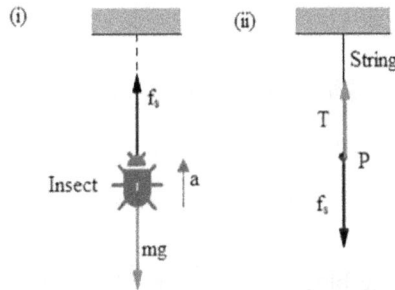

Force equation on string:
At the point of contact P of the string, static friction acts down, which pulls the string so that a tension T is induced. So, the net force acting at point P is

$$T - f_s = 0 \tag{5.36}$$

From equations (5.35) and (5.36), $T = f_s = m(g + a)$
Tension in the string is caused by the friction between string and insect. Ans.

Example 14 A block of mass m is in equilibrium on an inclined plane of angle of inclination θ. If $\mu = $ coefficient of kinetic friction between the block and plane, find the minimum force F parallel to the inclined plane, required to slide the block

(a) (i) up (ii) down along the line of greatest slope, assuming that the angle of inclination is greater than the angle of repose. (c) If the angle of inclination is smaller than the angle of repose find the minimum force F parallel to the inclined plane required to pull the block up and down.

Solution

(a) Force acting on the block are $mg\downarrow$, $N\nwarrow$, $F\nearrow$ and $f_{s_{max}}\swarrow$.

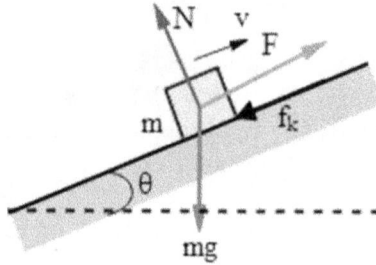

Force equation:

$$\sum F_x = F - mg\sin\theta - f_k = ma_x = 0 \qquad (5.37)$$

$$\sum F_y = N - mg\cos\theta = ma_y = 0 \qquad (5.38)$$

Law of kinetic friction:

$$f_k = \mu N \qquad (5.39)$$

Using the last three equations, we have

$$N = mg\sin\theta \text{ and } f = \mu mg\cos\theta$$

Substituting the values of N and f_k in equation (5.37),

$$F = mg(\sin\theta + \mu\cos\theta) \text{ Ans.}$$

(b) Force acting on the block are $mg\downarrow$, $N\nwarrow$, $F\nearrow$ and $f_{s_{max}}\nearrow$.

Force equation:

$$\sum F_x = F - mg\sin\theta + f_k = ma_x = 0 \qquad (5.40)$$

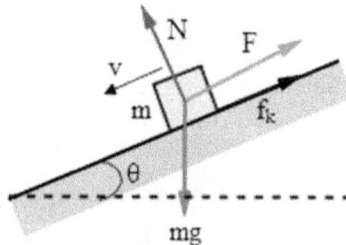

Putting the obtained values

$$N = mg \sin \theta \text{ and } f_k = \mu\, mg \cos \theta$$

in equation (5.40),

$$F = mg(\sin \theta - \mu \cos \theta) \text{ Ans.}$$

(c) If the angle of inclination is smaller than the angle of repose, the block needs to be pushed down the slant by the force F.

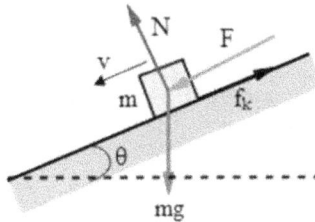

Force equation:

$$\sum F_x = F + mg \sin \theta - f_k = ma_x = 0 \tag{5.41}$$

By putting

$$N = mg \sin \theta \text{ and } f_k = \mu_k mg \cos \theta$$

in equation (5.41),

$$F = mg(\mu \cos \theta - \sin \theta) \text{ Ans.}$$

N.B: For equilibrium of the blocks, $mg(\sin \theta + \mu \cos \theta) > F > mg|\sin \theta - \mu \cos \theta|$. We note following points regarding angle of friction and angle of repose.

1. If a block of mass m does not slide on a stationary inclined plane of angle of inclination θ, $f = mg \sin \theta$ (but not $\mu_s mg \cos \theta$).
2. $f_{s_{max}} = mg \sin \theta_0 = \mu mg \cos \theta_0$, where $\theta_0 = $ angle of repose.
3. Angle of friction $\phi = \tan^{-1}(mg) = $ angle of repose θ_0.
4. When $\theta < \theta_0$, $f = f_s (\leqslant \mu_s N)$, where $N = mg \cos \theta$.
5. When $\theta > \theta_0$, $f = f_k (= \mu_k N)$, where $N = mg \cos \theta$.
6. At $\theta = \theta_0$, friction changes from static to kinetic.

Problem 1 A block is projected horizontally with a velocity v_0 from a point P of an inclined plane having an angle of inclination θ. The coefficient of kinetic friction between the block and inclined plane is μ_k, find the angle made by the velocity vector of the block with horizontal when its speed will be equal to v at Q.

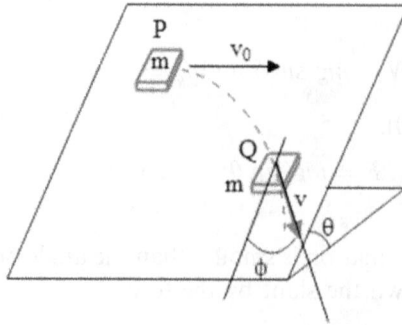

Solution

Free-body diagram:

When the block slides down, kinetic friction f_k acts up along the slant. The forces on the block are $mg\downarrow$, $f_k \searrow$ and $N \nwarrow$.

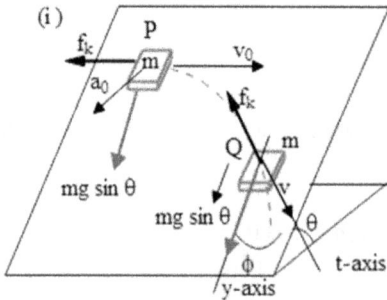

Normal reaction N and gravity mg are not shown here

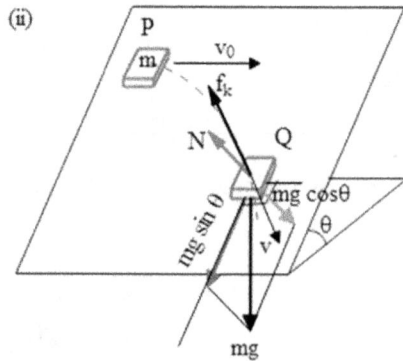

Force equation:

Referring to the free-body diagrams, the net force acting on the block down the inclined plane along the y-axis is

$$\sum F_y = mg \sin \theta - f_k \cos \phi = ma_y \tag{5.42}$$

The net force acting on the block down the inclined plane along the line of motion is

$$\sum F_t = mg \sin \theta \cos \phi - f_k = ma_t \tag{5.43}$$

The net force acting on the block normal to the inclined plane is

$$\sum F_n = -mg \cos \theta + N = ma_n \tag{5.44}$$

Kinematics:

We can see that velocity along the y-axis is

$$v_y = v \cos \phi \tag{5.45}$$

As the inclined plane is non-accelerating, the acceleration normal to the inclined plane is

$$a_n = 0 \tag{5.46}$$

Law of kinetic friction:

$$f_k = \mu_k N \tag{5.47}$$

Using equations (5.44) and (5.46), we have

$$N = mg \cos \theta \tag{5.48}$$

Substituting N from equation (5.48) in equation (5.47), we have

$$f_k = \mu_k mg \cos \theta \tag{5.49}$$

Then, substituting the above value of f_k from equation (5.49) in equation (5.42), we have

$$mg \sin \theta - \mu_k mg \cos \theta \cos \phi = ma_y$$

$$\Rightarrow a_y = g(\sin \theta - \mu_k \cos \theta \cos \phi) \tag{5.50}$$

Then, substituting the above value of f_k from equation (5.49) in equation (5.43), we have

$$mg \sin \theta \cos \phi - \mu_k mg \cos \theta = ma_t$$

$$a_t = g(\sin \theta \cos \phi - \mu_k \cos \theta) \tag{5.51}$$

Putting $\mu_k = \tan \theta$ in the last two equations, we obtain the following two equations:

$$a_y = g \sin \theta (1 - \cos \phi) \tag{5.52}$$

$$a_t = g \sin \theta (\cos \phi - 1) \tag{5.53}$$

From the last two equations, we have

$$a_t = -a_y$$

$$\Rightarrow \frac{dv}{dt} = -\frac{dv_y}{dt}$$

$$\Rightarrow dv = -dv_y$$

Integrating both sides,

$$\int_{v_o}^{v} dv = -\int_{0}^{v_y} dv_y$$

$$\Rightarrow v - v_0 = -(v_y - 0)$$

$$\Rightarrow v + v_y = v_0 \tag{5.54}$$

By using equations (5.45) and (5.54), we have

$$v + v \cos \phi = v_0$$

$$\Rightarrow v(1 + \cos \phi) = v_0$$

$$\Rightarrow \phi = \cos^{-1}(v_0/v - 1)$$

Putting $v_0/v = \eta$, we have

$$\phi = \cos^{-1}(\eta - 1) \text{ Ans.}$$

Problem 2 A man is moving with an acceleration a by pulling a string which passes over a smooth pulley at B. The string is connected with a block of mass m at A. If at the given position the angle made by the string ABC is 90°, find the tension in the string. Assume that μ_k = coefficient of kinetic friction between the block and the ground.

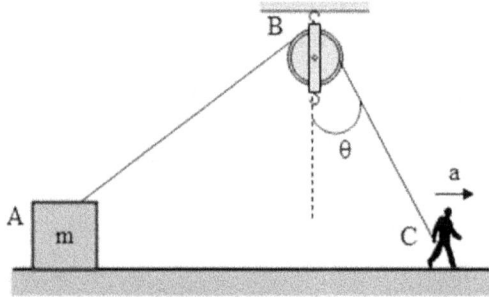

Solution

Free-body diagram

When the block slides kinetic friction f_k acts on the block to the left. So, the forces acting on the block are $mg\downarrow$, $f_k \leftarrow$ and $N\uparrow$.

Force equation:

Referring to the free-body diagram

$$\sum F_x = T \cos \theta - f_k = ma_x = ma' \tag{5.55}$$

$$\Rightarrow -mg + T \cos \theta + N = ma_y \tag{5.56}$$

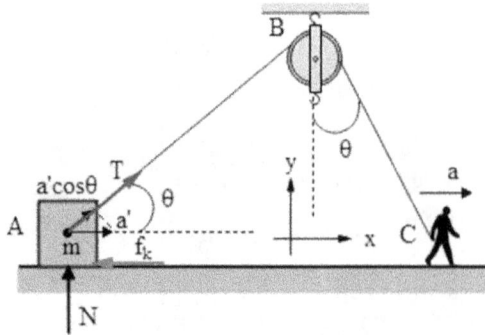

Kinematics:

The acceleration of the block in the vertical given as

$$a_y = 0 \tag{5.57}$$

The component of acceleration of the block and man along the string will be equal to each other because the string is inextensible.

$$a' \cos \theta = a \sin \theta$$

$$\Rightarrow a' = a \tan \theta \tag{5.58}$$

Using the above force equations, we have

$$f_k = T \cos \theta - ma \tan \theta \tag{5.59}$$

$$N = mg - T \sin \theta \tag{5.60}$$

Law of kinetic friction:

$$f_k = \mu_k N \tag{5.61}$$

Using the last three equations, we have

$$T \cos \theta - ma \tan \theta = \mu_k(mg - T \sin \theta)$$

$$\Rightarrow T(\cos \theta + \mu_k \sin \theta) = m(\mu_k g + a \tan \theta)$$

$$\Rightarrow T = \frac{m(\mu_k g + a \tan \theta)}{(\cos \theta + \mu_k \sin \theta)} \quad \text{Ans.}$$

Problem 3 A man is pulling a block A of mass m while walking with an acceleration a, by holding an inextensible string at C. A bird B of mass m' is walking on the rope with an acceleration a_0 relative to the string. Find the tension of the segment AB and BC of the string. Take necessary assumptions and notations.

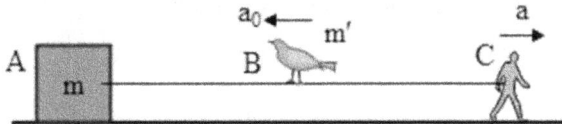

Solution

Free-body diagram:

When the block slides, kinetic friction f_k acts on the block to the left. The forces acting on the block are $mg\downarrow$, $f_k\leftarrow$ and $N\uparrow$. The force acting on the bird is static friction f_s, say, as it walks along the string; equal and opposite static friction also acts on the string at B to the right. So, the tensions in the portion AB and BC of the string will be different, assumed as T_1 and T_2, respectively. For the sake of simplicity, we assume the sag of the string due to the weight of the bird is very small so that we can assume that the string is nearly straight.

Force equation:

Referring to the free-body diagram the net horizontal force acting on the block is

$$\sum F_x = T_1 - f_k = ma_x = ma \tag{5.62}$$

The net vertical force acting on the block is

$$\Rightarrow -mg + N = ma_y = 0 \tag{5.63}$$

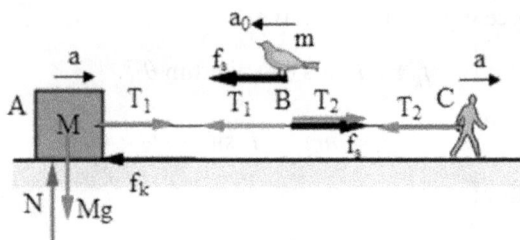

The kinetic friction is

$$f_k = \mu_k N = \mu_k mg \tag{5.64}$$

Using the last three equations,

$$T_1 = m(\mu_k g + a) \text{ Ans.}$$

The net horizontal force acting on the bird is

$$f_s = ma_b = m(a_0 - a) \tag{5.65}$$

The net horizontal force acting on the element of the string at B is

$$T_2 + f_s = T_1 \tag{5.66}$$

From the last two equations,

$$T_2 + m(a_0 - a) = T_1 \tag{5.67}$$

Putting the value of T_1 in equation (5.67), we have

$$T_2 + m(a_0 - a) = m(\mu_k g + a)$$

$$\Rightarrow T_2 = m(\mu_k g + 2a - a_0) \text{ Ans.}$$

Problem 4 A block is at rest on a rough inclined plane, if the force F acts on the block applied horizontally parallel to the inclined plane, find the value of F to slide the block μ_s = coefficient of static friction.

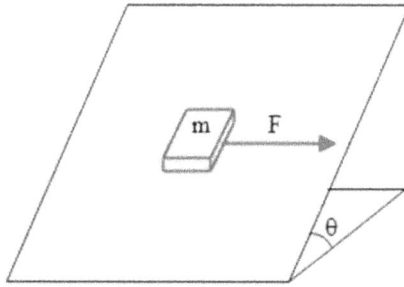

Solution

Referring to the free-body diagram, the block is in equilibrium under the action of the coplanar forces f_s ↘, $mg \sin \theta$ ↗, $F \rightarrow$ and the normal reaction N ↖.

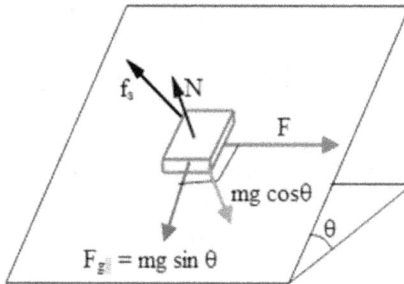

Force equation on the block:

The components of weight mg of the block parallel and normal to the inclined plane are $mg \sin \theta$ and $mg \cos \theta$, respectively. Let us assume an arbitrary direction of static friction f_s that must act in the inclined plane, as shown in the figure below. The net force acting on the block parallel to the inclined plane is

$$\sum F_{\parallel} = \vec{f_s} + \vec{F}_{gr\parallel} + \vec{F} = m\vec{a_1} = \vec{0}$$

$$\Rightarrow f_s = |(\vec{F} + F_{gr\parallel})|,$$

where $F_{gr\parallel} = mg \sin \theta$

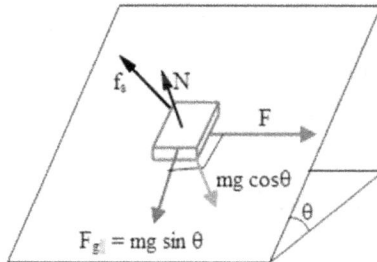

$$\Rightarrow f_s = \sqrt{F^2 + m^2g^2 \sin^2 \theta} \tag{5.68}$$

$$\sum F_n = N - mg \cos \theta = ma_n = 0$$

$$\Rightarrow N = mg \cos \theta \tag{5.69}$$

Law of f_s:

$$f_s \leqslant \mu_s N \tag{5.70}$$

Substituting f_s from equation (5.68) and N from equation (5.69) in equation (5.70),

$$F^2 + m^2g^2 \sin^2 \theta \leqslant \mu_s^2 m^2g^2 \cos^2 \theta$$

$$F \leqslant mg\sqrt{\mu_s^2 \cos^2 \theta - \sin^2 \theta} \quad \text{Ans.}$$

N.B: The critical value of F depends on the direction of its action.

Problem 5 A block of mass m_1 is placed on a block of mass m_2. The coefficient of static friction between the blocks is μ_s and the ground is smooth. If a horizontal force F acts on the block (a) m_1 (b) m_2. If no relative sliding occurs between the blocks find the maximum value of F.

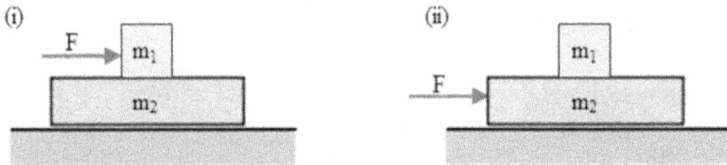

Solution

(a) *Free-body diagram:*
   Let us assume that the friction acting on m_1 is (to left)$f_s \leftarrow$, on m_2 is (to right) $f_s \rightarrow$; and the downward weight (m_1g) of the upper block and the normal reaction N (up) acting on it.
   Force equation on m_1:

$$\sum F = F - f_s = m_1 a_1 \tag{5.71}$$

   Force equation on m_2:

$$\sum F = f_s = m_2 a_2 \tag{5.72}$$

   Kinematics:
   Since f_s is present, there is no relative sliding;

$$a_1 = a_2 \tag{5.73}$$

Substituting a_1 from equation (5.71) and a_2 from equation (5.72) in equation (5.73),

$$\frac{F - f_s}{m_1} = \frac{f_s}{m_2}$$

$$\Rightarrow f_s = \frac{m_2 F}{m_1 + m_2} \tag{5.74}$$

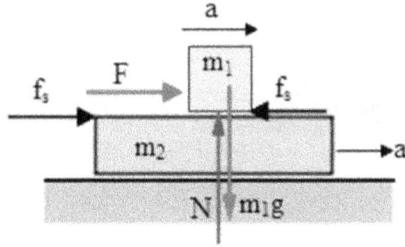

Law of static friction:

$$f_s \leqslant \mu_s N, \text{ where } N = m_1 g$$

$$\Rightarrow f_s \leqslant \mu_s m_1 g \tag{5.75}$$

From the last two equations, we have

$$\frac{m_2 F}{m_1 + m_2} \leqslant \mu_s m_1 g$$

$$\Rightarrow F \leqslant \mu_s m_1 g \left(1 + \frac{m_1}{m_2} \right)$$

So, the maximum force for no relative sliding between the blocks is

$$F = \mu_s m_1 g \left(1 + \frac{m_1}{m_2} \right) \text{ Ans.}$$

(b) *Free-body diagram*:

Let us assume that the friction acting on m_2 is (to left) $f_s \leftarrow$, on m_1 is (to right) $f_s \rightarrow$; the free-body diagram is drawn without showing their weights and normal reactions.

Force equation on m_1:

$$\sum F = f_s = m_1 a_1 \tag{5.76}$$

Force equation on m_2:

$$\sum F = F - f_s = m_2 a_2 \tag{5.77}$$

Kinematics:

Since f_s is present, there is no relative sliding;

$$a_1 = a_2 \qquad (5.78)$$

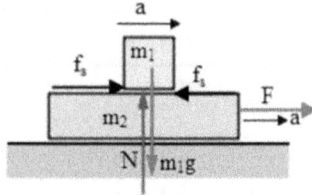

Substituting a_1 from equation (5.76) and a_2 from equation (5.77) in equation (5.78),

$$\frac{f_s}{m_1} = \frac{F - f_s}{m_2}$$

$$\Rightarrow f_s = \frac{m_1 F}{m_1 + m_2} \qquad (5.79)$$

Law of static friction:

$$f_s \leqslant \mu_s N, \text{ where } N = m_1 g$$

$$\Rightarrow f_s \leqslant \mu_s m_1 g \qquad (5.80)$$

From the last two equations, we have

$$\frac{m_1 F}{m_1 + m_2} \leqslant \mu_s m_1 g$$

$$\Rightarrow F \leqslant \mu_s (m_1 + m_2) g$$

So, the maximum force for no relative sliding between the blocks is

$$\Rightarrow F = \mu_s (m_1 + m_2) g \text{ Ans.}$$

N.B.:

1. When m_2 is stationary (put $m_2 \rightarrow \infty$), we have $f_s = F$ and $F \leqslant \mu_s m_1 g$. This is a special case.
2. The static friction f_s may favor or oppose the acceleration of m_1 and m_2 relative to the ground; but it must oppose the relative acceleration of the blocks (measured in the absence of friction), that is what is known as tendency of relative sliding.

Problem 6 A block of mass m_1 is placed on a block of mass m_2. The coefficient of static friction between the blocks is μ and the ground is smooth. If one end of a horizontal string is connected with the block m_1 and the other end of the string is connected with the hanging mass m_3, and if no relative sliding occurs between the blocks, find the maximum value of m_3.

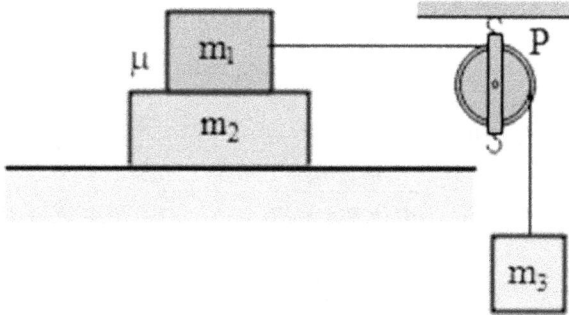

Solution

As there is no relative sliding between the blocks, they move as a combined unit. The acceleration of the combination of two blocks and the hanging mass have same magnitude because the string is inextensible. Let the tension in the string be T.

Force equation on $m_1 + m_2$:

$$\sum F_x = T = (m_1 + m_2)a \tag{5.81}$$

Force equation on m_3:

$$\sum F_y = m_3 g - T = m_3 a \tag{5.82}$$

Force equation on m_2:

$$\sum F_x = f_s = m_2 a \tag{5.83}$$

Solving the last two equations (5.81) and (5.82), the acceleration is

$$a = \frac{m_3 g}{m_1 + m_2 + m_3} \tag{5.84}$$

Using the last two equations,

$$f_s = m_2\left(\frac{m_3 g}{m_1 + m_2 + m_3}\right) = \frac{m_2 m_3 g}{m_1 + m_2 + m_3} \tag{5.85}$$

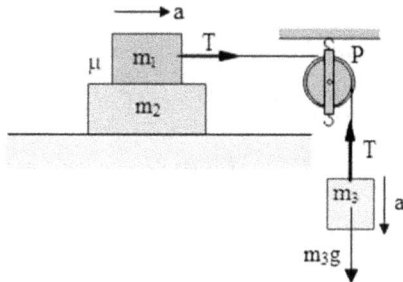

Law of static friction:

$$f_s \leqslant \mu_s N, \text{ where } N = m_1 g$$

$$\Rightarrow f_s \leqslant \mu_s m_1 g \tag{5.86}$$

Using the last two equations,

$$\frac{m_2 m_3 g}{m_1 + m_2 + m_3} \leqslant \mu_s m_1 g$$

$$\Rightarrow \frac{m_2 m_3}{m_1 + m_2 + m_3} \leqslant \mu_s m_1$$

$$\Rightarrow m_3 \leqslant \frac{m_1 + m_2}{\left(\frac{m_2}{m_1} - \mu_s\right)} \text{ Ans.}$$

Problem 7 A block of mass m is placed on another block of mass M which is kept on smooth ground. The upper block is pulled by a force F at an angle θ with the horizontal. If $\mu=$ coefficient of friction between the block and ground, find (a) the value of F at the given angle of orientation, (b) angle at which F will be minimum and (c) minimum possible value of F so that m does not slip with M.

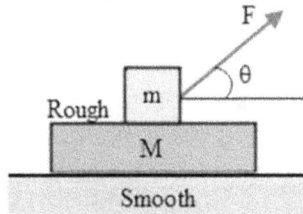

Solution

(a) For the force F to be maximum at the given angle, the limiting friction $f_{s_{max}}$ must prevail between the blocks. Referring to the free-body diagram, the forces acting on the block m are $F \nearrow$, $mg\downarrow$, $N\uparrow$ and $f_{s_{max}} \leftarrow$; the horizontal force acting on the lower block is $f_{s_{max}} \rightarrow$ at the verge of sliding of m relative to M.

Force equation on $(M + m)$:

$$\sum F_x = F \cos \theta = (M + m)a$$

$$\Rightarrow a = \frac{F \cos \theta}{M + m} \tag{5.87}$$

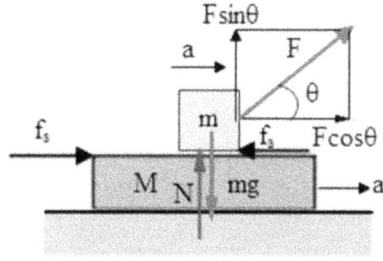

Force equation on m:

$$\sum F_y = F \sin \theta + N - mg = ma_y = 0$$

$$N = mg - F \sin \theta \tag{5.88}$$

Force equation on M:

$$f_s = Ma \tag{5.89}$$

Law of static friction:

$$f_{s_{max}} = \mu N \tag{5.90}$$

From equation (5.87) in equation (5.89),

$$f_s = Ma = \frac{MF \cos \theta}{M + m} \tag{5.91}$$

From equation (5.88) in equation (5.90),

$$f_s = \mu(mg - F \sin \theta) \tag{5.92}$$

From equation (5.91) in equation (5.92),

$$\frac{MF \cos \theta}{M + m} = \mu(mg - F \sin \theta)$$

$$\Rightarrow \left(\frac{M \cos \theta}{M + m} + \mu \sin \theta \right) F = \mu mg$$

$$\Rightarrow F = \frac{\mu mg}{\frac{M \cos \theta}{M + m} + \mu \sin \theta} \quad \text{Ans.}$$

(b) For F will be minimum, the denominator of the expression of F must be maximum.

$$\Rightarrow \frac{d}{d\theta} \left(\frac{M \cos \theta}{M + m} + \mu \sin \theta \right) = 0$$

$$\Rightarrow -\frac{M \sin \theta}{M + m} + \mu \cos \theta = 0$$

$$\Rightarrow \tan \theta = \left(\frac{M+m}{M}\right)\mu$$

$$\Rightarrow \theta = \tan^{-1}\left(\frac{M+m}{M}\right)\mu \quad \text{Ans.}$$

(c) Putting $\cos \theta = \dfrac{M}{\sqrt{\{\mu(M+m)\}^2 + M^2}}$, $\sin \theta = \dfrac{m}{\sqrt{\{\mu(M+m)\}^2 + M^2}}$ in the expression of F, and simplifying the expressions, we have

$$\Rightarrow F_{\min} = \frac{\mu mg}{\sqrt{\left(\frac{M}{M+m}\right)^2 + \mu^2}} \quad \text{Ans.}$$

N.B.:

1. You can also maximize the denominator by trigonometric method. By using the theorem that if

$$f(\theta) = a \cos \theta + b \sin \theta,$$

its maximum value is given as

$$f(\theta)_{\max} = \sqrt{a^2 + b^2}$$

2. If the block is massive, putting $M \to \infty$, we have

$$F = \frac{\mu mg}{\cos \theta + \mu \sin \theta}$$

3. If you take $\theta = 0$ and write $F_{\min} = \mu mg$, it will be wrong.

Problem 8 Let us assume that a block of mass m is placed on an inclined plane of angle of inclination θ. Let us pull the block up at an angle β with the line of greatest slope of the inclined plane. (a) Find the force required to slide the block slowly if angle of friction is ϕ. (b) What is the minimum possible value of the force F at the proper value of the angle β?

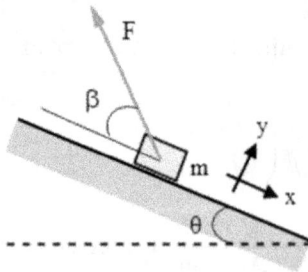

Solution

(a) Referring to the free-body diagram, the forces acting on the body are applied force $F \nwarrow mg\downarrow$, $f_k \searrow$ and $N \nearrow$.

Force equation of m:

Resolving forces parallel and perpendicular to the inclined plane, we have

$$\sum F_x = -F \cos \beta + mg \sin \theta + f_k = ma_x \tag{5.93}$$

and

$$\sum F_y = N + F \sin \beta - mg \cos \theta = ma_y \tag{5.94}$$

Kinematics:

Since, the body moves slowly along the inclined plane; $v = 0$. So, we have

$$a_x = 0 \tag{5.95}$$

Because the body does not move $\perp$ to the inclined plane, we have

$$a_y = 0 \tag{5.96}$$

Substituting a_x from equation (5.95) in equation (5.93) and a_y from equation (5.96) in equation (5.94), we have

$$f_s = F \cos \beta - mg \sin \theta \tag{5.97}$$

and

$$N = mg \cos \theta - F \sin \beta \tag{5.98}$$

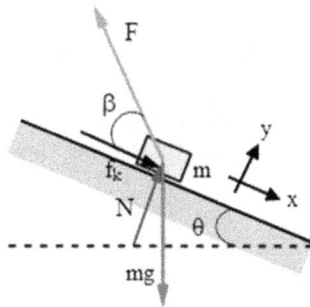

Law of static friction:

$$f_k = \mu N \tag{5.99}$$

Substituting f_s from equation (5.97) and N from equation (5.98) in equation (5.99), we have

$$F \cos \beta - mg \sin \theta = \mu(mg \cos \theta - F \sin \beta)$$

$$\Rightarrow F = mg\frac{\sin \theta + \mu \cos \theta}{\mu \sin \beta + \cos \beta}, \text{ where } \mu = \tan \phi$$

$$\Rightarrow F = mg\frac{\sin(\theta + \phi)}{\cos(\beta - \phi)} \text{ Ans.}$$

Hence, the force is

$$F = mg\frac{\sin \theta + \mu \cos \theta}{\mu \sin \beta + \cos \beta} \text{ Ans.}$$

(b) The force is minimum when the denominator is maximum. The maximum value of $\mu \sin \beta + \cos \beta$ is equal to $\sqrt{1 + \mu^2}$; so, the minimum value of F is

$$F = mg\frac{\sin \theta + \mu \cos \theta}{\sqrt{1 + \mu^2}} \text{ Ans.}$$

N.B.:
1. You should not understand this force as the minimum force for upward and downward motion of the block due to a force that acts parallel to the line of the greatest slope.
2. By putting $\theta = 0$ for the horizontal surface, we can get the familiar answer of

$$F_{min} = \frac{\mu mg}{\sqrt{1 + \mu^2}}.$$

Problem 9 A block of mass m is placed on an inclined plane μ_s = coefficient of static friction between the block and inclined plane. If the inclination is gradually increased, the block is at the verge of sliding at the angle of repose θ_0. If this angle is maintained and the block is given a gentle tap, it starts sliding. Find the acceleration of the block.

Solution

(a) Referring to the free-body diagram, at the verge of sliding, maximum static friction $f_{s_{max}}$ ↗ acts up along the inclined plane. Apart from this, normal reaction N ↖ and weight mg↓ act on the body.

Force equation:

As the kinetic friction comes into play just after giving a gentle tap, the body will slide down with an acceleration

$$a = g(\sin \theta_0 - \mu_k \cos \theta_0) \tag{5.100}$$

since $\tan \theta_0 = \mu_s$, we have

$$\sin \theta_0 = \frac{1}{\sqrt{1 + \mu_s^2}} \tag{5.101}$$

$$\cos \theta_0 = \frac{\mu_s}{\sqrt{1 + \mu_s^2}} \tag{5.102}$$

Using equations (5.100), (5.101) and (5.102), we have

$$a = g\left\{ \frac{1}{\sqrt{1 + \mu_s^2}} - \mu_k\left(\frac{\mu_s}{\sqrt{1 + \mu_s^2}} \right) \right\}$$

$$\Rightarrow a = \left(\frac{1 - \mu_s \mu_k}{\sqrt{1 + \mu_s^2}} \right) g \quad \text{Ans.}$$

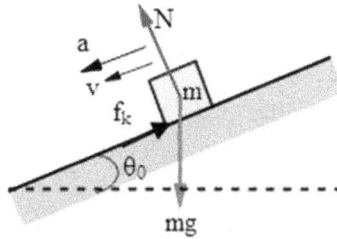

Problem 10 A block 1 is kept on a block 2 and this combination is placed on an inclined plane. If $\mu_1 =$ coefficient of friction between blocks 1 and 2, and $\mu_2 =$ coefficient of friction between block 1 and the inclined plane, find the (a) condition if block 1 does not slide on block 2 when block 2 slides on the inclined plane after releasing the system of two blocks (b) acceleration of the blocks and the friction between the blocks if the inclined plane is smooth.

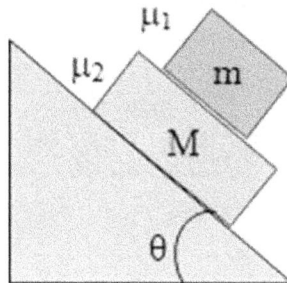

Solution

(a) If block 1 does not slide on block 2 when block 2 slides on the inclined plane, let both blocks will move with the same acceleration, a. Referring to the free-body diagram, let us write the force equation on $(M + m)$:

$$\sum F_x = (M + m)g \sin \theta - f_k = (M + m)a \qquad (5.103)$$

Law of kinetic friction between block 2 and the inclined plane:

$$f_k = \mu_2(M + m)g \cos \theta \qquad (5.104)$$

Using equations (5.103) and (5.104)

$$a = g(\sin \theta - \mu_2 \cos \theta) \qquad (5.105)$$

Force equation on m:

$$\sum F_x = mg \sin \theta - f_{s1} = ma \qquad (5.106)$$

$$N_1 = mg \cos \theta \qquad (5.107)$$

Using equations (5.105) and (5.106), we have

$$mg \sin \theta - f_{s1} = mg(\sin \theta - \mu_2 \cos \theta)$$

$$\Rightarrow mg \sin \theta - f_{s1} = mg(\sin \theta - \mu_2 \cos \theta)$$

$$\Rightarrow f_{s1} = \mu_2 mg \cos \theta \qquad (5.108)$$

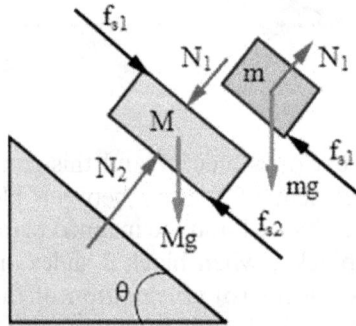

Law of static friction between blocks 1 and 2:

$$f_{s1} \leqslant \mu_1 N_1 \qquad (5.109)$$

Using equations (5.107), (5.108) and (5.109), we have

$$\mu_2 mg \cos \theta \leqslant \mu_1(mg \cos \theta)$$

$$\Rightarrow \mu_2 \leqslant \mu_1 \text{ Ans.}$$

(b) If the inclined plane is smooth, $\mu_2 = 0$; since $\mu_2 \leqslant \mu_1$ block 1 does not slide on block 2. This means that both blocks will move with an acceleration given as

$$a = g \sin \theta \text{ Ans.}$$

Putting $\mu_2 = 0$ in equation (5.109), we have

$$f_{s1} = \mu_2 mg \cos \theta = 0 \text{ Ans.}$$

Problem 11 In cases (i) and (ii), a block of mass m does not slide relative to an accelerating trolley car. If μ_s = coefficient of static friction between the block and accelerating frame, in each case, find the accelerations.

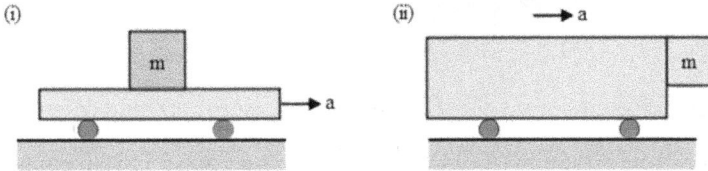

Solution

In figure (i), the forces acting on m are $mg\downarrow$, $f_s \rightarrow$, $N\uparrow$
Force equations:

$$\sum F_x = f_s = ma_x = ma \tag{5.110}$$

$$\sum F_y = N - mg = ma_y = 0 \tag{5.111}$$

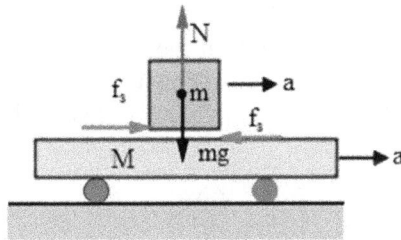

Law of f_s:

$$f_s \leqslant \mu_s N \tag{5.112}$$

Substituting f_s from equation (5.110) and N from equation (5.111) in equation (5.112),

$$ma \leqslant \mu_s mg$$

$$\Rightarrow a \leqslant \mu_s g \text{ Ans.}$$

In figure (ii), the forces acting on m are $mg\downarrow$, $f_s \uparrow$ and $N\rightarrow$.

Force equations:

$$\sum F_x = N = ma_x = ma \tag{5.113}$$

$$\sum F_y = f_s - mg = ma_y = 0 \tag{5.114}$$

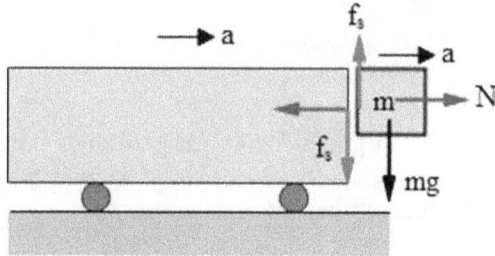

Law of f_s:

$$f_s \leqslant \mu_s N \tag{5.115}$$

Substituting f_s from equation (5.114) and N from equation (5.113) in equation (5.115),

$$mg \leqslant \mu_s ma$$

$$a \geqslant \frac{g}{\mu_s} \text{ Ans.}$$

N.B.: In the first case f_s accelerates the block and normal reaction counteracts gravity, but in the second case f_s counteracts gravity and normal reaction accelerates the block. You can also do the above example by pseudo-force method.

Problem 12 A block of mass m is kept on the slanted surface of a prismatic wedge which moves with an acceleration A, as shown in the figure below. If $\mu_s < \tan \theta$, find the (a) maximum and (b) minimum value of A such that the block will tend to slide up.

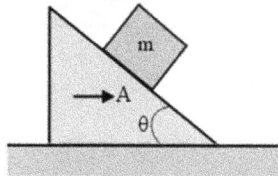

Solution
Method 1 (Ground frame method):

(a) Referring to the free-body diagram, the forces acting on the block are $mg \downarrow$, $f_s \searrow$ (if the block tends to slide up) $N \nearrow$.

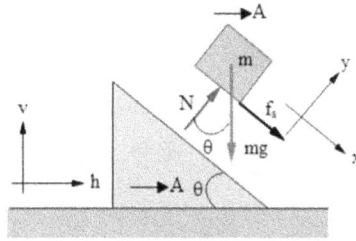

Force equation:

Let us write $F = ma$ in the horizontal and vertical directions resolving the forces and accelerations in those directions.

$$\sum F_{\text{hor}} = N \sin \theta + f_s \cos \theta = ma_x = mA \tag{5.116}$$

$$\sum F_{\text{vert}} = (N \cos \theta - mg \sin \theta - f_s \sin \theta) = ma_y = 0 \tag{5.117}$$

Law of f_s:

$$f_s \leqslant \mu_s N \tag{5.118}$$

Solving equations (5.116) and (5.117) find N and f_s in terms of A and substitution in equation (5.118) to obtain.

$$\Rightarrow A \leqslant g\left(\frac{\sin \theta + \mu \cos \theta}{\cos \theta - \mu \sin \theta}\right).$$

Method 2 (Ground frame method):

Let us write $F = ma$ in the x- and y-directions (as shown in the free-body diagrams) by resolving the forces and accelerations in those directions.

Force equation:

$$\sum F_x = mg \sin \theta + f_s = ma_x = mA \cos \theta \tag{5.119}$$

$$\sum F_y = (N - mg \cos \theta) = ma_y = mA \sin \theta \tag{5.120}$$

Law of f_s:

$$f_s \leqslant \mu_s N \tag{5.121}$$

Solving equations (5.119) and (5.120) find N and f_s in terms of A and substitution in equation (5.121) to obtain.

$$\Rightarrow A \leqslant g\left(\frac{\sin \theta + \mu \cos \theta}{\cos \theta - \mu \sin \theta}\right).$$

Method 3 (Pseudo-force method):

Imposing pseudo-force in addition to the real forces, write the force equations relative to the accelerating frame. So, in $F = ma$, $a =$ acceleration of the block relative to the wedge.

Since, the block is at rest relative to the accelerating frame, a pseudo force mA acts to the left of the block in addition to all real forces, as shown in the free-body diagrams.

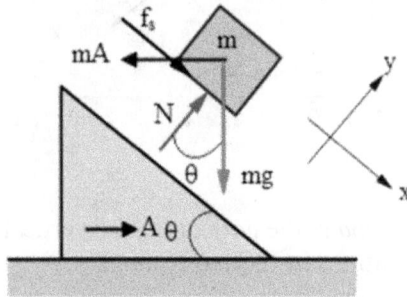

Since the block does not slip with the wedge, $a = 0$; so, both of its components must be zero, which can be given as

Kinematics: $a_x = a_y = 0$

Force equation:

$$\sum F_x = f_s + mg \sin \theta - mg \cos \theta = 0 \qquad (5.122)$$

$$\sum F_y = N - ma \sin \theta - mg \cos \theta = 0 \qquad (5.123)$$

Law of f_s:

$$f_s \leqslant \mu_s N \qquad (5.124)$$

Substituting f_s from equation (5.122) and N from equation (5.123) in equation (5.124),

$$m(a \cos \theta - g \sin \theta) \leqslant \mu m(a \sin \theta + g \cos \theta)$$

$$\Rightarrow a \leqslant g \left(\frac{\sin \theta + \mu \cos \theta}{\cos \theta - \mu \sin \theta} \right) \text{ Ans.}$$

(b) If the block tends to slide down, static friction acts up along the plane and the directions of other forces remain the same.

Method 1 (ground frame method):
Referring to the free-body diagram the forces acting on the block are $mg\downarrow$, $f_s \nwarrow$ (if the block tends to slide up) $N \nearrow$.
Force equation:

$$\sum F_h = N \sin \theta - f_s \cos \theta = ma_h = mA \qquad (5.125)$$

$$\sum F_v = (N \cos \theta - mg \sin \theta + f_s \sin \theta) = ma_y = 0 \qquad (5.126)$$

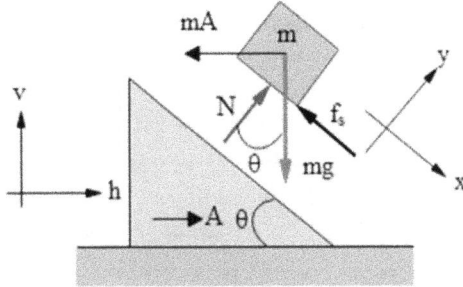

Law of f_s:

$$f_s \leqslant \mu_s N \qquad (5.127)$$

Solving equations (5.125) and (5.126) find N and f_s in terms of A and substitution in equation (5.127) to obtain.

$$mg \sin \theta - mA \cos \theta \leqslant \mu_s(mg \cos \theta + mA \sin \theta)$$

$$\Rightarrow A \geqslant g\left(\frac{\sin \theta - \mu \cos \theta}{\cos \theta + \mu \sin \theta}\right) \text{ Ans.}$$

Method 2 (Ground frame method):
Force equation:

$$\sum F_x = mg \sin \theta - f_s = ma_x = mA \cos \theta \qquad (5.128)$$

$$\sum F_y = (N - mg \cos \theta) = ma_y = mA \sin \theta \qquad (5.129)$$

Law of f_s:

$$f_s \leqslant \mu_s N \qquad (5.130)$$

Solving equations (5.128) and (5.129) find N and f_s in terms of A and substitution in equation (5.130) to obtain.

$$\Rightarrow A \geqslant g\left(\frac{\sin \theta - \mu \cos \theta}{\cos \theta + \mu \sin \theta}\right) \text{ Ans.}$$

N.B: So, the range of the acceleration can be given as

$$g\left(\frac{\sin \theta - \mu \cos \theta}{\cos \theta + \mu \sin \theta}\right) < A < g\left(\frac{\sin \theta + \mu \cos \theta}{\cos \theta - \mu \sin \theta}\right).$$

Problem 13 A block is given a velocity $\vec{v}_1 = v_0\hat{i}$ on a horizontal plate which moves with a velocity (a) $\vec{v}_2 = 2v_0\hat{i}$ (b) $\vec{v}_2 = -v_0\hat{j}$ as shown in figures (i) and (ii), respectively. If the coefficient of kinetic friction between the particle and plane is μ, find the acceleration the block in each case at the given instant. Assume that the xy-plane is horizontal.

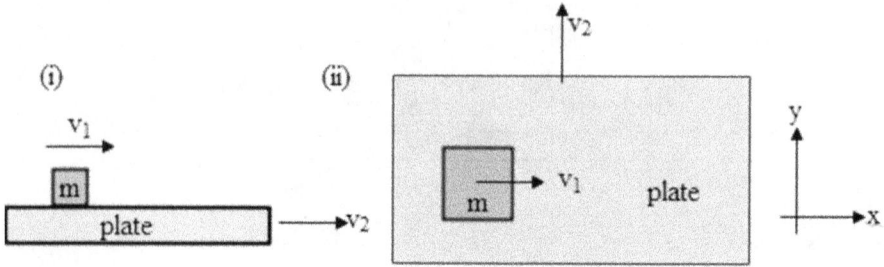

Solution

(a) The velocity of block relative to the horizontal plate is

$$\vec{v}_{12} = \vec{v}_1 - \vec{v}_2 = v_0\hat{i} - 2v_0\hat{i} = -v_0\hat{i}$$

Since, $\vec{f}_k$ opposes $\vec{v}_{12}$, we have

$$\vec{f}_k = \mu N\hat{i} = \mu_1 mg\hat{i}$$

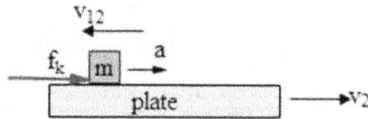

So, the acceleration of the block is

$$\vec{a}_1 = \vec{f}_1 / m = \mu g\hat{i} \quad \text{Ans.}$$

The velocity of block relative to the horizontal plate is

$$\vec{v}_{12} = \vec{v}_1 - \vec{v}_2 = v_0\hat{i} - v_0\hat{j}$$

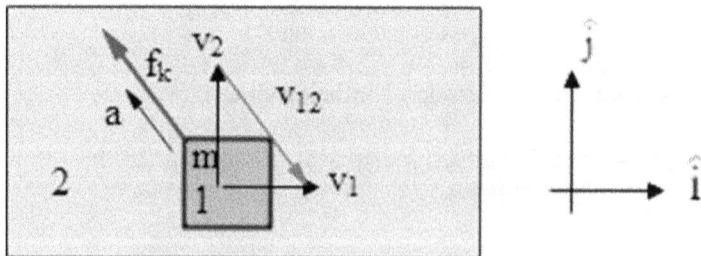

Since, $\vec{f_k}$ opposes $\vec{v}_{12}$, we have

$$\vec{f_k} = \frac{\mu mg}{\sqrt{2}}(-\hat{i} + \hat{j})$$

So, the acceleration of the block is

$$\vec{a} = \vec{f_k}/m = \frac{\mu g}{\sqrt{2}}(-\hat{i} + \hat{j}) \text{ Ans.}$$

Problem 14 A block of mass m is projected at $t = 0$ with a velocity v_0 on an elevator which starts moving vertically with a constant acceleration a. Find the (a) acceleration of the block at $t = 0$ and (b) time after which the horizontal and vertical components of the velocity of the block become equal in magnitude (c) speed and (d) distance covered by the elevator when the block stops sliding on the base of the elevator.

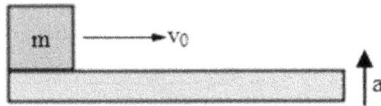

Solution

(a) Referring to the free-body diagram, the forces acting on the block are $mg\downarrow$, $N\uparrow$ and $f_k\leftarrow$.
    Force equations:

$$\sum F_x = -f_k = ma_x \tag{5.131}$$

$$\sum F_y = N - mg = ma_y = ma \tag{5.132}$$

Law of f_k:

$$f_k = \mu N \tag{5.133}$$

Substituting N from equation (5.132) and f_k from equation (5.131) in equation (5.133),

$$a_x = -\mu(g + a) \text{ Ans.}$$

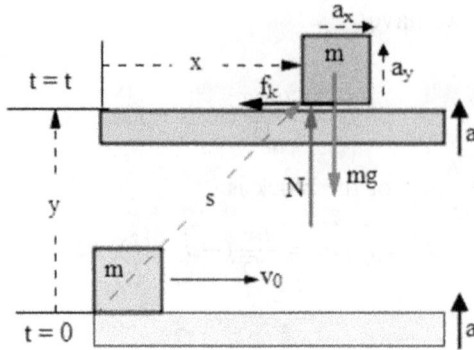

(b) Kinematics:

$$v_x = v_{0_x} + a_x t = v_0 - \mu(g + a)t \tag{5.134}$$

$$v_y = v_{0_y} + a_y t = 0 + at \tag{5.135}$$

Since, $v_x = v_y$ (given), using equation (5.134) and equation (5.135),

$$v_0 - \mu(g + a)t = at$$

$$t = \frac{v_0}{\mu g + a(\mu + 1)}$$

(c) The horizontal velocity of the block is zero as it stops sliding. The velocity of the block is equal to the velocity of the plank which is vertically upward, given as

$$v = at = \frac{v_0 a}{\mu g + a(1 + \mu)}$$

(d) The distance covered by the block is

$$s = \sqrt{x^2 + y^2}, \tag{5.136}$$

where $= x = \frac{v_0^2}{2\mu g}$ and $y = \frac{v_0^2 a}{2[\mu g + a(1 + \mu)]^2}$.

$$\Rightarrow s = \sqrt{\left(\frac{v_0^2}{2\mu g}\right)^2 + \left[\frac{v_0^2 a}{2\{\mu g + a(1 + \mu)\}^2}\right]^2}$$

$$= \frac{v_0^2}{2\mu g}\sqrt{1 + \frac{a^2\mu^2 g^2}{\{\mu g + a(1 + \mu)\}^4}} \quad \text{Ans.}$$

N.B: In this problem, the magnitude of kinetic friction is

$$f_k = \mu_k N \text{ (but not } \mu_k mg)$$

Problem 15 A block is projected up along the line of greatest slope of an inclined plane having an angle of inclination θ and coefficient of kinetic friction μ_k. The velocity of projection is v_0. Find the (a) time of ascent, (c) time of descent, (d) ratio of time of ascent and time of descent, and (e) velocity with which the block returns back to the point of projection. Assume that the angle of inclination is greater than the angle of friction.

Solution

(a) Referring to the free-body diagram, while going up, f_k points down the plane; furthermore, the block experiences $mg\downarrow$ and $N\nearrow$.

Force equation:

Resolving all the forces acting on the block parallel and perpendicular to the inclined plane as shown in the free-body diagram in the figure below, let us apply Newton's laws by assuming the x-direction (down the slant) and y-direction (up and normal to the inclined plane),

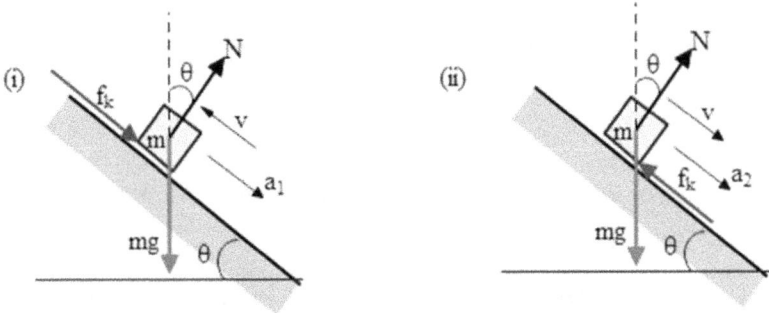

$$\sum F_x = mg \sin \theta + f_k = ma_1 \tag{5.137}$$

$$\sum F_y = N - mg \cos \theta = 0 \tag{5.138}$$

Law of kinetic friction:

$$f_k = \mu_k N \tag{5.139}$$

Using the above equations, we have

$$a_1 = g(\sin \theta + \mu_k \cos \theta) \tag{5.140}$$

Kinematics:

As the body slows down, let after a time t, it comes to rest instantaneously $v = u + at$, where $v = 0$, $a = g(\sin \theta + \mu_k \cos \theta)$ and $u = -v_0$.

This gives the time of ascent as

$$t = \frac{v_0}{g(\sin \theta + \mu_k \cos \theta)} \quad \text{Ans.}$$

(b) Let the body moves through a distance x before coming to rest.
Applying the formula

$$v^2 = u^2 + 2as,$$

where $v = 0$, $u = v_0$, $s = -x$ and $a = a_1 = -g(\sin \theta + \mu_k \cos \theta)$

$$\Rightarrow x = \frac{v_0^2}{2g(\sin \theta + \mu_k \cos \theta)} \quad \text{Ans.}$$

(c) The body will have to cover this distance in its return journey (descent) with an acceleration

$$a_2 = g(\sin \theta - \mu_k \cos \theta)$$

Because the kinetic friction changes its direction from down to up along the plane. The time of descent is given as

$$t' = \sqrt{2x/a_2}$$

Putting the obtained values of a_2 and x, and simplifying the factors, we have

$$t' = \frac{v_0}{g\sqrt{\sin^2 \theta - \mu_k^2 \cos^2 \theta}} \quad \text{Ans.}$$

(d) The ratio of time of ascent and time of descent is

$$t/t' = \left(\frac{v_0}{g(\sin \theta + \mu_k \cos \theta)} \right) \Big/ \left(\frac{v_0}{g\sqrt{\sin^2 \theta - \mu_k^2 \cos^2 \theta}} \right)$$

$$\Rightarrow t/t' = \sqrt{\left(\frac{\sin \theta - \mu_k \cos \theta}{\sin \theta + \mu_k \cos \theta} \right)} \quad \text{Ans.}$$

(e) Speed with which the body passes through the point of its projection and the speed of projection is given as

$$v = \sqrt{2a_2 x} = \sqrt{2g(\sin \theta - \mu_k \cos \theta)\left(\frac{v_0^2}{2g(\sin \theta + \mu_k \cos \theta)} \right)}$$

$$\Rightarrow v = v_0 \sqrt{\left(\frac{\sin \theta - \mu_k \cos \theta}{\sin \theta + \mu_k \cos \theta} \right)} \quad \text{Ans.}$$

N.B: The time of ascent is less than the time of descent because it ascends with greater downward acceleration along the line of motion than that during its descent.

Problem 16 A block of mass m_1 is placed on another block of mass m_2. The blocks have initial velocities v_{0_1} and v_{0_2}, respectively. If $v_{0_1} > v_{0_2}$ and $\mu_k =$ coefficient of kinetic friction between the blocks, find the (a) acceleration of the blocks, (b) variation of velocities of the blocks versus time, (c) time of relative sliding, (d) distance of relative sliding, (e) common velocity and acceleration of the system $(m_1 + m_2)$ after the relative sliding stops, and (f) individual distances covered by the bodies till relative sliding stops.

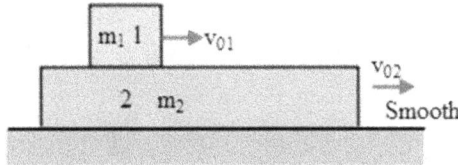

Solution

(a) *Free-body diagram*:

Initially, block 1 moves faster than block 2. So, the block 1 moves with a velocity $(v_0)_{12} = (v_{0_1} - v_{0_2})$ towards the right relative to block 2, the kinetic friction on block 1 is directed to the left and the same amount of friction acts on block 2, towards the right. Furthermore, we have shown the weights $m_1 g$ and $m_2 g$. Let the normal contact forces acting on the top and bottom blocks be N and N', respectively.

Force equation:

$$\text{For } m_1: \quad \sum F_x = -f_x = m_1 a_1 \tag{5.141}$$

$$\text{For } m_2: \quad \sum F_y = f_k = m_2 a_2 \tag{5.142}$$

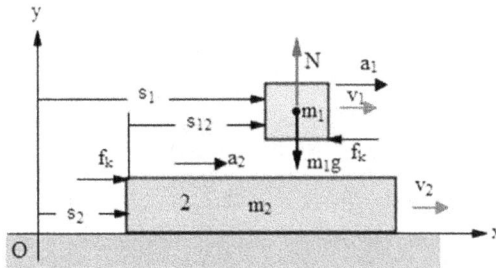

Law of kinetic friction:

$$f_k = \mu_k N, \quad \text{where } N = mg$$

$$\Rightarrow f_k = \mu_k m_1 g \tag{5.143}$$

Substituting f_k from equation (5.143) in equations (5.141) and (5.142), we have

$$a_1 = -\mu_k g \text{ and } a_2 = \mu_k \frac{m_1}{m_2} g \text{ Ans.}$$

(b) *Kinematics*:

Applying kinematical equation for m_1, we have

$$v = u + at \qquad (5.144)$$

Then by substituting $v = v_1$, $u = v_{0_1}$ and $a = -\mu_k g$, we have

$$v_1 = v_{0_1} - \mu_k gt \quad \text{Ans.}$$

Similarly, by substituting $v = v_2$, $u = v_{0_2}$ and $a = \mu_k \frac{m_1}{m_2} g$, we have

$$v_2 = v_{0_2} + \mu_k \frac{m_1}{m_2} gt \quad \text{Ans.}$$

(c) When the relative sliding will stop, their relative velocity will be zero;

$$v_1 = v_2$$

$$\Rightarrow v_{0_1} - \mu_k gt = v_{0_2} + \mu_k \frac{m_1}{m_2} gt$$

$$\Rightarrow t = \frac{v_{0_1} - v_{0_2}}{\mu_k g \left(1 + \frac{m_1}{m_2} \right)} \quad \text{Ans.}$$

(d) The velocity of the lower block at time t is

$$v_2 = v_{0_2} + \mu_k \frac{m_1}{m_2} gt \qquad (5.145)$$

Putting the obtained value of time t in equation (5.145), we have

$$v_2 = v_{0_2} + \mu_k \frac{m_1}{m_2} g \left(\frac{v_{0_1} - v_{0_2}}{\mu_k g \left(1 + \frac{m_1}{m_2} \right)} \right)$$

After simplifying the expressions, we have

$$v_1 = v_2 = v = \frac{m_1 v_{0_1} + m_2 v_{0_2}}{m_1 + m_2} \quad \text{Ans.}$$

(e) The distance of relative sliding is given by the formula

$$2\vec{a}_{12} \cdot \vec{s}_{12} = v_{12}^2 - u_{12}^2 \qquad (5.146)$$

When the relative sliding ceases, $v_{12} = 0$. The acceleration of 1 relative to 2 is

$$\vec{a}_{12} = \vec{a}_1 - \vec{a}_2$$

$$= -\mu_k g \hat{i} - \mu_k \frac{m_1}{m_2} g \hat{i}$$

$$\Rightarrow \vec{a}_{12} = -\mu_k\left(1 + \frac{m_1}{m_2}\right)g\hat{i} \qquad (5.147)$$

The displacement of 1 relative to 2 is

$$\vec{s}_{12} = l\hat{i}, \text{ say}; \qquad (5.148)$$

Using the last three equations,

$$s_{12} = l = \frac{|v_{0_1} - v_{0_2}|^2}{2\mu_k g\left(1 + \frac{m_1}{m_2}\right)} \quad \text{Ans.}$$

(f) The distances covered by body 1 till relative sliding stops can be given by

$$2\vec{a}_1 \cdot \vec{s}_1 = v_1^2 - u_1^2$$

$$\Rightarrow 2(-\mu_k g\hat{i} \cdot s_1\hat{i}) = \left(\frac{m_1 v_{0_1} + m_2 v_{0_2}}{m_1 + m_2}\right)^2 - v_{01}^2$$

$$\Rightarrow 2\mu_k g \cdot s_1 = v_{01}^2 - \left(\frac{m_1 v_{0_1} + m_2 v_{0_2}}{m_1 + m_2}\right)^2$$

$$\Rightarrow s_1 = \frac{v_{01}^2 - \left(\frac{m_1 v_{0_1} + m_2 v_{0_2}}{m_1 + m_2}\right)^2}{2\mu_k g} \quad \text{Ans.}$$

The individual distances covered by body 2 till relative sliding stops can be given by

$$2\vec{a}_2 \cdot \vec{s}_2 = v^2 - u_2^2$$

$$\Rightarrow 2\left(\mu_k \frac{m_1}{m_2}g\hat{i}\right) \cdot (s_2\hat{i}) = \left(\frac{m_1 v_{0_1} + m_2 v_{0_2}}{m_1 + m_2}\right)^2 - v_{02}^2$$

$$\Rightarrow 2\mu_k \frac{m_1}{m_2}gs_2 = \left(\frac{m_1 v_{0_1} + m_2 v_{0_2}}{m_1 + m_2}\right)^2 - v_{02}^2$$

$$\Rightarrow s_2 = \frac{m_2}{2\mu_k m_1 g}\left\{\left(\frac{m_1 v_{0_1} + m_2 v_{0_2}}{m_1 + m_2}\right)^2 - v_{01}^2\right\} \quad \text{Ans.}$$

N.B.:

1. For $v_{0_1} > v_{0_2}$, m_1 slows down and m_2 speeds up till their velocities are equal at time t_0, which is given as

$$t_0 = \frac{\left|\vec{v}_{0_1} - \vec{v}_{0_2}\right|}{\left(1 + \frac{m_1}{m_2}\right)\mu_k g} \quad \text{in general.}$$

2. When the relative sliding stops, f_k will disappear.
3. Since, there is no external force excluding friction acting on the blocks so as to generate a tendency of relative sliding, no static friction will prevail. As a result, the bodies will move as a combined block with a velocity

$$v = \frac{\left|m_1\vec{v}_{0_1} + m_2\vec{v}_{0_2}\right|}{m_1 + m_2}.$$

You can also find this expression by conserving the linear momentum of the blocks. The velocity–time and acceleration–time graphs for the blocks are shown as follows:

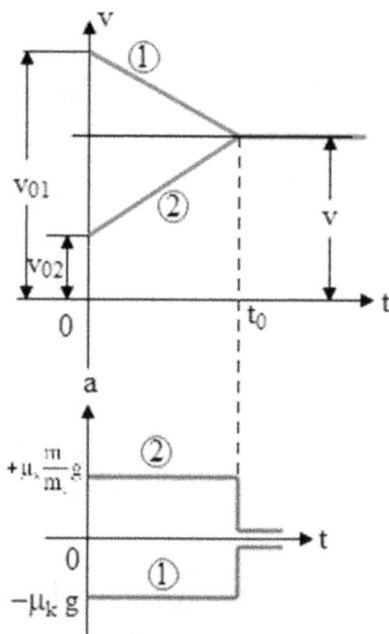

Problem 17 A block of mass m is gently loaded onto a conveyor belt at A when the belt moves with a constant speed v_0. If the coefficient of kinetic friction between the belt and block is μ, find the (a) time after which the block stops sliding with the belt; (b) time after which the block reaches the pulley, B, if (i) $v_0 < \sqrt{2\mu g l}$

(ii) $v_0 > \sqrt{2\mu g l}$, (c) variation of distance that the (i) block moves relative to ground, (ii) block slides relative to the belt, (iii) belt moves relative to ground, if l is large; (d) the distance covered by the (i) block relative to the belt (ii) belt (iii) block relative to the ground; (e) critical velocity of the block; (f) draw the variation of velocity and acceleration of the block relative to the ground for finite l and $v_0 > \sqrt{2\mu g l}$.

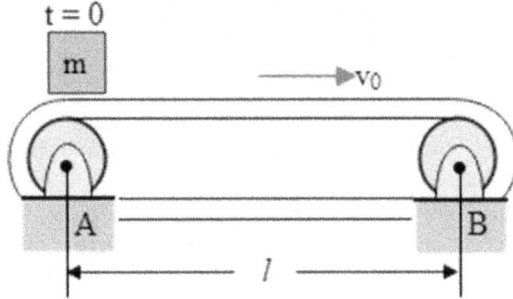

Solution

(a) *Free-body diagram*:

Initially the block slides to the left relative to the belt; so, f_k acts forward on the block. The other forces are $N\uparrow$ and $mg\downarrow$.

Force equation:

$$f_k = ma \tag{5.149}$$

$$N - mg = 0 \tag{5.150}$$

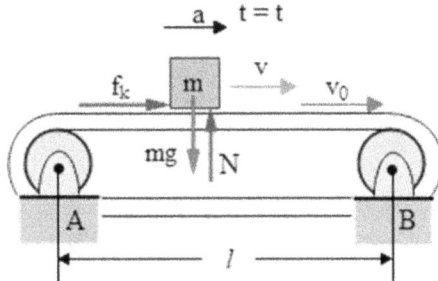

Law of f_k:

$$f_k = \mu N \tag{5.151}$$

Substituting f_k from equation (5.149) and N from equation (5.150) in equation (5.151),

$$ma = \mu mg$$

$$\Rightarrow a = \mu g$$

Kinematics:

Since $v = u + at$, putting $u = 0$, $a = \mu g$, we have

$$v = \mu g t$$

At $t = t_0$, let $v = v_0$; then the block stops sliding relative to the belt and moves with a velocity v_0 after

$$t_0 = \frac{v_0}{\mu g} \text{ Ans.}$$

(b) (i) The distance covered by the block till it stops sliding is

$$x = \frac{1}{2} a t_0^2 = \frac{1}{2} (\mu g) \left(\frac{v_0}{\mu g} \right)^2 = \frac{v_0^2}{2 \mu g}$$

If $x = \frac{v_0^2}{2 \mu g} = l$, $v_0 = \sqrt{2 \mu g l}$; If $v_0 < \sqrt{2 \mu g l}$, $x < l$. Then, the block will move the rest of the distance,

$$x' = l - x = l - \frac{v_0^2}{2 \mu g}$$

with a velocity equal to the velocity of the belt as it stops sliding after a time $t_0 = \frac{v_0}{\mu g}$. Then, the extra time taken by the block to reach pulley B is given as

$$t' = \frac{x'}{v_0} = \frac{l - \frac{v_0^2}{2 \mu g}}{v_0} = \frac{l}{v_0} - \frac{v_0}{2 \mu g}$$

So, the total time taken by the block to travel from A to B is

$$T = t + t' = \frac{x'}{v_0} = \frac{v_0}{\mu g} + \frac{l}{v_0} - \frac{v_0}{2 \mu g}$$

$$\Rightarrow T = \frac{v_0}{2 \mu g} + \frac{l}{v_0} \text{ Ans.}$$

(ii) If $v_0 > \sqrt{2 \mu g l}$, the block will keep on sliding on the belt till it reaches the end B. Since the block moves with a constant acceleration $a = \mu g$ from A to B, it will take a time T to cover the distance $AB = l$ given as

$$l = \frac{1}{2} a T^2$$

$$\Rightarrow T = \sqrt{2l/a},$$

where $a = \mu g$. Then, we have $T = \sqrt{2l/\mu g}$ Ans.

(c)

(i) If the belt is too long, the variation of distance covered relative to the ground is given as

$$s = \frac{\mu g t^2}{2} \text{ for } t \leqslant \frac{v_0}{\mu g} \text{ and } s = v_0 t \text{ for } t > \frac{v_0}{\mu g}$$

(ii) If the belt is too long, the variation of distance covered relative to the ground is given as

$$s_{rel} = -v_0 t + \frac{\mu g t^2}{2} \text{ for } t \leqslant \frac{v_0}{\mu g} \text{ and } s = 0 \text{ for } t > \frac{v_0}{\mu g}$$

(d)

(i) The distance covered by the block relative to the belt till it stops sliding relative to the belt is

$$s_{rel} = -v_0 t + \frac{\mu g t^2}{2} = -v_0 \left(\frac{v_0}{\mu g} \right) + \frac{\mu g}{2} \left(\frac{v_0}{\mu g} \right)^2 = -\frac{v_0^2}{2\mu g} \text{ (to left) Ans.}$$

(ii) The distance covered by the belt relative to the ground till the block stops sliding relative to the belt is

$$s_b = v_0 t = v_0 \left(\frac{v_0}{\mu g} \right)^2 = \frac{v_0^2}{\mu g} \text{ Ans.}$$

(iii) So, the distance covered by the block relative to the ground till the block stops sliding relative to the belt is

$$\vec{s} = \vec{s}_{rel} + \vec{s}_b = -\frac{v_0^2}{2\mu g} \hat{i} + \frac{v_0^2}{\mu g} \hat{i} = \frac{v_0^2}{2\mu g} \hat{i} \text{ (to right) Ans.}$$

(e) If this distance is equal to $AB = l$, then

$$\frac{v_0^2}{2\mu g} = l$$

So, for the block just reaching point B when it just stops sliding with the belt, the critical velocity is of the belt is

$$v_0 = \sqrt{2\mu g l}$$

(f) The variation of velocity and acceleration of the block relative to the ground can be given as follows;

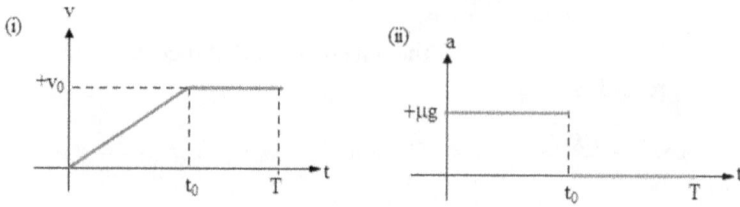

$$v = \mu g t \text{ and } a = \mu g \text{ for } t \leqslant \frac{v_0}{\mu g}$$

$$v = v_0 \text{ and } a = 0 \text{ for } t = T - t_0 = T - \frac{v_0}{\mu g}$$

Problem 18 A block of mass m is projected at $t = 0$ with a horizontal velocity v_0 on a stationary plank which starts moving with a constant horizontal acceleration a_0. If μ_k = coefficient of kinetic friction between the block and plank, find the (a) acceleration of the block, (b) time t_0 after with the block comes to rest relative to the plank, (c) variation of acceleration and velocity of the block with time if (i) $a_0 < \mu_s g$ (ii) $a_0 > \mu_s g$. (i), (ii) $a = -\mu_k g; t < t_0; t > t_0$ $a' = +\mu_k g; t > t_0$.

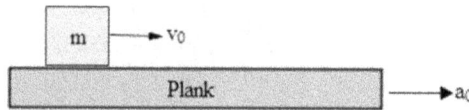

Solution

(a) The velocity of the block relative to the plank is

$$\vec{v}_{12} = \vec{v}_1 - \vec{v}_2 = v_0 \hat{i} - 0\hat{i} = v_0 \hat{i}$$

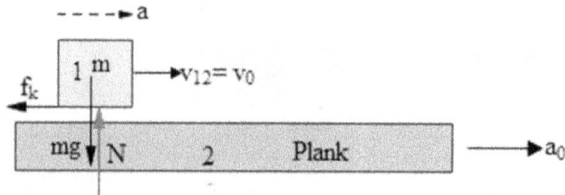

Hence, f_k acts backward on the block. The other forces acting on the block are $N\uparrow$, $mg\downarrow$.

Newton's force equations:

$$-f_k = ma \qquad\qquad (5.152)$$

$$N = mg \qquad\qquad (5.153)$$

Law of kinetic friction f_k:

$$f_k = \mu_k mg \tag{5.154}$$

Using the above equations, the acceleration of the block is

$$a = -\mu_k g \quad \text{Ans.}$$

(b) Kinematics:

The velocity of the block after a time t is

$$v_1 = v_0 = \mu_k g t \tag{5.155}$$

The velocity of the plank after a time t is

$$v_2 = a_0 t \tag{5.156}$$

If $v_1 = v_2$ at $t = t_0$ using equations (5.155) and (5.156), we have

$$t_0 = \frac{v_0}{(\mu_k g + a_0)} \quad \text{Ans.}$$

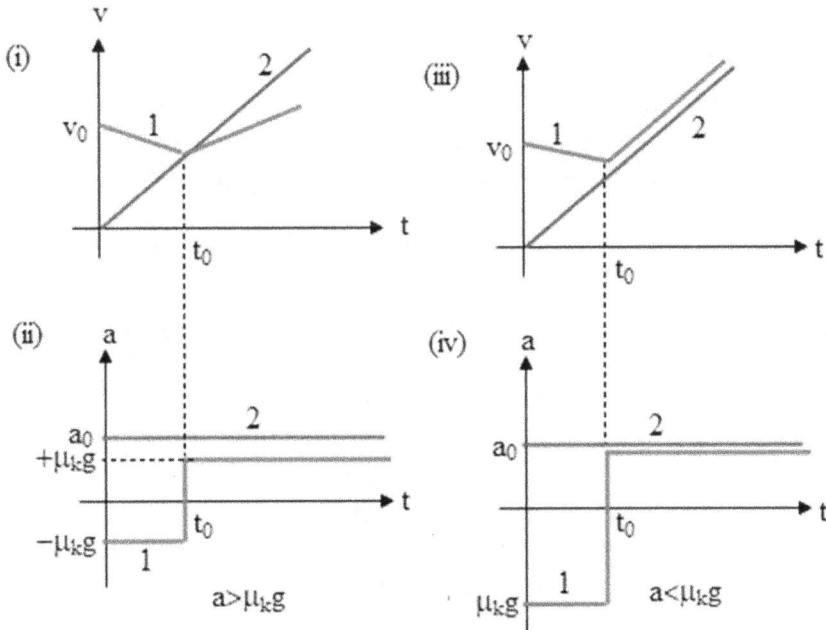

(c) If (i) $a < \mu_s g$, the kinetic friction acts on the block to the left till it slides on the plank. So, the acceleration of the block is

$$a = f_k / m = -\mu_k g \text{ (to left)}; \quad t < t_0$$

The velocity of the block is

$$v_{\text{block}} = v_0 - \mu_k g t$$

Since, there will be no relative sliding after the block stops relative sliding, the acceleration of the block is equal to that of the plank. So, we can write

$$a = a_0; \, t > t_0$$

If (ii) the block moves to the right relative to the plank, kinetic friction acts on the block to the left till it slides on the plank. After the relative sliding ceases, let us assume that static friction comes into play, which can have the maximum value of $(f_s)_{\text{max}} = \mu_s mg$ acting to the right on the block; so, the block can have a maximum acceleration

$$a = f_s/m = +\mu_s g \text{ (to right)}$$

Since the plank has an acceleration a_0 greater than $\mu_s g$, the acceleration of the block relative to the plank will point to the left. So, the block will start sliding to the left relative to the plank and a kinetic friction acts on the block to the right. Then, we can write

$$v_{block} = +\mu_s g t; \, t > t_0.$$

N.B.: In the above example:
1. The kinetic friction f_{k_1} acting on the block opposes v_{12}.
2. If $a_0 > \mu_s g$, the block will slide relative to the plank and $f = f_k$ (to right) after $t = t_0$.
3. If $a < \mu_s g$, the block will not slide relative to the plank and $f = f_s$ (to right).

Problem 19 A block 1 is placed on a block 2. Horizontal forces F_1 and F_2 act on the blocks, respectively. Find the (a) magnitude and direction of static friction acting on each block (b) condition so that the static friction acting on block 1 will act to the right.

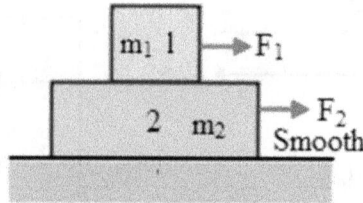

Solution

(a) The external forces F_1 and F_2 act on bodies m_1 and m_2, respectively, and the ground is smooth. Assume that f_s is forward on m_1 and backward on m_2. Referring to the free-body diagram, applying Newton's second law on the blocks, we have the following equations:

$$F_x = F_1 + f_s = m_1 a \qquad (5.157)$$

$$F_y = N - m_1 g = 0 \qquad (5.158)$$

$$F_x = F_2 - f_s = m_2 a \qquad (5.159)$$

Solving equations (5.157) and (5.158),

$$f_s = (F_2/m_2 - F_1/m_1)\left(\frac{m_1 m_2}{m_1 + m_2}\right) \text{ Ans.}$$

If $F_2/m_2 > F_1/m_1$, f_s is +ve; if $F_1/m_1 = F_2/m_2$, $f_s = 0$ and if $F_2/m_2 < F_1/m_1$, f_s is −ve (and we have to alter the direction of f_s).

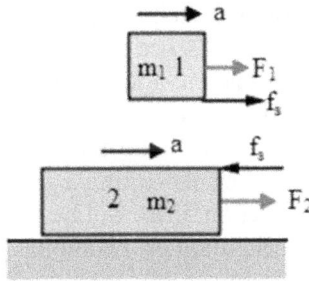

(b) The law of static friction:

$$f_s \leqslant \mu_s N \qquad (5.160)$$

Using equations (5.157) and (5.160), the limiting friction is given as

$$f_s \leqslant \mu_s mg \qquad (5.161)$$

Putting the value of f_s in the last equation, we have

$$f_s = (F_2/m_2 - F_1/m_1)\left(\frac{m_1 m_2}{m_1 + m_2}\right) \leqslant \mu_s m_1 g$$

$$\Rightarrow (F_2/m_2 - F_1/m_1) \leqslant \mu_s g\left(\frac{m_1 + m_2}{m_2}\right) \text{ Ans.}$$

N.B.: The physical meaning of the expression

$$f_s = (F_2/m_2 - F_1/m_1)\left(\frac{m_1 m_2}{m_1 + m_2}\right)$$

can be explained by putting

$$F_1/m_1 = a_1, \quad F_2/m_2 = a_2, \quad \left(\frac{m_1 m_2}{m_1 + m_2}\right) = m_{\text{reduced}}$$

we obtain the expression, given as

$$\vec{f_s} = m_{\text{reduced}}(a_2 - a_1)\hat{i} = m_{\text{reduced}}\vec{a}_{21}$$

This tells us that the static friction will prevail between blocks 1 and 2 if there is a relative acceleration between them in the absence of friction between them. This is what we call *tendency of relative sliding*, which is one of the conditions for the static friction to exist. So, if you intend to find the static friction between two blocks, first of all, mentally eliminate the friction between the blocks. Then draw the free-body diagram and find the acceleration of each block parallel to the ground. Now, you can find the relative accelerations (in the absence of friction between the blocks); this may not be equal to the actual relative acceleration in the presence of friction. Then you can calculate the friction required to prevent relative sliding between the blocks. Then you can compare the required friction with the maximum available friction or limiting friction between the blocks by using the formula $(f_s)_{\max} = \mu_s m_1 g$. If the required friction is less than the limiting friction, there will be NO relative sliding and both the blocks will move with equal acceleration as a combined unit, which is given as

$$\vec{a} = \frac{\vec{F_1} + \vec{F_2}}{m_1 + m_2}$$

If the required friction is greater than the limiting friction, there will be a relative sliding between the blocks; so, both blocks will have different accelerations because a kinetic friction will come into play.

Problem 20 Blocks of mass 3 and 6 kg are acted upon by horizontal forces, as shown in the following figures. Find the friction acting on each block and accelerations of the blocks for each case. Assume that the coefficients of static and kinetic friction between the blocks are 0.5 and 0.45, respectively.

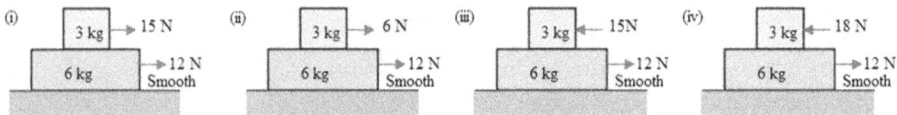

Solution

Figure (i): Referring to the last problem, the friction required to prevent relative sliding between the blocks is

$$f_s = (F_2/m_2 - F_1/m_1)\left(\frac{m_1 m_2}{m_1 + m_2}\right)$$

In figure (i), putting the given values,

$$f_s = (12/6 - 15/3)\left(\frac{3 \times 6}{3 + 6}\right) = -2N$$

The maximum available friction or limiting friction between the blocks is

$$(f_s)_{\max} = \mu_s m_1 g = (0.5)(3)(10) = 15 \text{ N}$$

The required friction is 2 N which is less than the limiting friction (15 N); so, there will be no relative sliding between the blocks; eventually, both the blocks will move with equal acceleration as a combined unit which is given as

$$\vec{a} = \frac{\vec{F_1} + \vec{F_2}}{m_1 + m_2}$$

$$\Rightarrow \vec{a} = \frac{15\hat{i} + 12\hat{i}}{3 + 6} = 3\hat{i} \text{ m s}^{-2}$$

So, the friction will be static of magnitude 2 N that acts between the blocks. Negative sign (-2 N) signifies that the direction of friction is opposite to the assumed direction (to the right) on the upper block. In other words, static friction of magnitude 2 N acts to the right on the upper block and it acts to the left on the lower block. Ans.

Figure (ii): In figure (ii), putting the given values,

$$f_s = (12/6 - 6/3)\left(\frac{3 \times 6}{3 + 6}\right) = 0 \text{ N}$$

The maximum available friction or limiting friction between the blocks is

$$(f_s)_{\max} = \mu_s m_1 g = (0.5)(3)(10) = 15 \text{ N}$$

The required friction is 0 N which less than the limiting friction (15 N); so, there will be no relative sliding between the blocks; eventually, both the blocks will move with equal acceleration as a combined unit which is given as

$$\vec{a} = \frac{\vec{F_1} + \vec{F_2}}{m_1 + m_2}$$

$$\Rightarrow \vec{a} = \frac{6\hat{i} + 12\hat{i}}{3 + 6} = 2\hat{i} \text{ m s}^{-2}$$

So, the zero-friction acts between the blocks and each of them has an acceleration of 2 m s^{-2} to the right. Ans.

Figure (iii): In figure (iii), putting the given values,

$$f_s = \{12/6 - (-15/3)\}\left(\frac{3 \times 6}{3 + 6}\right) = 14 \text{ N}$$

The maximum available friction or limiting friction between the blocks is

$$(f_s)_{\max} = \mu_s m_1 g = (0.5)(3)(10) = 15 \text{ N}$$

The required friction is 14 N which less than the limiting friction (15 N); so, there will be no relative sliding between the blocks; eventually, both the blocks will move with equal acceleration as a combined unit, which is given as

$$\vec{a} = \frac{\vec{F_1} + \vec{F_2}}{m_1 + m_2}$$

$$\Rightarrow \vec{a} = \frac{15\hat{i} + 12\hat{i}}{3 + 6} = 3\hat{i} \text{ m s}^{-2}$$

Figure (iv): So, the friction will be static of magnitude 2 N that acts between the blocks. Positive sign (+14 N) signifies that the direction of friction is the same as the assumed direction (to the right) on the upper block. In other words, static friction of magnitude 14 N acts to the right on the upper block and it acts to the left on the lower block. Ans.

Figure (iv): In figure (iv), putting the given values,

$$f_s = \{12/6 - (-18/3)\}\left(\frac{3 \times 6}{3 + 6}\right) = 16 \text{ N}$$

The maximum available friction or limiting friction between the blocks is

$$(f_s)_{\max} = \mu_s m_1 g = (0.5)(3)(10) = 15 \text{ N}$$

The required friction is 16 N which greater than the limiting friction (15 N); so, kinetic friction comes into play, which is given as

$$f_k = \mu_k m_1 g = (0.45)(3)(10) = 13.5 \text{ N}$$

Let us assume that kinetic friction acts on the upper block to the right; so, the same amount of kinetic friction will act on the lower block to the left, as shown in the following figure.

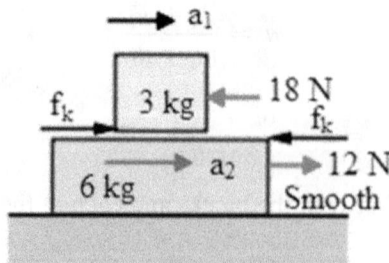

Applying Newton's second law and by putting $f_k = 13.5$ N, the net horizontal force acting on the upper block is

$$F_x = -18 + f_k = -18 + \mu_k m_1 g = -18 + (0.45)(3)(10) = -18 + 13.5 = -4.5 \text{ N}$$

So, the acceleration of the upper block is

$$a_1 = -4.5/3 = -1.5 \text{ m s}^{-2}$$

A negative answer means that the upper block will accelerate to the left.
  By putting $f_k = 13.5$ N the net horizontal force acting on the lower block is

$$F_x = 12 - f_k = 12 - \mu_k m_1 g = 12 - (0.45)(3)(10) = 12 - 13.5 = -1.5 \text{ N}$$

So, the acceleration of the lower block is

$$a_2 = -1.5/3 = -0.5 \text{ m s}^{-2}$$

Negative answer signifies that the lower block will accelerate to the left. Eventually, both the blocks will move with unequal acceleration. Ans

Problem 21 (Block pulled by a horizontal force)
  When you push the block with a time-varying horizontal force $F = \alpha t$, (a) find the variation of (i) friction (ii) acceleration (iii) velocity of the block, with time. (b) Is it always true to say that 'when the applied horizontal force overcomes limiting friction the block will move'? Explain.
  Solution

  (a) (i) At time t_0, let the applied force F be just be equal to $f_{s_{max}}$. Then, the time after which the block will begin to slide is

$$t_0 = \frac{f_{s_{max}}}{\alpha} = \frac{\mu_{mg}}{\alpha}.$$

  During this time, $a = 0$; so, the friction acting on the block is

$$f_s = F = \alpha t; \ t \leqslant t_0.$$

  After time t_0, the block will slip; so, the friction acting on the block is equal to kinetic friction given as $f_k = \mu_k mg; \ t > t_0$. Ans.
  (ii) Before the time t_0, block still remains stationary. So, $a = 0$ for $t \leqslant t_0$. For $t > t_0$, the body starts sliding and the kinetic friction $f_k = \mu_k mg$ opposes the motion of the block. Then, the net force acting on the block is,

$$F_{net} = F - f_k = \alpha t - \mu_k mg$$

So, the acceleration will be given as

$$a = \frac{\alpha t - f_k}{m} = \frac{\alpha t - \mu_k mg}{m} = \frac{\alpha t}{m} - \mu_k g \quad t > t_0 \text{ Ans.}$$

(iii) Then, the speed of the body is zero for $t \leqslant t_0$ because the block remains stationary in this time interval. For $t > t_0$, the body moves with a velocity,

$$v = \int a \, dt = \frac{\alpha t^2}{2m} - \mu_k gt; \quad t > t_0; \quad v = 0; \quad t < t_0 \text{ Ans.}$$

(b) No, it is true when the ground is fixed or moves with constant velocity. In an accelerating reference frame, the given statement is not valid.

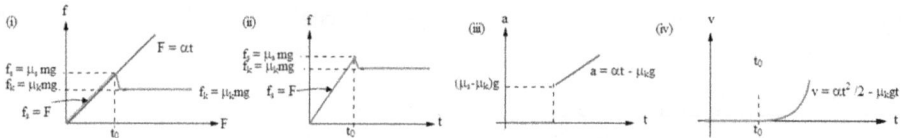

Problem 22 A horizontal force F acts on a cubical wedge of mass m such that the wedge accelerates on the horizontal surface. Find the minimum force F so as to prevent the sliding of the block of mass m on the wedge.

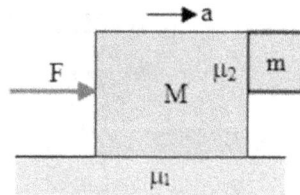

Solution

In figure (i), the forces acting on $M + m$ are $(M + m)g\downarrow$, $f_k \leftarrow$ and $N\uparrow$.
Force equations:

$$\sum F_x = F - f_k = (M + m)a_x = (M + m)a \tag{5.162}$$

$$\sum F_y = N - (M + m)g = ma_y = 0 \tag{5.163}$$

Law of f_k:

$$f_k = \mu_k N \tag{5.164}$$

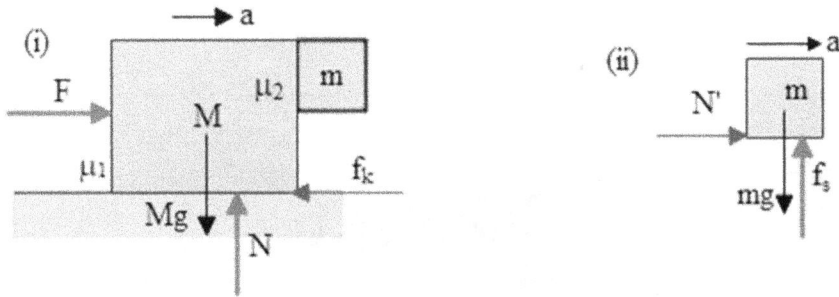

In figure (ii), the forces acting on m are $mg\downarrow$, $f_s \uparrow$, $N'\rightarrow$, $N'\uparrow$

Force equations:

$$\sum F_x = N' = ma_x = ma \tag{5.165}$$

$$\sum F_y = mg - f_s = ma_y = 0 \tag{5.166}$$

Law of f_s:

$$f_s \leqslant \mu_s N' \tag{5.167}$$

Using equations (5.162), (5.163) and (5.164),

$$a = F/(M + m) - \mu_k g \tag{5.168}$$

Using equations (5.165), (5.166) and (5.167),

$$mg \leqslant \mu_s ma$$

$$\Rightarrow a \geqslant \frac{g}{\mu_s} \tag{5.169}$$

Using equations (5.168) and (5.169),

$$F/(M + m) - \mu_k g \geqslant \frac{g}{\mu_s}$$

$$\Rightarrow F \geqslant (M + m)\left(\mu_k + \frac{1}{\mu_s}\right)g \quad \text{Ans.}$$

IOP Publishing

Problems and Solutions in Particle Mechanics

Pradeep Kumar Sharma

Chapter 6

Dynamics of circular motion

6.1 Cause of curvilinear motion: centripetal force

When a particle moves in a curve its speed may remain constant but it must change the direction of motion (velocity $\vec{v}$). In chapter 3, you learnt that a particle has to accelerate towards the center of curvature due to the change in its direction of motion. This means that the particle must experience an inward normal acceleration a_n which is also called centripetal acceleration. The net force that produces a centripetal acceleration is called centripetal force $F_{C.P.}$.

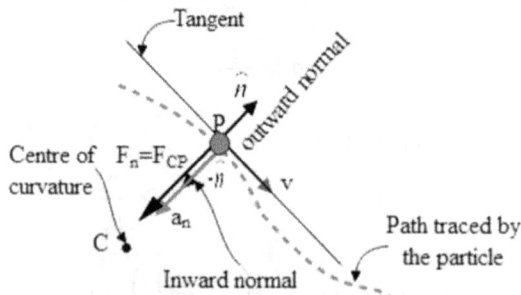

The particle's acceleration points towards the centre of curvature when it moves with a uniform speed in a curve.

Applying Newton's second law in normal n-axis,

$$F_{C.P}(=F_n) = ma_{C.P} = ma_n,$$

where $a_n = a_{cp} = v^2/r$; $r =$ radius of curvature of the path followed by the particle.

The centripetal force is not a new kind of force. It is the new effect of a force that changes the direction of velocity of the particle.

doi:10.1088/978-0-7503-6442-3ch6

The force that causes a particle to change its direction of motion is called centripetal force. It may be a component of the net force F acting on the particle.

However, the tangential component of the net force F along the tangential direction is responsible for changing the magnitude of the velocity of the particle. Applying Newton's second law in normal t-axis,

$$F_t = ma_t,$$

where $a_t = dv/dt = v\,dv/ds$.

Example 1 Discuss the presence of centripetal force in the following examples of circular motion.

(i) Satellites revolving around Earth.

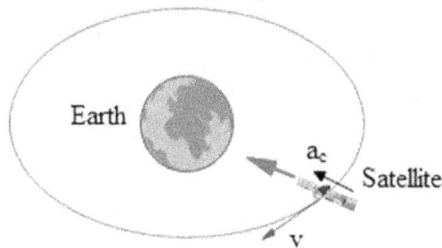

(ii) Electrons revolving around atoms.

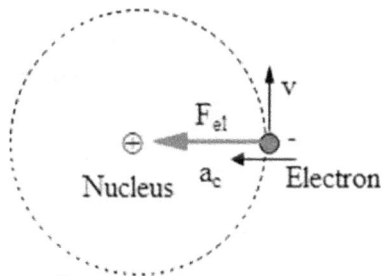

(iii) A vehicle negotiating a plane curve.

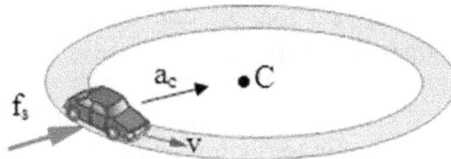

(iv) Revolving a stone in horizontal plane by a string.

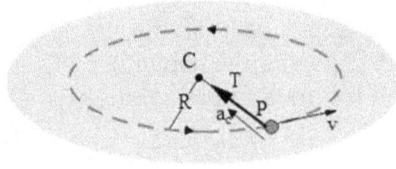

Solution

   (i) The Earth pulls the satellites towards its center by its own gravity. The orbits of Earth-satellites are circular in the equatorial plane. This means that a satellite accelerates towards the center of its circular path by gravitational force. So, in this case, 'gravitational force' is the centripetal force. Here, tangential force is zero.
  (ii) In an atom, electrons are pulled by the electrostatic (Coulombic) force offered by the nucleus. So, in this case, electrostatic force is the cause of centripetal acceleration for an electron. In other words, electrostatic force is the centripetal force in this case.
 (iii) When we negotiate a plane curve with a uniform speed, static friction pushes the vehicle towards the center of curvature of the curved path. So, 'friction' is the centripetal force for the vehicle. The friction produces necessary centripetal acceleration.
  (iv) When we revolve a stone by a string on a smooth horizontal surface, the string force (tension) pulls the stone towards the center of the circle. So, the 'tension' is the 'centripetal force'. So, the tension generates a centripetal acceleration of the stone.

N.B.: The net force which produces centripetal acceleration to change the direction of motion of a particle is known as centripetal force denoted as $\vec{F_c}$; $\vec{F_c} = m\vec{a_c}$, where a_c is the centripetal acceleration and m = mass of the particle. It is not a new type of force, rather it is a new effect of forces.

Example 2 Tension, normal reaction, friction and gravitation can change the direction of motion of a particle. Explain.
   Solution
   When a body is whirled in a horizontal circle by a string, tension is the centripetal force that pulls the particle continuously towards the center of the circle. We have explained it in the last example.
   When a motorcyclist moves in a cage, the normal reaction N pushes the motorcycle towards the center of the circular path changing the direction of motion of the motorcycle.

When you ride a bicycle in a plane curve, the static friction pushes the motorcycle towards the center of curvature of the path. Thus, f_s can also change the direction of velocity, as explained in the last example.

In an Earth–satellite system gravitational force of Earth pulls the satellites towards Earth's center providing necessary centripetal acceleration.

Example 3 An apple of mass m is projected with a speed v_0 at an angle θ with horizontal in gravity. Find the (a) tangential force, (b) centripetal force, (c) tangential acceleration, (d) centripetal acceleration, and (e) radius of curvature, at the given instant.

Solution

(a) The component of gravitational force (weight of the body) along the tangent, that is, $mg \sin \theta$ provides necessary tangential acceleration increasing or decreasing the speed of the projectile. Ans.

(b) The component of mg perpendicular to $\vec{v}$, that is, $mg \cos \theta$ is the centripetal force providing centripetal acceleration v^2/r, where $r =$ radius of curvature at the given position of the particle, or given point of the curve. Ans.

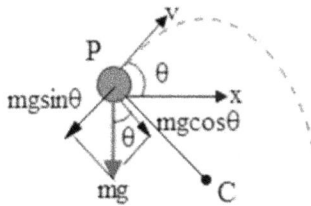

(c) The tangential acceleration is

$$a_t = F_t/m = mg \sin \theta/m = g \sin \theta \ \text{Ans.}$$

(d) The normal or radial or centripetal acceleration is

$$a_n = F_n/m = mg \cos \theta/m = g \cos \theta \ \text{Ans.}$$

(e) The radius of curvature at the point of projection is

$$r = v^2/a_{cp} = v_0^2/g \cos \theta \ \text{Ans.}$$

N.B.: In projectile motion, a constant force, that is, weight of the body $(m\vec{g})$ changes the direction of motion, making the path parabolic. However, in uniform circular motion the centripetal acceleration of the particle has constant magnitude while it changes its direction continuously. Hence, the centripetal force is not a constant force.

6.2 Uniform circular motion

When a particle moves in a circular path its motion is called circular motion. If the speed remains constant, it is called uniform circular motion. Since $|\vec{v}| =$ constant, the tangential acceleration of the particle is

$$a_t = \frac{d\,|\vec{v}|}{dt} = 0.$$

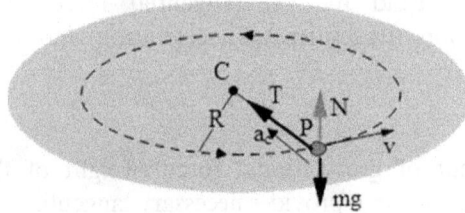

A smooth ball of mass m moves in a horizontal circle of radius R by the centripetal force $F_c = F_n$.

Then, the total acceleration of the particle is always centripetal (directed towards the center of the circle). The net force, that is, centripetal force causing the centripetal acceleration is

$$F_{net} = F_n = ma_n,$$

where $a_n = \frac{v^2}{R}$; $R =$ radius of the circle,

$$\Rightarrow F = \frac{mv^2}{R}$$

In uniform circular motion, any force such as tension, normal reaction, friction, or any field-force (such as gravity, electrostatic and magnetic forces) can act as centripetal forces. In uniform circular motion, the net force, that is, F_n is always perpendicular to the velocity $\vec{v}$ of the particle. Please note that F_n is the centripetal force but not mv^2/R; rather F_n is equal to mass times centripetal acceleration, that is, mv^2/R.

Example 4 A stone P moves in a circle of radius R by an inextensible string. (a) If the breaking strength of the string is equal to double the weight of the string, find the maximum velocity of the stone. (b) What happens when the velocity exceeds the maximum value. Put $R = 1/5$ m and $g = 10$ m s^{-2}.
  Solution
  (a) *Kinematics*:
  The stone is pulled by the tension T of the string towards the center of the circle. The tension generates a centripetal acceleration

$$a_{cp} = v^2/R \tag{6.1}$$

Since the horizontal plane does not accelerate up

$$a_y = 0 \qquad (6.2)$$

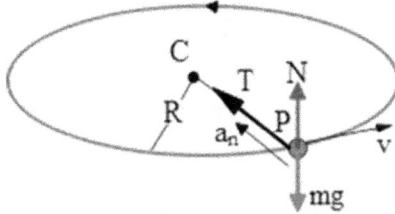

Force equation:

$$F_{cp}(=T) = ma_{cp} \qquad (6.3)$$

Since the normal reaction of the ground and the weight of the stone act vertically, the net vertical force

$$F_v = N - mg = ma_v \qquad (6.4)$$

Since N and mg do not contribute in centripetal force, equations (6.2) and (6.3) are dummy equations. So, using the other two equations, we have

$$T = mv^2/R \qquad (6.5)$$

The maximum tension is

$$T = 2mg \qquad (6.6)$$

Using the last wo equations,

$$mv^2/R = 2mg$$

$$\Rightarrow v = \sqrt{2gR} = \sqrt{2 \times 10 \times 1/5} = 2 \text{ m s}^{-1} \text{ Ans.}$$

(b) If the string breaks, the net force acting on the stone will be zero. Then the stone will move with a constant velocity along the tangent leaving the circle at the moment of breaking of the string.

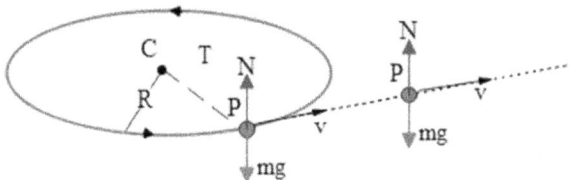

6.3 Non-uniform circular motion

When the speed of a particle moving in a circle changes, the motion is called non-uniform circular motion. Then, the tangential acceleration is

$$a_t = \frac{d\,|\vec{v}|}{dt} = \frac{|\vec{v}|\,d\,|\vec{v}|}{ds}$$

The force causing the tangential acceleration is

$$F_t = ma_t.$$

Then, the net force acting on the particle is the vector sum of $\vec{F_t}$ and $\vec{F_n}$.

$$\vec{F} = \vec{F_t} + \vec{F_n}$$

$$\Rightarrow F = \sqrt{F_t^2 + F_n^2} = \sqrt{(ma_t)^2 + (ma_n)^2}$$

$$\Rightarrow F = m\sqrt{a_t^2 + a_n^2}$$

(i)

(ii)

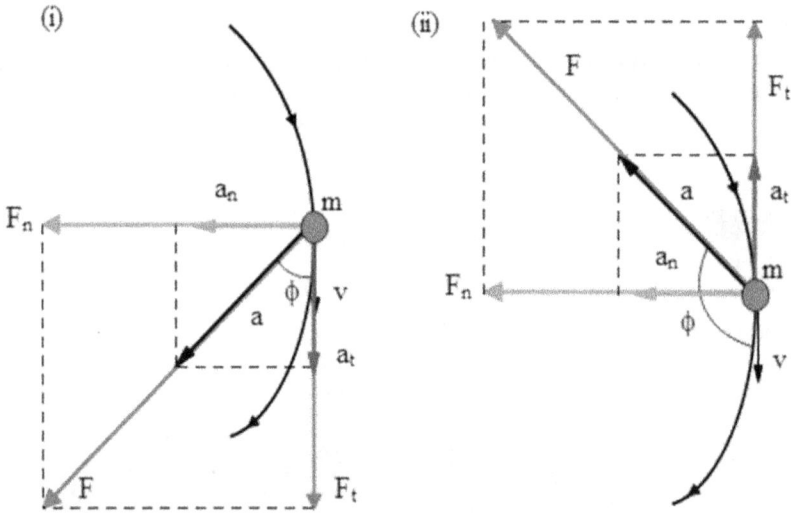

Vector diagram of centripetal acceleration, tangntial acceleration and total acceleration; figure (i) - speeding up; figure (ii)-slowing down (retardation)

The direction $\vec{F}$ relative to $\vec{v}$ is

$$\phi = \tan^{-1}\left(\frac{a_n}{a_t}\right),$$

where $a_n = \frac{v^2}{R}$ and $a_t = \frac{|\vec{v}|\,d\,|\vec{v}|}{ds} = \frac{d\,|\vec{v}|}{dt}$

N.B.: When a particle moves in a curve experiencing a net force, the component of this force along the line of motion (tangent) changes the speed of the particle and the radial component of the net force changes the direction of motion of the particle. The centripetal force is,

$$F_n = \frac{mv^2}{R},$$

where R = radius of curvature of the path. The tangential force is

$$F_t = ma_t = \frac{mdv}{dt} = m\frac{vdv}{ds}.$$

When the particle speeds up, the tangential acceleration favors the velocity; while the particle is slowing down, tangential acceleration opposes the velocity. So, the angle between the total force and velocity is acute in the former case and obtuse in the latter case.

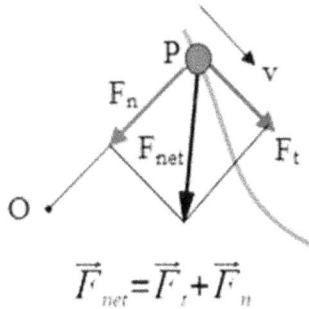

$$\vec{F}_{net} = \vec{F}_t + \vec{F}_n$$

Example 5 (Vertical circle)

A pendulum bob is released with a downward velocity v_0 when the string is just taut and horizontal. Find the (a) velocity of the bob and (b) tension in the string as the function of the angle θ made by the string with the horizontal.

Solution

(a) *Free-body diagram*:

At any angular position θ, the bob experiences $mg\downarrow$ and as shown in the figure.

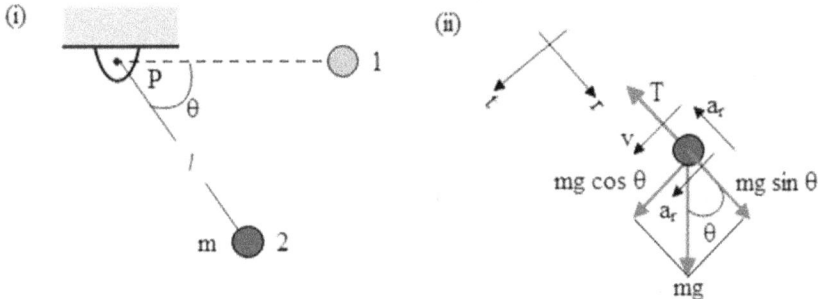

Force equation:

$$\sum F_r = T - mg \sin \theta = ma_r$$

$$\Rightarrow T = m(g \sin \theta + a_r) \tag{6.7}$$

The net tangential force acting on the bob is

$$\sum F_t = mg \cos \theta = ma_t$$

$$\Rightarrow a_t = g \cos \theta \tag{6.8}$$

Kinematics:

The tangential acceleration of the bob is

$$a_t = \frac{vdv}{ld\theta} \tag{6.9}$$

The radial acceleration of the bob is

$$a_r = \frac{v^2}{l} \tag{6.10}$$

Using equations (6.8) and (6.9),

$$vdv = gl \cos \theta \, d\theta \tag{6.11}$$

As the bob describes an angle θ, its speed increases from 0 to v. So, integrating both sides of equation (6.11)

$$\int_{v_0}^{v} vdv = gl \int_{0}^{\theta} \cos \theta d\theta$$

$$\Rightarrow v = \sqrt{v_0^2 + 2gl \sin \theta} \quad \text{Ans.}$$

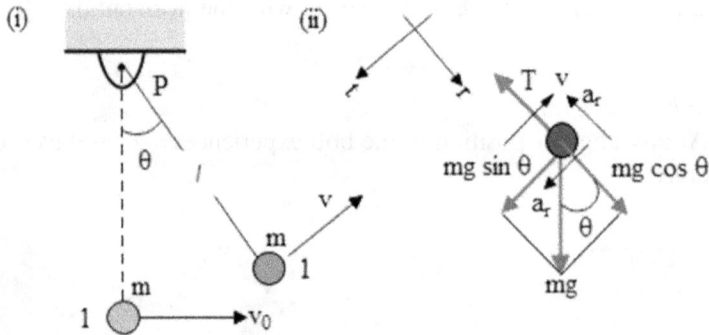

(b) From equations (6.7) and (6.10)

$$T = m\left(g \sin \theta + \frac{v^2}{l}\right) \tag{6.12}$$

Putting $v = \sqrt{v_0^2 + 2gl \sin \theta}$ in the last equation,

$$T = m\left(3g \sin \theta + \frac{v_0^2}{l}\right) \text{ Ans.}$$

Example 6 (Vertical circle-projected from the lowest point)

A pendulum bob is given a horizontal velocity v_0 when the bob is at the lowest position. Find the (a) tension in the string and (b) acceleration of the bob as the function of the angle θ made by the string with vertical.

Solution

(a) *Free-body diagram*:

At any angular position θ, the bob experiences $mg{\downarrow}$ as shown in figure.
Force equation:

$$\sum F_r = T - mg \cos \theta = ma_r$$

$$\Rightarrow T = m(g \cos \theta + a_r) \tag{6.13}$$

The net tangential force acting on the bob is

$$\sum F_t = mg \sin \theta = ma_t$$

$$\Rightarrow a_t = g \sin \theta \tag{6.14}$$

Kinematics:

The tangential acceleration of the bob is

$$a_t = \frac{v dv}{l d\theta} \tag{6.15}$$

The radial acceleration of the bob is

$$a_r = \frac{v^2}{l} \tag{6.16}$$

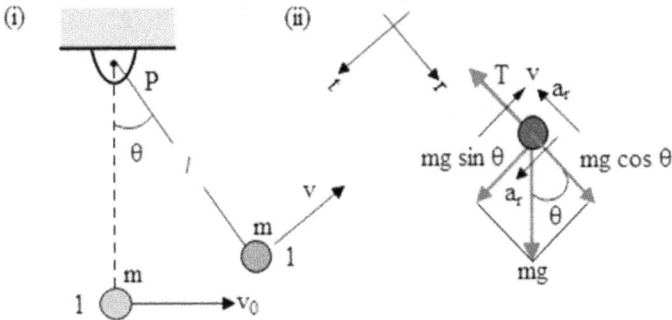

Using equations (6.14) and (6.15),

$$vdv = gl \sin \theta \, d\theta \tag{6.17}$$

As the bob describes an angle θ, its speed increases from 0 to v. So, integrating both sides of equation (6.17)

$$\int_{v_0}^{v} vdv = gl \int_{0}^{\theta} \sin \theta d\theta$$

$$\Rightarrow v = \sqrt{v_0^2 + 2gl(1 - \cos \theta)} \quad \text{Ans.}$$

From equations (6.13) and (6.16)

$$T = m\left(g \cos \theta + \frac{v^2}{l}\right) \tag{6.18}$$

Putting $\Rightarrow v = \sqrt{v_0^2 + 2gl(1 - \cos \theta)}$ in the last equation,

$$T = m\left((3 \cos \theta - 2)g + \frac{v_0^2}{l}\right) \quad \text{Ans.}$$

Example 7 The speed v of a particle in a circular motion increases with time as
$$v = kt.$$

(a) At what speed of the particle,
(b) after what time, will the centripetal force be half of the net force acting on the particle?

Solution

(a) It is given that the magnitude of centripetal force = half of the magnitude of total force

$$\Rightarrow F/2 = m\sqrt{a_t^2 + a_n^2}/2 = ma_n$$

$$\Rightarrow a_t = \sqrt{3} \, a_n \tag{6.19}$$

The centripetal acceleration is

$$a_n = \frac{v^2}{R} \tag{6.20}$$

The speed of the particle is

$$v = kt$$

$$\Rightarrow \frac{dv}{dt} = a_t = k \tag{6.21}$$

Putting a_n from equation (6.20) and a_t from equation (6.21) in equation (6.19),

$$k = \sqrt{3}\,v^2/R$$

$$\Rightarrow v = \sqrt{kR/\sqrt{3}}\quad \text{Ans.}$$

(b) Putting $v = kt$, we have

$$kt = \sqrt{kR/\sqrt{3}}$$

$$\Rightarrow t = \sqrt{R/k\sqrt{3}}\quad \text{Ans.}$$

6.4 Centrifugal force and rotating frame

When you spin about a vertical axis or stand on a rotating frame, each surrounding object seems to have an additional relative acceleration which is radially outward. This is called centrifugal acceleration, which can be given as $a_{C.F.} = \omega^2 \vec{r}$, where $\vec{r} =$ radius vector of the object from the axis of rotation.

So, you need to imagine a force acting on a particle to provide the centrifugal acceleration; this radially outward force is called a centrifugal force.

$$\vec{F}_{C.F.} = m\vec{a}_{C.F.}, \quad \text{where } a_{C.F.} = \omega^2 \vec{r}$$

$$\Rightarrow \vec{F}_{C.F.} = m\omega^2 \vec{r}$$

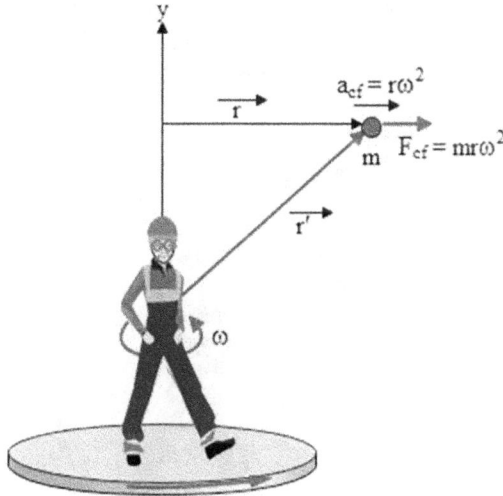

An observer (boy) standing on the rotating frame (disc/table) needs to impose a centrifugal force F_{cf} that can produce a centrifugal acceleration a_{cf} of an object of mass m (it may be stationary or moving relative to ground).

This force is a pseudo-force because it arises from the rotation of the reference frame but not caused by the interaction with any material bodies.

Let us summarize the following points from the previous discussions:

1. As the centrifugal force exists in the rotating (accelerating) frames, it is a pseudo-force. So, centrifugal force has no action–reaction pair.
2. $F = m\omega^2 \vec{r}$, $\omega=$ angular velocity of the rotating frame.
3. Centrifugal force increases linearly with the radial distance r, whereas generally real force decreases with distance.
4. Centrifugal force is directly proportional to the mass of the particle like gravitational force. Then, can you think that the gravitational force is the same sort of a pseudo-force?
5. Centripetal force are real forces (arising from interactions), whereas centrifugal forces are imaginary or pseudo-force (arising from the rotation of the reference frames).
6. The centrifugal force acting on a particle remains unaltered, if the direction of spin (angular velocity or rotation) is the reverse.
7. The centrifugal force acts on a particle from the axis of rotation of the reference frame. In this way,

    N.B: In general, the last two points are valid for centripetal forces. the particle need not move in a circle to experience a centrifugal force.

Example 8 You are standing at the center of a turntable that is spinning about its vertical axis passing through your body with an angular speed ω_0. Suppose that a mosquito revolves in a circular path of radius R around you in a horizontal plane. If the angular velocity of the mosquito about the vertical axis is ω, find the (i) centrifugal force (ii) centripetal force. In this example, (i) can you say that centrifugal force counteracts centripetal force? Explain. (ii) Do the above forces change their directions if we reverse the directions of ω_0 and ω.

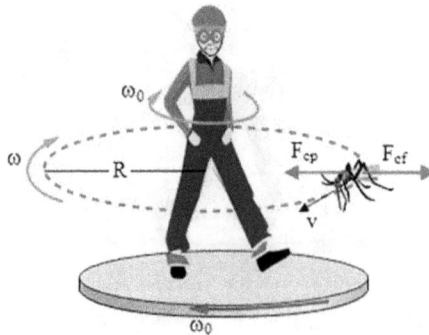

Solution

(i) The centrifugal force acing on the mosquito is directed radially outward by its definition and its magnitude is given as,

$$F_{C.F.} = mR\omega_0^2 \text{ (but not } mR\omega^2) \text{ Ans.}$$

(ii) The centripetal force acting on the mosquito is directed inward (towards the center of the circular path) by its definition and its magnitude is given as,

$$F_{C.F} = mR\omega^2 \text{ (but not } mR\omega_0^2) \text{ Ans.}$$

$F_{C.F.}$ depends on the perpendicular distance of the particle from the axis of rotation of the observer and the angular speed of the rotating frame (observer), whereas the $F_{C.P.}$ depends on the center of the circular path or radius of curvature of the path of the particle and its angular velocity relative to the center of curvature. If you say 'centrifugal and centripetal forces cancel', it does not make any sense because centripetal force acts on the object when viewed from the ground or an inertial frame (in general), whereas the centrifugal force is observed to act when viewed from a non-inertial frame. So, the question of equality is meaningless.

Example 9 Explain the physical significance of centrifugal force when you are sitting by the side of the window of a car while negotiating a plane curve.

  Solution

When you are sitting by the side of a window of a car while it negotiates a curve, you feel as if you are pushed away from the center of the curve. This is known as centrifugal force as it gives a physical effect of being centrifuged (pushed radially away). Since you are sitting in an accelerating frame, you think that you are not accelerating towards the center of the circle, rather you are centrifuged. But the actual situation can be viewed from the ground frame. The actual thing is that 'you are moving in a circular path' by the inward reaction force (pressing force) offered by the inner wall of the car on you. You may feel as if you are pressed against the walls of the vehicle. Then, obeying Newton's third law, the wall pushes you, towards the center of the circle with a force of equal magnitude N, say. In fact, the reaction force given by the wall of the car is centripetal force and in the rotating frames '$\vec{F}_{pseudo} = -m\vec{a}_{C.P}$' is the centrifugal force which never exists in the inertial frame. In this case the statement that 'centripetal and centrifugal forces balance each other' does not make any sense.

Example 10 Suppose that you are observing a block which remains stationary on a rotating disk. Can you use the phrase centripetal force is equal to centrifugal force to solve the problem?

Solution

No; relative to the rotating frame the block is stationary; so, the $a = 0$ then, the net force is

$$F_{net} = -f_s + F_{C.F} = 0$$

$$\Rightarrow F_{C.F} = f_s.$$

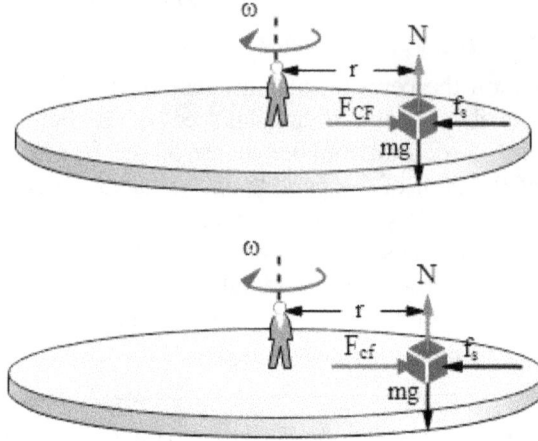

The centrifugal force is equal in magnitude and opposite in direction to static friction f_s.

In the ground (inertial) frame, the body moves in uniform circle motion. The net force acting on it is static friction f_s. No centrifugal force acts on the body when observed from the ground frame. Then, relative to the ground frame, the acceleration of the body is centripetal, that is,

$$a = \frac{v^2}{R}.$$

So, the static friction is

$$f_s = ma_r = \frac{mv^2}{R}.$$

The centripetal and centrifugal force do not act on an object when viewed from the same reference frame.

Problem 1 A simple pendulum of mass m is suspended from the roof of a trolley car that moves with a velocity v in a circle of radius R. If the string makes an angle θ with the vertical, find the velocity v.

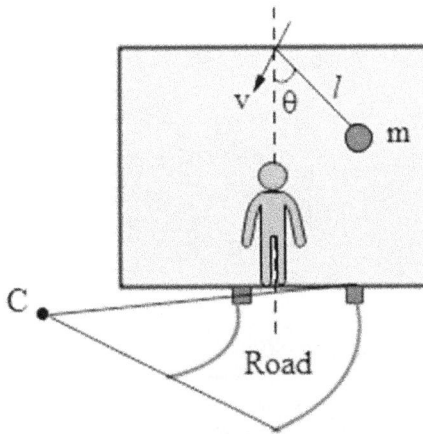

Solution

(a) Let us solve this problem by writing force equations relative to the rotating frame. Referring to the free-body diagram, the forces acting on the bob are, centrifugal force $F_{C.F.} = m(R + l \sin \theta)\omega^2 \rightarrow$, weight $mg\downarrow$ and tension $T \nwarrow$.

   Force equations:

   The net radial force acting on the bob is

$$F_r = F_{C.F.} - T \sin \theta = 0$$
$$\Rightarrow T \sin \theta = mr\omega^2$$

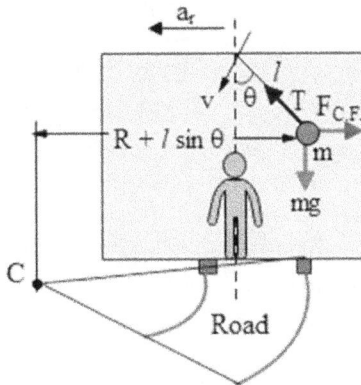

Putting $r = (l \sin \theta + R)$, we have

$$\Rightarrow T \sin \theta = (l \sin \theta + R)\omega^2 \qquad (6.22)$$

The net vertical force acting on the bob is

$$F_y = T \cos \theta - mg = ma_y = 0$$

$$T \cos \theta = mg \qquad (6.23)$$

Using the last two equations,

$$\frac{T \sin \theta}{T \cos \theta} = \frac{(l \sin \theta + R)\omega^2}{mg}$$

$$\Rightarrow \tan \theta = \frac{(l \sin \theta + R)\omega^2}{g}$$

$$\Rightarrow \omega = \sqrt{\frac{g \tan \theta}{R + l \sin \theta}}$$

So, the velocity of the vehicle is

$$v = R\omega = R\sqrt{\frac{g \tan \theta}{R + l \sin \theta}} \quad \text{Ans.}$$

Problem 2 A block of mass m is constrained to move in the diametrical groove made on the rotating disc under the action of a spring of stiffness k and friction. At any length r of the spring, the block will slip towards the center of the disc if the disc is stationary. The coefficient of static friction between block and the rotating disc is μ_s. Find the maximum and minimum angular velocity of the disc so that the block will not slip along the groove at the radial distance r. Assume that the undeformed length of the spring is l.

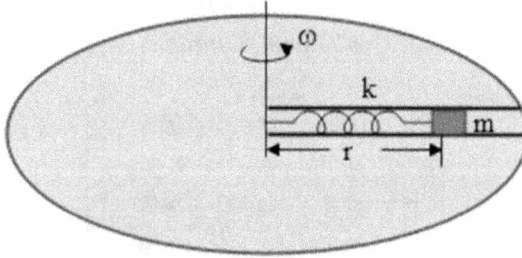

Solution
Referring to the free-body diagram made relative to the rotating frame, the forces acting on the bob are, centrifugal force $F_{C.F.} = mr\omega^2 \rightarrow$, $F_{sp} = kx \leftarrow$; $x = r - l$, weight $mg\downarrow$, $N\uparrow$, limiting friction $f_s \rightleftarrows$.

(a) For maximum angular speed the block will tend to slip radially away. So, the static friction will point radially inwards. The net radial force acting on the block is

$$-f_s - kx = ma_r \tag{6.24}$$

$$F_y = N - mg = ma_y = 0 \tag{6.25}$$

Law of static friction:

$$f_s \leqslant \mu N \qquad (6.26)$$

The centripetal acceleration of the block relative to ground is given as

$$a_r = -r\omega^2 \qquad (6.27)$$

So, using equations (6.24) and (6.27), we have

$$f_s = mr\omega^2 - kx \qquad (6.28)$$

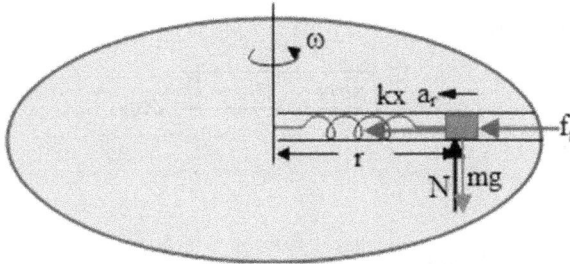

By using equations (6.25), (6.26) and (6.28), we have

$$(mr\omega^2 - kx) \leqslant \mu_s mg$$

$$\Rightarrow mr\omega^2 \leqslant \mu mg + kx,$$

where $x = (r - l)$, we have

$$\Rightarrow \omega \leqslant \sqrt{\frac{\mu mg + k(r - l)}{mr}} \quad \text{Ans.}$$

(b) For minimum angular speed the block will tend to slip radially inward. So, the static friction will point radially outwards. The net radial force acting on the block is

$$f_s - kx = ma_r \qquad (6.29)$$

Using equations (6.27) and (6.29), we have

$$f_s = kx - mr\omega^2 \qquad (6.30)$$

By using equations (6.25), (6.26) and (6.30), we have

$$(kx - mr\omega^2) \leqslant \mu_s mg$$

$$\Rightarrow mr\omega^2 \geqslant \mu mg - kx,$$

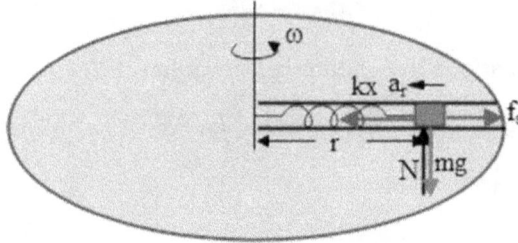

where $x = (r - l)$, we have

$$\Rightarrow \omega \geqslant \sqrt{\frac{\mu mg - k(r - l)}{mr}} \quad \text{Ans.}$$

Problem 3 (Conical pendulum)

Find the speed of a pendulum bob when it revolves in a horizontal circle about a vertical axis passing through the point of suspension of the pendulum. The length of the string is l and the angle of inclination of the string with the vertical is θ as shown in the figure. We call it 'conical pendulum' as the string generates a conical surface of revolution. Find the speed of the bob. If the breaking strength of the above example is equal to twice the weight of the bob, find the maximum possible angular frequency of revolution.

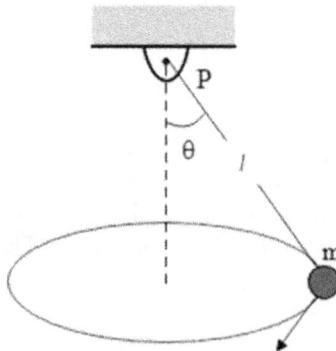

Solution

(a) *Free-body diagram*

The only forces acting on the bob are gravity force $mg\downarrow$ and tension $T\nwarrow$.

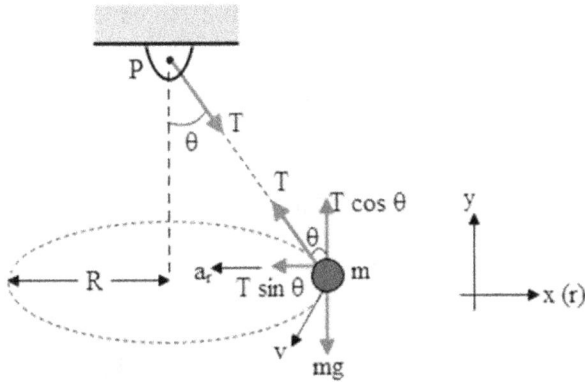

Force equation:

Resolving the forces along the radius, we have

$$\sum F_r = T \sin \theta = ma_r \tag{6.31}$$

Similarly, resolving the forces vertically, we have

$$\sum F_y = T \cos \theta - mg = ma_y \tag{6.32}$$

Kinematics:

As the bob does not change the 'y' coordinate, $a_y = 0$. Furthermore, $a_r = \frac{v^2}{R}$.

Substituting the accelerations a_r and a_y in equations (6.31) and (6.32), we have

$$T \sin \theta = m\frac{v^2}{R} \tag{6.33}$$

$$T \cos \theta = mg \tag{6.34}$$

Eliminating T in equations (6.33) and (6.34), we have

$$v = \sqrt{Rg \tan \theta} \tag{6.35}$$

Finally, substituting $R = l \sin \theta$ in equation (6.35), we have

$$v = \sqrt{\frac{gl}{\cos \theta}} \sin \theta \text{ Ans.} \tag{6.36}$$

(b) Putting $T = 2 \, mg$ in equation (6.36), we have

$$2mg \cos \theta = mg$$

$$\Rightarrow \cos \theta = 1/2$$

$$\Rightarrow \theta = 60°$$

Putting the obtained angle $\theta = 60°$ in the expression

$$v = \sqrt{\frac{gl}{\cos \theta}} \sin \theta,$$

we have

$$v = \sqrt{\frac{3gl}{2}} \quad \text{Ans.}$$

Problem 4 (Rotating loop)

A flexible loop of radius R and linear mass density μ assumes a circular shape when it spins in a smooth horizontal plane with an angular velocity w. (a) Find the tension in the loop. (b) If the breaking stress and density of the string of the loop are p and ρ, respectively, find the maximum velocity of any point of the loop.

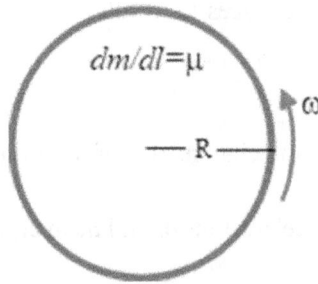

Solution

(a) *Free-body diagram*

Two tensions each of magnitude T act at the end points 1 and 2 of the elementary segment of mass $dm = 2\mu R d\theta$.

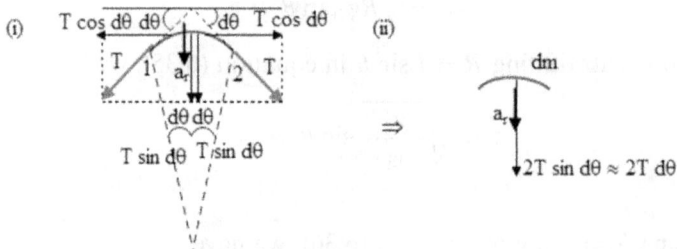

Force equation:

Since the tangential components of tensions are balanced, the net force acting on the segment is radially inward which is given as $dF = 2T \sin d\theta$. As $\sin \delta\theta \simeq \delta\theta$, we have $dF \simeq 2T \, d\theta$. This inward force pushes the elementary segment dm with an acceleration $a = R\omega^2$. Applying Newton's second law on 'dm', we have

$$dF = dm.\, a \tag{6.37}$$

Substituting $dF = 2Td\theta$, $dm = 2\mu Rd\theta$ and $a = a_r$, we have

$$2Td\theta = 2\mu\, Rd\theta\, 4 = dm\, a_r \tag{6.38}$$

Kinematics:

The radial acceleration of the element is

$$a_r = R\omega^2 \tag{6.39}$$

Using equations (6.37) and (6.38), we have

$$T = \mu R^2\omega^2 \text{ Ans.}$$

(b) The stress in the loop is

$$p = T/A = \mu R^2\omega^2/A,$$

where A = area of cross-section of the wire. Putting $\mu/A = \rho$, we have

$$p = \rho R^2\omega^2 = \rho v^2 \;(\because v = R\omega)$$

$$\Rightarrow v = \sqrt{\frac{p}{\rho}} \text{ Ans.}$$

Problem 5 A particle of mass m is moving with a constant (a) speed v (b) angular speed ω in a circular path in a smooth horizontal plane (plane of the paper) by a spring. If the natural length of the spring is l_0 and stiffness of the spring is k, find the elongation of the spring.

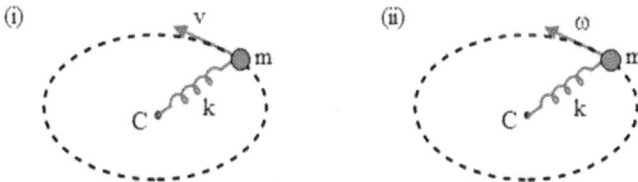

Solution

(a) *Free-body diagram*:

If the particle executes uniform circular motion, its speed must be uniform because, there is no tangential force to speed it up. Here the spring force kx acting on the particle is centripetal force caused by the elongation x of the spring.

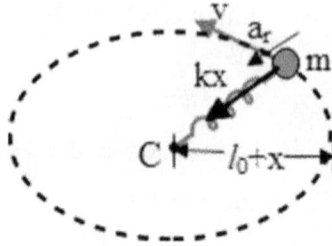

Force equation:

$$\sum F_r = F_{sp} = ma_r \tag{6.40}$$

Kinematics:

The radial acceleration is

$$a_r = \frac{v^2}{R} \quad (\text{where } R = \text{radius of circular path}) \tag{6.41}$$

Constraints equation:

The radius of the circular path is

$$R = l_0 + x \tag{6.42}$$

Hooke's law:

$$|F_{sp}| = kx \tag{6.43}$$

Using all four equations, we have

$$kx = m\frac{v^2}{l_0 + x} \tag{6.44}$$

$$\Rightarrow kx^2 + kl_0 x - mv^2 = 0$$

$$\Rightarrow x = \frac{\sqrt{k^2 l_0^2 + 4mv^2 k} - kl_0}{2k} \quad \text{Ans.}$$

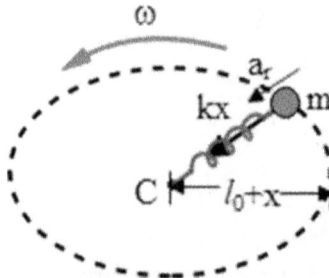

(b) The angular velocity of the particle is

$$\omega = \frac{v}{l_0 + x} \tag{6.45}$$

Using equations (6.44) and (6.45),

$$kx = m\frac{v^2}{l_0 + x} = m(l_0 + x)\omega^2$$

$$\Rightarrow kx - mx\omega^2 = ml_0\omega^2$$

$$\Rightarrow x = \frac{ml_0\omega^2}{k - m\omega^2}$$

The speed of the particle is

$$v = (l_0 + x)\omega = \left(l_0 + \frac{kl_0\omega^2}{k - m\omega^2}\right)\omega$$

$$\Rightarrow v = \frac{kl_0\omega}{k - m\omega^2} \quad \text{Ans.}$$

Problem 6 A block is placed near the edge of a disc of radius R which is rotating about the vertical axis passing through its center. The angular velocity of the disc is ω and the block (a) does not slip with the disc (b) is at the verge of slipping. For the case (b) use the coefficient of friction between the block and disc. Find (i) the total contact force received by the block (ii) the angle of inclination of the reaction force with vertical (iii) value of v_{max}. Assume that μ_s = coefficient of static friction between the block and disc.
Solution

(a)

(i) *Free-body diagram:* the forces acting on the block are $f_s \rightarrow$, $N\uparrow$ and weight $mg\downarrow$.
Force equation:

$$\sum F_r = f_s = ma_r \tag{6.46}$$

$$\sum F_v = N - mg = ma_y \tag{6.47}$$

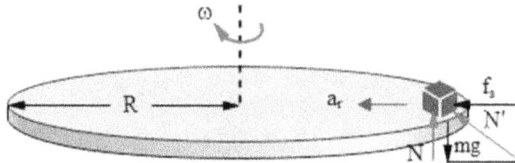

Kinematics:

$$a_r = R\omega^2 \tag{6.48}$$

$$a_y = 0 \tag{6.49}$$

Solving the above equations, we have

$$f_s = mR\omega^2 \tag{6.50}$$

$$N = mg \tag{6.51}$$

So, the net reaction (contact) force is

$$\vec{N'} = \vec{f_s} + \vec{N}$$

$$\left|\vec{N'}\right| = =\sqrt{f_s^2 + N^2} \tag{6.52}$$

Using the last three equations,
$$N' = m\sqrt{R^2\omega^4 + g^2} \quad \text{Ans.}$$

(ii) The angle of inclination of the reaction force with the vertical is

$$\beta = \tan^{-1}\left(\tfrac{f_s}{N}\right) = \tan^{-1}\left(\tfrac{mR\omega^2}{mg}\right)$$

$$= \tan^{-1}\left(\tfrac{R\omega^2}{g}\right) \quad \text{Ans.}$$

(iii) For the given angular velocity ω, the maximum velocity of the block is equal to $R\omega$. Ans.

(b)

(i) If the speed is maximum, the limiting friction will develop, which is given as

$$(f_s)_{\max} = \mu_s N = \mu_s mg \tag{6.53}$$

Then the net reaction force is

$$\left|\vec{N'}\right| = \sqrt{(f_s)_{\max}^2 + N^2} \tag{6.54}$$

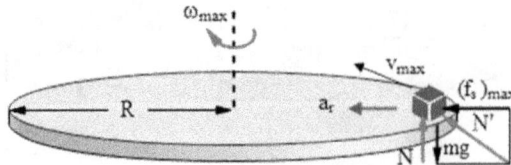

By using equations (6.51), (6.53) and (6.54), we have

$$N' = \sqrt{(\mu_s mg)^2 + (mg)^2}$$

$$\Rightarrow N' = (\sqrt{\mu_s^2 + 1})mg \quad \text{Ans.}$$

(ii) The angle of inclination of the reaction force with the vertical is

$$\beta = \tan^{-1}\left(\frac{f_s}{N}\right)$$

$$= \tan^{-1}\left(\frac{\mu mg}{mg}\right) = \tan^{-1}\mu = \phi_s \quad \text{Ans.}$$

(iii) For the maximum speed of the block for no relative sliding, the maximum or limiting friction provides the maximum centripetal acceleration, which is given as

$$(f_s)_{\max} = m(a_{cp})_{\max}.$$

where $(f_s)_{\max} = \mu_s N = \mu_s mg$ and $(a_{cp})_{\max} = v_m^2/R$; so, the maximum speed of the block is

$$v_{\max} = \sqrt{\mu_s gR} \quad \text{Ans.}$$

N.B.: In case (a) β is less than ϕ_s, but in case (b) $\beta = \phi_s$

Problem 7 (Banking of curve)
A vehicle moves in a banked curve of radius of curvature R with constant speed. The angle of banking and angle of static friction are equal to θ and $\varnothing$, respectively. Find the (a) maximum speed of the vehicle without skidding (b) speed of the vehicle so that it does not tend to slip.
Solution

(a) At the maximum speed of the vehicle, it tends to skid/slip away from the center of curvature along the plane and the friction will be maximum. Then, the vehicle experiences gravity $mg\downarrow$ and normal reaction $N(\nearrow)$ and $f_m(\searrow)$. If we intend to solve the problem in the reference frame of the vehicle, the centrifugal force acting on it is $F_{C.F.}(\leftarrow) = mA$.
Force equation:
The net force along the x-axis is

$$\sum F_x = mg\sin\theta + f_m - mA\cos\theta = ma_x = 0$$

$$f_m = m(A \cos \theta - g \sin \theta) \tag{6.55}$$

The net force along the x-axis,

$$\sum F_y = N - mg \cos \theta - mA \sin \theta = ma_y = 0$$

$$\Rightarrow N = m(A \sin \theta + g \cos \theta) \tag{6.56}$$

Kinematics:
As the vehicle negotiates a curve the centripetal acceleration A is given as

$$A = \frac{v^2}{R} \tag{6.57}$$

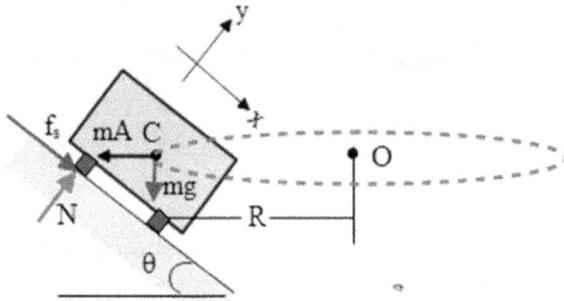

Putting A from equation (6.57) in equation (6.55), we have

$$N = \frac{mv^2}{R} \sin \theta + mg \cos \theta \tag{6.58}$$

Putting $a_y = 0$ from equation (6.58) in equation (6.56), we have

$$f_m = m\left(\frac{v^2}{R} \cos \theta - g \sin \theta\right) \tag{6.59}$$

Law of static friction:
  The limiting friction is given as

$$f_m = \mu N \tag{6.60}$$

Putting f_m and N from equations (6.58) and (6.59) in equation (6.60), we have

$$m\left(\frac{v^2}{R} \cos \theta - g \sin \theta\right) = \mu\left(\frac{mv^2}{R} \sin \theta + mg \cos \theta\right)$$

$$\Rightarrow m\frac{v^2}{R}(\cos \theta - \mu g \sin \theta) = (\sin \theta + \mu \cos \theta)mg$$

$$\Rightarrow v = \sqrt{\left(\frac{\sin \theta + \mu \cos \theta}{\cos \theta - \mu \sin \theta} \right) gR}$$

$$\Rightarrow v = \sqrt{\left(\frac{\sin \theta + \tan \phi \cos \theta}{\cos \theta - \tan \phi \sin \theta} \right) gR}$$

$$\Rightarrow v = \sqrt{\left(\frac{\tan \theta + \tan \phi}{1 - \tan \theta \tan \phi} \right) gR}$$

$$\Rightarrow v = \sqrt{gR \tan(\phi + \theta)} \quad \text{Ans.}$$

(b) If the vehicle neither tends to slip up nor down, static friction will be absent. Then, equation (6.55) can be modified as

$$f_s = m(A \cos \theta - g \sin \theta) = 0$$

$$\Rightarrow A = g \tan \theta \tag{6.61}$$

Using equations (6.57) and (6.61)

$$\frac{v^2}{R} = g \tan \theta$$

$$\Rightarrow v = \sqrt{gR \tan \theta} \quad \text{Ans.}$$

N.B.: If the vehicle moves with a speed greater than the speed $\sqrt{gR \tan \theta}$ it will tend to skid up the inclined plane (banked curve). Hence, frictional force will act downwards along the banked curve.

Problem 8 A disc of radius R rotates from rest about a vertical axis with a constant angular acceleration such that a block placed at the periphery of the disc moves with a tangential acceleration a as shown in the figure. If the coefficient of static friction between the block and disc is μ_s, find the (a) velocity of the block before it starts sliding relative to the disc, (b) total distance traversed by the block before sliding, (c) time taken by the block before sliding.
    Solution

(a) *Free-body diagram*:
    As the block moves in a circle, it experiences radial force F_r and tangential force F_t. and F_t are the components of static friction f_s.

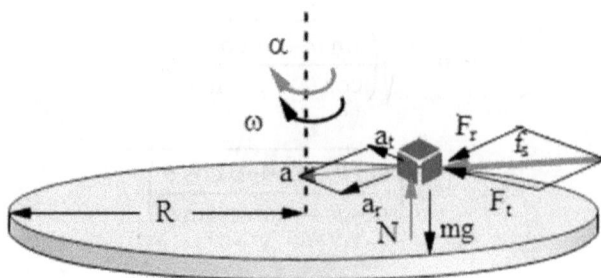

Force equation:

$$F_r = ma_r \qquad (6.62)$$

Since, $(a_t = a;$ given),

$$F_t = ma_t = ma \qquad (6.63)$$

$$\sum F_y = N - mg = ma_y \qquad (6.64)$$

Law of static friction:

$$f_s \leqslant \mu_s N \qquad (6.65)$$

Kinematics:

$$a_r = \frac{v^2}{R} \qquad (6.66)$$

Since the disc does not move vertically

$$a_y = 0 \qquad (6.67)$$

Vector addition of forces:

$$\sqrt{F_t^2 + F_r^2} \leqslant f_s \qquad (6.68)$$

From equations (6.62) and (6.66), we have

$$F_r = \frac{mv^2}{R}$$

From equations (6.64) and (6.67), we have

$$N = mg$$

Substituting $N = mg$ in equation (6.65), we have

$$f_s \leqslant \mu_s mg$$

Substituting F_t, F_r and f_s, we have

$$\frac{m^2 v^4}{R^2} + m^2 a^2 \leqslant \mu_s^2 m^2 g^2,$$

where $a = r\alpha$

$$\Rightarrow v \leqslant \sqrt{R\sqrt{\mu_s^2 g^2 - R^2\alpha^2}} \quad \text{Ans.}$$

(b) The tangential acceleration is

$$\frac{vdv}{ds} = a_t = R\alpha$$

Separating the variable,

$$vdv = R\alpha ds$$

Integrating both sides,

$$\int_0^{v_{max}} vdv = R\alpha \int_0^s ds$$

$$\Rightarrow s = v_{max}^2 / 2R\alpha$$

Putting the obtained value of v_{max} and simplifying the factors, the total distance traversed by the block before sliding is given as

$$\Rightarrow s = R\sqrt{\mu_s^2 g^2 - R^2\alpha^2} / 2R\alpha$$

$$= \sqrt{\mu_s^2 g^2 - R^2\alpha^2} / 2\alpha \quad \text{Ans.}$$

(c) The tangential acceleration is

$$\frac{dv}{dt} = a_t = R\alpha$$

Separating the variable,

$$dv = R\alpha dt$$

Integrating both sides,

$$\int_0^{v_{max}} dv = R\alpha \int_0^t dt$$

Then the time after which the block will begin to slip is

$$t = v_{max} / R\alpha$$

Putting the obtained value of v_{max} and simplifying the factors, the total distance traversed by the block before sliding is given as

$$t = \sqrt{R\sqrt{\mu_s^2 g^2 - R^2\alpha^2}} / R\alpha$$

$$\Rightarrow t = \sqrt{(1/R)\sqrt{\mu_s^2 g^2 - R^2\alpha^2}} / \alpha \quad \text{Ans.}$$

N.B.:

1. When $\alpha \neq 0$, f_s is neither tangential nor radial. The tangential component of f_s provides necessary tangential acceleration (or retardation) speeding or slowing the block. The radial component of f_s pushes the block towards the center of the disc providing necessary centripetal acceleration.

2. If $\alpha = 0$, the static friction f_s is radially inward.

Problem 9 A bead of mass m is given an initial velocity $v_0 = \sqrt{2gR}$ so that it moves along a rigid horizontal circular frame of radius R. (a) Write all equations related to the motion of the bead. (b) Find the angular distance covered by the bead when its speed becomes $v = \sqrt{gR}$. (c) What is the total number of turns revolved by the bead before it comes to rest. Assume $\mu =$ coefficient of kinetic friction between the bead and frame. Assume vertically up is $+y$ direction.

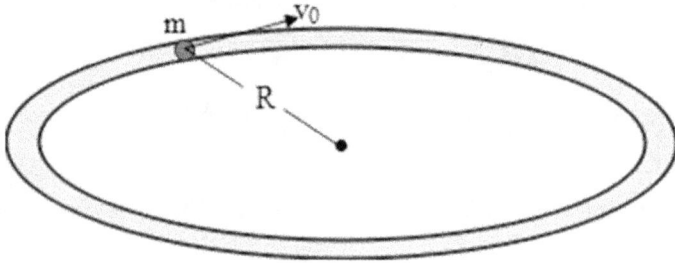

Solution

(a) *Free-body diagram*:

As the ring skids along the circular frame, kinetic friction f_k acts tangentially opposite to the velocity $\vec{v}$ of the ring. Apart from this, gravity $mg\downarrow$ and the reaction force N act so that the radial component of N, that is, N_r provides necessary centripetal acceleration a_n and the vertical component of N, that is, N_y counterbalances the gravity.

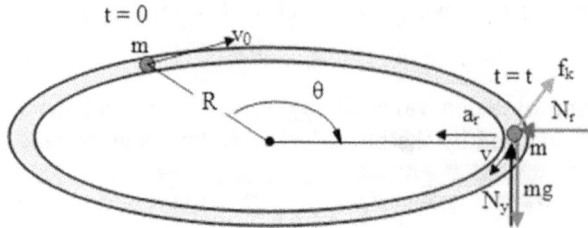

Vector addition:

Adding N_r and N_y, we have

$$N = \sqrt{N_r^2 + N_y^2} \tag{6.69}$$

Force equation:

Referring to the free-body diagram, we have the following equations:

$$\sum F_y = N_r = ma_r \tag{6.70}$$

$$\sum F_y = N_y - mg = ma_y \tag{6.71}$$

$$\sum F_t = f_k = ma_t \tag{6.72}$$

Law of kinetic friction:

$$f_k = \mu N \tag{6.73}$$

Kinematics:

$$a_r = \frac{v^2}{R} \tag{6.74}$$

$$a_y = 0 \tag{6.75}$$

$$a_t = -\frac{dv}{dt} = -\frac{v \, dv}{R \, d\theta} \quad \text{Ans.} \tag{6.76}$$

(b) Using equations (6.70) and (6.74),

$$N_r = ma_r = \frac{mv^2}{R} \tag{6.76}$$

Using equations (6.71) and (6.75),

$$N_y = mg \tag{6.77}$$

Using equations (6.69), (6.76) and (6.77),

$$N = \sqrt{\left(\frac{mv^2}{R}\right)^2 + (mg)^2} = m\sqrt{g^2 + \frac{v^4}{R^4}} \tag{6.78}$$

Using equations (6.73) and (6.78),

$$f_k = \mu N = \mu m \sqrt{g^2 + \frac{v^4}{R^4}} \tag{6.79}$$

Using equations (6.72) and (6.79),

$$ma_t = \mu m \sqrt{g^2 + \frac{v^4}{R^4}}$$

$$\Rightarrow a_t = \mu \sqrt{g^2 + \frac{v^4}{R^4}} \qquad (6.80)$$

Using equations (6.72) and (6.79),

$$a_t = -\frac{v dv}{R d\theta} = \mu \sqrt{g^2 + \frac{v^4}{R^4}}$$

$$\Rightarrow -\frac{v dv}{R d\theta} = \mu \sqrt{g^2 + \frac{v^4}{R^4}}$$

$$\Rightarrow -\frac{v dv}{d\theta} = \mu \sqrt{g^2 R^2 + v^4}$$

Separating the variables,

$$\frac{v dv}{\sqrt{g^2 R^2 + v^4}} = -\mu d\theta$$

As the bead describes an angular distance θ, its speed decreases from v_0 to v; so, integrating both sides, we have

$$\int_{v_0}^{v} \frac{v dv}{\sqrt{g^2 R^2 + v^4}} = -\mu \int_0^{\theta} d\theta = -\mu\theta \qquad (6.81)$$

If we put $v^2 = k \cdot v dv = dk$; putting these values and the limits of $k = v^2$ (from $2gR$ to gR) in the integration, we have

$$I = \frac{1}{2} \int_{2gR}^{gR} \frac{dk}{\sqrt{g^2 R^2 + k^2}} = \frac{1}{2} \ln\left(k + \sqrt{g^2 R^2 + k^2}\right)_{2gR}^{gR}$$

$$\Rightarrow I = \frac{1}{2} \ln\left\{\left(gR + \sqrt{g^2 R^2 + g^2 R^2}\right) - \ln\left(2gR + \sqrt{g^2 R^2 + 4g^2 R^2}\right)\right\}$$

$$\Rightarrow I = \frac{1}{2} \ln\left(\frac{gR + \sqrt{g^2 R^2 + g^2 R^2}}{2gR + \sqrt{g^2 R^2 + 4g^2 R^2}}\right) = -\frac{1}{2} \ln\left(\frac{2 + \sqrt{5}}{1 + \sqrt{2}}\right) \qquad (6.82)$$

Using equations (6.81) and (6.82),

$$-\frac{1}{2} \ln\left(\frac{2 + \sqrt{5}}{1 + \sqrt{2}}\right) = -\mu\theta$$

$$\Rightarrow \theta = \frac{1}{2} \ln \frac{2 + \sqrt{5}}{1 + \sqrt{2}} \quad \text{Ans.}$$

(c) Let the bead describe the total angular distance θ as its speed decreases from v_0 to 0; so, using equation (6.82), we have

$$I = \frac{1}{2} \int_{2gR}^{0} \frac{dk}{\sqrt{g^2 R^2 + k^2}} = \frac{1}{2} \ln\left(k + \sqrt{g^2 R^2 + k^2}\right)_{2gR}^{0}$$

$$\Rightarrow I = \frac{1}{2} \ln\left\{\left(0 + \sqrt{g^2 R^2 + 0}\right) - \ln\left(2gR + \sqrt{g^2 R^2 + 4g^2 R^2}\right)\right\}$$

$$\Rightarrow I = -\frac{1}{2} \ln(2 + \sqrt{5}) \tag{6.83}$$

Using equations (6.81) and (6.83),

$$-\frac{1}{2} \ln(\sqrt{5} + 2) = -\mu\theta$$

$$\Rightarrow \theta = \frac{1}{2\mu} \ln(2 + \sqrt{5}) \quad \text{Ans.}$$

Problem 10 A cylindrical cage of radius R is rotating about its vertical axis. A block of mass m is pressed against the wall of the cage by the centrifugal effect of the cage. If the block does not fall down, find the minimum angular velocity ω of rotation of the cage if its angular acceleration is α.

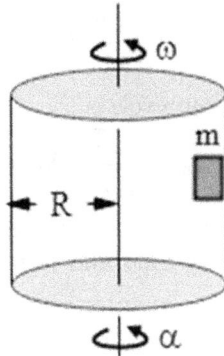

Solution

Free-body diagram:

The block experiences the forces $mg\downarrow$, $f\uparrow$ and $N\rightarrow$.

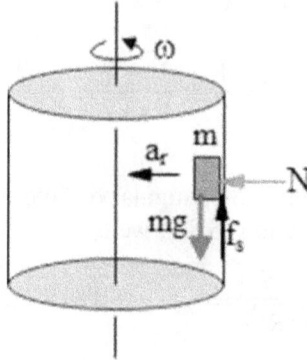

Force equation:

Referring to the free-body diagram, let us write the following equations:

$$\sum F_y = f - mg = ma_y \tag{6.84}$$

$$\sum F_r = N = ma_r \tag{6.85}$$

Law of static friction:

$$f \leqslant \mu_s N \tag{6.86}$$

Kinematics:

$$a_r = \frac{v^2}{R} \tag{6.87}$$

As the motorcylce does not fall,

$$a_y = 0 \tag{6.88}$$

Using the above equations, we have

$$v \geqslant \sqrt{\frac{gR}{\mu}} \tag{6.89}$$

So, the minimum speed of the motorcycle is

$$v_{\min} = \sqrt{\frac{gR}{\mu}} \quad \text{Ans.}$$

N.B: If $v < \sqrt{gR/\mu}$, the motorcycle slips down describing a helical path. In the above example, f_s balances gravity (mg) and N provides centripetal acceleration; $f_s = mg$ and $N = \frac{mv^2}{R}$. The above equations hold good even though the cage is stationary and a motorcyclist rides inside the cage in a horizontal circle.

Problem 11 A cylindrical cage of radius R is rotating about its vertical axis. A block of mass m is pressed against the wall of the cage by the centrifugal effect of the cage. If the block does not fall down, find the minimum angular velocity ω of rotation of the cage if its angular acceleration is α.

Solution

If the cylindrical cage has constant angular acceleration α, the radial component of resultant or net static friction f_s can provide the tangential acceleration and the vertical component of the net static friction balances the weight of the block. So, the static friction in this case is no longer vertical, as shown in the free-body diagram.

Free-body diagram:

The block experiences the forces $mg\downarrow$, $f_s \searrow$ and $N\rightarrow$.

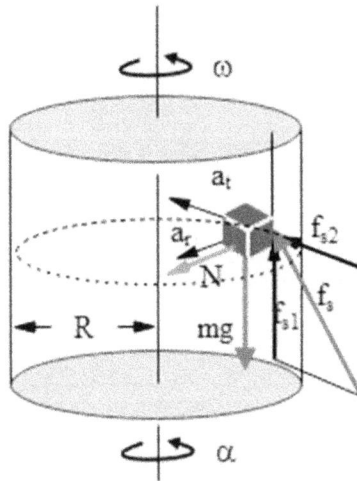

Force equation:

Referring to the free-body diagram, let us write the following equations:

The net vertical force is

$$\sum F_v = f_{s1} - mg = ma_v \tag{6.90}$$

The net radial force is

$$\sum F_r = N = ma_r \tag{6.91}$$

The net tangential force is

$$\sum F_t = f_{s2} = ma_t \tag{6.92}$$

Kinematics:

The centripetal (radial) acceleration of the block is

$$a_r = R\omega^2 \tag{6.93}$$

The tangential acceleration of the block is

$$a_t = R\alpha \tag{6.94}$$

As the motorcycle does not fall,

$$a_v = 0 \tag{6.95}$$

According to the law of vector addition, the net static friction is

$$f_s^2 = f_{s1}^2 + f_{s2}^2 \tag{6.96}$$

Law of static friction:

$$f_s \leqslant \mu_s N \tag{6.97}$$

From equations (6.90) and (6.95)

$$f_{s1} = mg \tag{6.98}$$

From equations (6.90) and (6.93)

$$N = mR\omega^2 \tag{6.99}$$

From equations (6.92) and (6.94)

$$f_{s2} = mR\alpha \tag{6.100}$$

From equations (6.98), (6.99) and (6.100)

$$f_s^2 = (mg)^2 + (mR\alpha)^2$$

$$\Rightarrow f_s = \sqrt{(mg)^2 + (mR\alpha)^2}$$

$$\Rightarrow f_s = m\sqrt{g^2 + R^2\alpha^2} \tag{6.101}$$

From equations (6.97), (6.99) and (6.101)

$$m\sqrt{g^2 + R^2\alpha^2} \leqslant \mu_s(mR\omega^2)$$

$$\omega \geqslant \sqrt{\sqrt{g^2 + R^2\alpha^2}/\mu_s R} \quad \text{Ans.}$$

N.B.: If you put $\alpha = 0$, we have

$$\omega \geqslant \sqrt{g/\mu_s R}$$

So, the velocity of the block is given as

$$v(=R\omega) \geqslant \sqrt{\frac{gR}{\mu_s}}$$

So, the minimum speed of the motorcycle is

$$v_{\min} = \sqrt{\frac{gR}{\mu_s}}$$

as obtained in the last example.

N.B.: If, ω is less than $\sqrt{\dfrac{\sqrt{R^2\alpha^2 + g^2}}{\mu_s R}}$ the motorcycle slips down describing a helical path.

Problem 12 A bead of mass m slides over a vertical string due to friction between it and the string. A block of mass M is connected with the other end of the string which passes over the smooth pulley. If the block is given a horizontal velocity v at a time t, say, and the acceleration of the bead relative to the string is downward, at this instant, find the (a) tension in the string, (b) frictional force between the bead and string, (c) acceleration of the string, (d) acceleration of the bead, and (e) acceleration of the block.

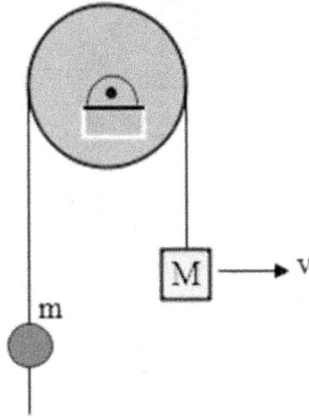

Solution

(a) In the free-body diagram, the tension in the string is T (assumed). Since the bead slides and accelerates down with a given relative acceleration a_r, kinetic friction f, say, acts upon the bead and a kinetic friction of the same magnitude acts down on the string at the contact point A. Let the acceleration of the contact point A of the bead and string be a (up); since the string is inextensible, the acceleration of point C of the string is a (down). As the block M turns around C with an angular velocity ω_{BC} relative to the point C, by applying the concept of relative angular velocity, we have

$$\omega_{BC} = \frac{v}{l} \tag{6.102}$$

Then, the acceleration of block B relative to C is

$$a_{BC} = l\omega_{BC}^2 = l\left(\frac{v}{l}\right)^2 = \frac{v^2}{l} \tag{6.103}$$

The acceleration of block M relative to the ground is

$$\vec{a}^{\,\prime} = \vec{a_B} = \vec{a_{BC}} + \vec{a_C} = \frac{v^2}{l}\hat{j} - a\hat{j} = \left(\frac{v^2}{l} - a\right)\hat{j}$$

$$\Rightarrow a' = \left(\frac{v^2}{l} - a\right) \tag{6.104}$$

The acceleration of the bead relative to the ground is

$$\vec{a_1} = a\hat{j} - a_r\hat{j} = (a - a_r)\hat{j}$$

$$a_1 = (a_r - a) \tag{6.105}$$

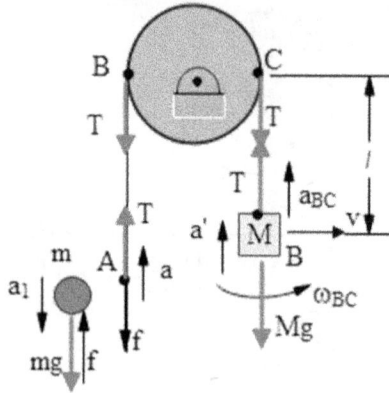

Force equations on bead:

$$mg - f = ma_1 \tag{6.106}$$

Force equations on string at A:
As the string is massless

$$f - T = 0 \tag{6.107}$$

Force equations on the block:

$$-Mg + T = ma' \tag{6.108}$$

Using equations (6.104) and (6.108)

$$T - Mg = M\left(\frac{v^2}{l} - a\right)$$

$$T - Mg = M\left(\frac{v^2}{l} - a\right) \tag{6.109}$$

Using equations (6.105),(6.106) and (6.107)

$$mg - T = m(a_r - a) \tag{6.110}$$

Solving the last two equations,

$$T = \frac{Mm}{M + m}\left(2g + \frac{v^2}{l} - a_r\right) \text{ Ans.}$$

(b) The frictional force is

$$f = T = \frac{Mm}{M + m}\left(2g + \frac{v^2}{l} - a_r\right) \text{ Ans.}$$

(c) Putting the obtained value of T in equation (6.110), the acceleration of the string is

$$a = \frac{M - m}{M + m}\left(g + \frac{v^2}{l}\right) \text{ Ans.}$$

(d) Putting the obtained value of a in equation (6.105), the acceleration of the bead is

$$a_1 = a_r - \frac{M - m}{M + m}\left(g + \frac{v^2}{l}\right) \text{ Ans.}$$

(e) Putting the obtained value of a in equation (6.104), the acceleration of the block is

$$a' = \left(\frac{v^2}{l} - a\right) = \frac{v^2}{l} - \frac{M - m}{M + m}\left(g + \frac{v^2}{l}\right)$$

$$\Rightarrow a' = \frac{2m}{M + m}\frac{v^2}{l} - \frac{M - m}{M + m}g \text{ Ans.}$$

N.B.: If you put $a_r = 0$ for $v = 0$, the obtained expression will be reduced to the familiar equations for tension of an ideal pulley–particle system, given as

$$T = \frac{2Mm}{M + m}g.$$

Problem 13 A block of mass m_1 placed at the edge of a rotating turntable of radius R. A string connecting blocks m_1 and m_2 after passing over a smooth hole made at the center C of the table, as shown in the figure below. If the block does not slide relative to the rotating table, find the maximum angular velocity of the disc.

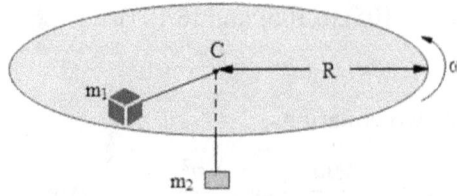

Solution

Free-body diagram:

To prevent relative sliding, static friction will prevail. For maximum angular velocity the block must tend to slip radially away due to the centrifugal effect. So, the friction will be directed radially inward. The forces on m_1 are $f_s \nearrow$ (assumed), $T \nearrow$, $N\uparrow$, $m_1 g\downarrow$. The forces on m_2 are $m_2 g\downarrow$ and $T\uparrow$, as shown in figure.

Force equation for m_1:

$$\sum F_y = N - m_1 g = m_1 a_{1y} \qquad (6.111)$$

$$\sum F_r = f_s + T = m_1 a_r \qquad (6.112)$$

Force equation for m_2:

$$\sum F_y = T - m_2 g = m_2 a_{2y} \qquad (6.113)$$

Law of static friction:

$$f_s \leqslant \mu_s N \qquad (6.114)$$

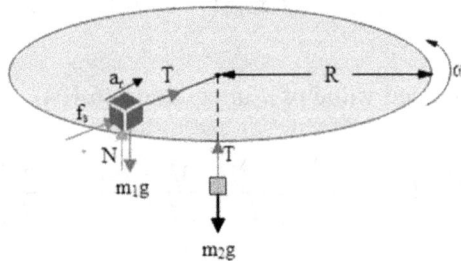

Kinematics:

Since the block does not lose contact with the rotating disc and the disc does not accelerate up, the vertical acceleration of the block is

$$a_{1y} = 0 \qquad (6.115)$$

$$a_r = \frac{v^2}{R} \qquad (6.116)$$

$$a_{2y} = 0 \qquad (6.117)$$

Solving the above equation, we have

$$f_s = m_1 R\omega^2 - m_2 g \qquad (6.118)$$

$$N = m_1 g \qquad (6.119)$$

Using equations (6.114), (6.117) and (6.119),

$$m_1 R\omega^2 - m_2 g \leqslant \mu_s m_1 g$$

$$\Rightarrow \omega_{max} = \sqrt{\left(\frac{m_2 + \mu_s m_1}{m_1}\right) g/R} \quad \text{Ans.}$$

N.B.: For minimum angular speed, the block will tend to slip radially inward; so, friction will act radially outward on the block; following the above procedure, the minimum angular velocity of rotation of the disc to avoid relative sliding of the particle can be given as

$$\omega_{min} = \sqrt{\left(\frac{m_2 - \mu m_1}{m_1}\right) \frac{g}{R}}$$

1. When $\omega = \sqrt{\frac{m_2}{m_1} \frac{g}{R}}$, $f = 0$ because the particle neither tends to slide radially outward not tends to slide radially inward.
2. When $\omega < \sqrt{\frac{m_2}{m_1} \frac{g}{R}}$, f_s is directed radially outward because the particle tends to slide radially inward.
3. When $\omega > \sqrt{\frac{m_2}{m_1} \frac{g}{R}}$, f_s is directed radially inward because the particle tends to slide radially outward.

Problem 14 Two blocks of mass m_1 and m_2 being connected with a light inextensible string are placed on the diameter of a rotating disc. The coefficient of static friction between block m_1 and disc is μ and the other block is smooth. If the block does not slide relative to the rotating table, find the maximum angular velocity of the disc.

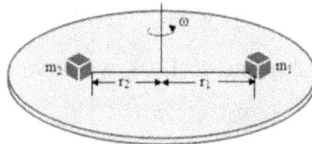

Free-body diagram:
To prevent relative sliding, static friction will prevail. For maximum angular velocity the block will have to tend to slip radially away due to the centrifugal effect. So, the friction will be directed radially inward. The forces on m_1 are $f_s \rightarrow$ (assumed), $T \leftarrow$, $N_1 \uparrow$, $m_1 g \downarrow$. The force on m_2 are $m_2 g \downarrow$ and $N_2 \uparrow$, $T \rightarrow$, as shown in figure.

Force equation for m_1:

$$\sum F_y = N_1 - m_1 g = m_1 a_{1y} \tag{6.120}$$

$$\sum F_r = f_s + T = m_1 a_{1r} \tag{6.121}$$

Force equation for m_2:

$$\sum F_r = T = m_2 a_{2r} \tag{6.122}$$

$$\sum F_y = N_2 - m_2 g = m_2 a_{2y} \tag{6.124}$$

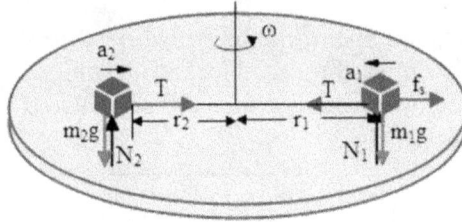

Law of static friction:

$$f_s \leqslant \mu_s N_1 \tag{6.125}$$

Using equations (6.121) and (6.122),

$$f_s = m_1 a_{1r} - m_2 a_{2r} \tag{6.126}$$

Kinematics:

Since the block does not lose contact with the rotating disc and the disc does not accelerate up, the vertical acceleration of the block is

$$a_{1y} = 0 \tag{6.127}$$

$$a_{1r} = r_1 \omega^2 \tag{6.128}$$

$$a_{2y} = 0 \tag{6.129}$$

$$a_{2r} = r_2 \omega^2 \tag{6.130}$$

Using equations (6.126), (6.129) and (6.130), we have

$$f_s = m_1 r_1 \omega^2 - m_2 r_2 \omega^2 = (m_1 r_1 - m_2 r_2)\omega^2 \tag{6.131}$$

Using equations (6.120) and (6.129), we have

$$N_1 = m_1 g \tag{6.132}$$

Using equations (6.125), (6.131) and (6.132),

$$(m_1 r_1 - m_2 r_2)\omega^2 \leqslant \mu_s m_1 g$$

$$\Rightarrow \omega \leqslant \sqrt{\frac{\mu_s m_1 g}{(m_1 r_1 - m_2 r_2)}}$$

$$\Rightarrow \omega_{max} = \sqrt{\frac{\mu_s m_1 g}{(m_1 r_1 - m_2 r_2)}} \quad \text{Ans.}$$

N.B.: In this problem $m_1 r_1 - m_2 r_2$ must be positive for the friction to act inward. This means that for this condition, the body m_1 will tend to slip outward because the centrifugal force acting on m_1 is greater than that of m_2. If $m_1 r_1 - m_2 r_2$ is zero, the centrifugal force acting on m_1 is equal to that on m_2; so, body m_1 will not tend to slip and eventually the friction will disappear. If $m_1 r_1 - m_2 r_2$ is negative, the centrifugal force acting on m_2 is greater than that on m_1; so, body m_1 will tend to slip inwards and eventually the friction will point radially outward.

Problem 15 A coin of mass m is placed on the top of a fixed sphere of radius R. If the coin is given a horizontal velocity $v = \sqrt{2gR}$, find the (a) acceleration of the coin, (b) radius of curvature of the trajectory of the coin (c) maximum horizontal distance covered by the coin.

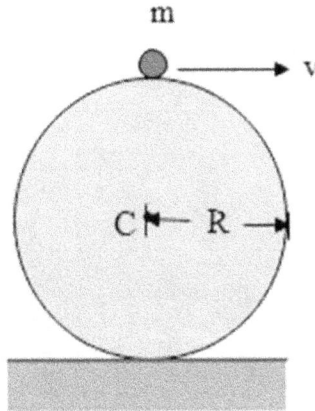

Solution

(a) Let N = normal reaction offered by the sphere on the coin which is vertically up. So, the net force acting on the coin is

$$F_{C.P.} = mg - N = ma_r$$

$$\Rightarrow N = mg - \frac{mv^2}{l}$$

Putting $v = \sqrt{2gl}$, we have

$$N = -mg$$

A negative sign signifies that the normal reaction is downward, which is impossible because the sphere cannot pull the coin down; it can only push it up. So, the coin must lose contact ($N = 0$) with the sphere and the only force acting on the coin is the weight of the coin. So, the acceleration of the coin is equal to acceleration due to gravity. Ans.

(b) The radius of curvature of the path of the coin at the top is

$$r = v^2/a_{cp}$$

The centripetal acceleration is

$$a_{cp} = F_{cp}/m = mg/m = g\downarrow$$

Using the last two equations, we have

$$r = v^2/a_{cp} = v^2/g$$

Putting $v = \sqrt{2gR}$, we have

$$r = v^2/a_{cp} = 2gR/g = 2R \text{ Ans.}$$

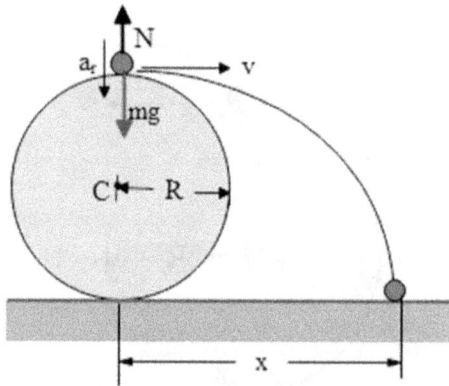

(c) Just after the projection, the coin will leave the spherical surface at the top with the given velocity $v = \sqrt{2gl}$. It will reach the ground after traversing a vertical distance $2R$. The time taken by the coin to reach the ground is given as

$$y = -2R = \frac{1}{2}(-g)t^2$$

$$\Rightarrow t = \sqrt{2(2R)/g} = 2\sqrt{R/g}$$

So, the horizontal distance traversed by the coin till it hits the ground is

$$x = vt = \left(\sqrt{2gR}\right)\left(2\sqrt{R/g}\right)$$

$$\Rightarrow x = (2\sqrt{2})R \text{ Ans.}$$

Problem 16 A pendulum bob is released from rest when the string is just taut and horizontal. Find the (a) tension in the string, (b) acceleration of the bob as the function of the angle θ made by the string with horizontal, (c) the position where the bob moves with a horizontal acceleration and find that horizontal acceleration, and (d) the position where the vertical component of the velocity of the bob is maximum.

Solution

(a) *Free-body diagram*:

At any angular position θ, the bob experiences $mg\downarrow$ and tension T (directed radially inward), as shown in figure.

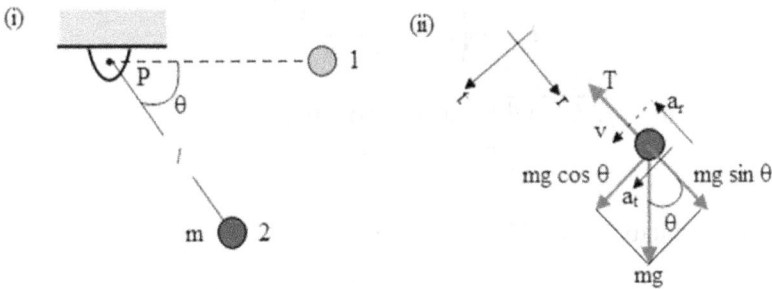

Force equation:

$$\sum F_r = T - mg \sin \theta = ma_r$$

$$\Rightarrow T = m(g \sin \theta + a_r) \tag{6.133}$$

The net tangential force acting on the bob is

$$\sum F_t = mg \cos \theta = ma_t$$

$$\Rightarrow a_t = g \cos \theta \tag{6.134}$$

Kinematics:

The tangential acceleration of the bob is

$$a_t = \frac{v\,dv}{l\,d\theta} \tag{6.135}$$

The radial acceleration of the bob is

$$a_r = \frac{v^2}{l} \tag{6.136}$$

Addition of acceleration:

The net acceleration of the bob is

$$a = \sqrt{a_t^2 + a_r^2} \tag{6.137}$$

Using equations (6.134) and (6.135),

$$v\,dv = gl \cos \theta \, d\theta \qquad (6.138)$$

As the bob describes an angle θ, its speed increases from 0 to v. So, integrating both sides of equation (6.136)

$$\int_0^v v\,dv = gl \int_0^\theta \cos \theta \, d\theta$$

$$\Rightarrow v = \sqrt{2gl \sin \theta} \qquad (6.139)$$

From equations (6.133) and (6.136)

$$T = m\left(g \sin \theta + \frac{v^2}{l}\right) \qquad (6.140)$$

Putting $v = \sqrt{2gl \sin \theta}$ in the last equation,

$$T = 3mg \sin \theta \quad \text{Ans.}$$

(b) From equations (6.136) and (6.138)

$$a_r = 2g \sin \theta$$

From equation (6.134),

$$a_t = g \cos \theta$$

Putting the values of a_t and a_r in equation (6.137), we have

$$a = \sqrt{(g \cos \theta)^2 + (2g \sin \theta)^2} \qquad (6.141)$$

$$\Rightarrow a = g\sqrt{1 + 3 \sin^2 \theta} \quad \text{Ans.}$$

(c) When the bob moves with a horizontal acceleration, its vertical acceleration is zero. The vertical component of the acceleration of the bob is

$$a_y = a_r \sin \theta - a_t \cos \theta = (2g \sin \theta)\sin \theta - (\cos \theta)(\cos \theta) = 0$$

$$\Rightarrow \theta = \tan^{-1}\left(\frac{1}{\sqrt{2}}\right)$$

Putting the obtained angle in the expression

$$a = g\sqrt{1 + 3 \sin^2 \theta},$$

we have

$$a = a_x = \sqrt{2}g \quad \text{Ans.}$$

(d) The vertical velocity is

$$v_y = v \cos \theta = (\sqrt{2gl \sin \theta})\cos \theta$$

For it to be is maximum,

$$\frac{dv_y}{d\theta} = \frac{d}{d\theta}\left\{(\sqrt{2gl \sin \theta})\cos \theta\right\} = 0$$

$$\Rightarrow \theta = \tan^{-1}\left(\frac{1}{\sqrt{2}}\right)$$

N.B.:
1. When v_y is maximum $a_y = 0$; so, directly you can write the result of (c)

$$\Rightarrow \theta = \tan^{-1}\left(\frac{1}{\sqrt{2}}\right)$$

2. The variation of magnitude and direction of the total acceleration vector of the bob is shown in the following figure. At $\theta = 0, a = g\downarrow$; at $\theta = \tan^{-1}\frac{1}{\sqrt{2}}, a = \sqrt{2}g\leftarrow$ and $\theta = 90°, \vec{a} = 2g\uparrow$.

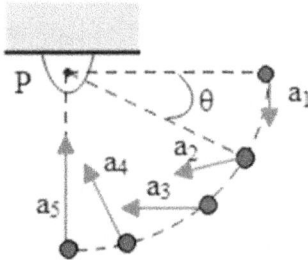

We can see that the magnitude of total acceleration vector gradually increases and the acceleration vector turns clockwise from vertically down to vertically up in the right side quadrant while the body is coming down.

Problem 17 The bob of mass m is released from the angular position θ such that the magnitude of its acceleration at its initial and lowest position are equal, find the (a) value of θ, and (b) ratio of tensions in the string at the initial position 1 and lowest position 2 of the bob.

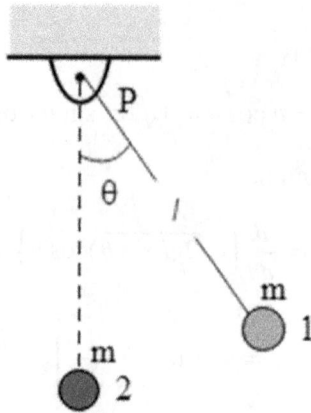

Solution

(a) As per the given condition

$$a_1 = a_2 \tag{6.142}$$

The initial acceleration (at the angular position θ) is

$$a_1 = g \sin \theta \tag{6.143}$$

The final acceleration (at the lowest position) is

$$a_2 = \frac{v^2}{l} \tag{6.144}$$

Referring to example 6, the velocity at the lowest position is

$$v = \sqrt{2gl(1 - \cos \theta)} \tag{6.145}$$

From the last two equations,

$$a_2 = 2g(1 - \cos \theta) \tag{6.146}$$

Substituting a_1 from equation (6.143) and a_2 from equation (6.146) in equation (6.142),

$$g \sin \theta = 2g(1 - \cos \theta)$$

$$\Rightarrow \sin \theta = 2(1 - \cos \theta)$$

$$\Rightarrow 5 \cos^2 \theta - 8 \cos \theta + 3 = 0$$

$$\Rightarrow \cos \theta = \frac{3}{5}$$

$$\Rightarrow \theta = \cos^{-1} \frac{3}{5} \quad \text{Ans.}$$

(b) The net radial force is

$$F_r = T_1 - mg \cos \theta = ma_r \tag{6.147}$$

Putting $a_r = \frac{v^2}{l} = 0$ (because initially the bob is at rest), we have

$$T_1 = mg \cos \theta \tag{6.148}$$

The tensions in the string at the lowest position of the bob is

$$T = m\left(g + \frac{v^2}{l}\right)$$

Putting $v = \sqrt{2gl(1 - \cos \theta)}$, we have

$$T_2 = mg(3 - 2 \cos \theta) \tag{6.149}$$

From the last two equations,

$$T_1/T_2 = \cos \theta/(3 - 2 \cos \theta)$$

Putting $\cos \theta = \frac{3}{5}$, we have

$$T_1/T_2 = (3/5)/\{3 - 2(3/5)\} = 1/3 \text{ Ans.}$$

Problem 18 A coin is given a downward velocity v tangentially at an angular position θ on a cylindrical surface, as shown in the figure below. The cylindrical surface moves up with a constant acceleration A. If the coefficient of friction between the coin and surface is μ, find the total acceleration of the coin.

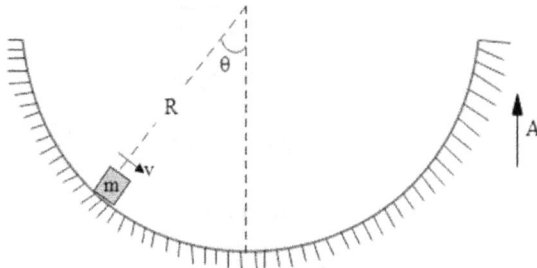

Solution

Free-body diagram:

As the coin slides down, friction is kinetic $f_k \uparrow$ and acts tangentially up. Furthermore, the coin experiences $mg \downarrow$ and $N \nearrow$ as shown in the figure. We intend to write the force equations from the accelerating cylinder. So, we need to impose pseudo-force $mA \downarrow$.

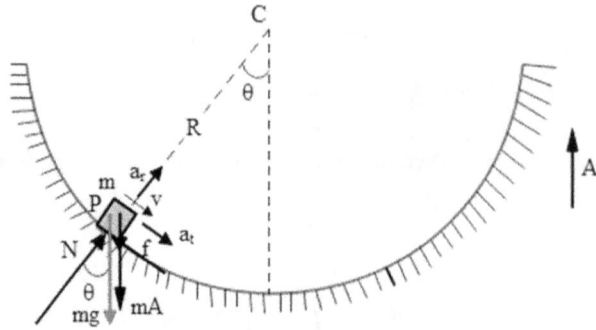

Force equations:

The net force along radial direction is

$$\sum F_r = N - m(g + A)\cos\theta = (ma_{PC})_r \qquad (6.150)$$

The net force along tangential direction is

$$\sum F_t = m(g + A)\sin\theta - f_k = ma_t \qquad (6.151)$$

Law of kinetic friction:

$$f_k = \mu N \qquad (6.152)$$

Kinematics:

$$(a_{PC})_r = \frac{mv^2}{R} \qquad (6.153)$$

Substituting f_k from equation (6.152), $(a_{PC})_r$ from equation (6.153) in equation (6.151), we have

$$(a_{PC})_t = (g + A)\sin\theta - \frac{\mu N}{m} \qquad (6.154)$$

Now, substituting N from equation (6.150) in equation (6.154), we have

$$(a_{PC})_t = (g + A)(\sin\theta - \mu\cos\theta) - \frac{\mu v^2}{R}$$

The acceleration of the coin relative to ground is

$$\vec{a_P} = \vec{a_{PC}} + \vec{a_C} = (\vec{a_{PC}})_t + (\vec{a_{PC}})_r + \vec{a_C}$$

So, the acceleration of P in the radial direction is

$$(\vec{a_P})_r = (\vec{a_{PC}})_r + (\vec{a_C})_r = \left(\frac{v^2}{R} + g\cos\theta\right)\hat{r}$$

The acceleration of P in the radial direction is

$$(\vec{a_P})_t = (\vec{a_{PC}})_t + (\vec{a_C})_t$$

$$(\overrightarrow{a_P})_t = \left[(g + A)(\sin \theta - \mu \cos \theta) - \frac{\mu v^2}{R} - (A \sin \theta) \right] \hat{\theta}$$

So, the acceleration of P relative to the ground is

$$(\overrightarrow{a_P}) = (\overrightarrow{a_P})_t + (\overrightarrow{a_P})_r \Rightarrow \overrightarrow{a_P}$$

$$= \left(\frac{v^2}{R} + g \cos \theta \right) \hat{r} + \left((g + A)(\sin \theta - \mu \cos \theta) - \frac{\mu v^2}{R} - (A \sin \theta) \right) \hat{\theta} \text{ Ans.}$$

N.B.: If we reverse the velocity of the block relative to the cylinder, the magnitude and direction of a_r remains unchanged, but, the friction gets reversed. So, there is a change in the tangential acceleration.

Problem 19 A disc is made to rotate about the vertical axis with a constant angular speed ω. A block of mass m can slide in a diametrical groove made on the disc. The block is attached to one end of a spring of stiffness k. The other end of the spring is connected with the center O of the disc. (a) Write the differential equation for the radial motion of the disc. (b) Find the total reaction force acting on the block by the disc. (c) Write the differential equation in the radial direction when (i) the bead is smooth, (ii) static friction is present, (iii) the bead is smooth and it does not slide on the groove, (iv) kinetic friction is present so also the spring is elongated and the block moves radially in the groove, and (v) the bead is smooth and the spring is absent.
Solution
Method 1 (Pseudo-force method)

(a) When observed from the rotating disc we have outward pseudo-force called centripetal force,

$$\overrightarrow{F_{C.F.}} = mr\omega^2 \hat{r} \tag{6.155}$$

Furthermore, we have two horizontal normal forces; one is horizontal or transverse normal reaction offered by the wall of the groove on the block,

$$\overrightarrow{F_\theta} = F_\theta \hat{\theta} \tag{6.156}$$

The other reaction force is vertical normal reaction N given by the disc to the block that balances the weight mg of the block.

$$\overrightarrow{F_y} = (N - mg)\hat{j} \tag{6.157}$$

Apart from this, relative to the disc, another transverse pseudo-force called Coriolis force comes into play, which is given as

$$\overrightarrow{F_{\text{Coriolis}}} = 2m\overrightarrow{v_r} \times \overrightarrow{\omega} = -2mv_r\omega\hat{\theta} \tag{6.158}$$

acts opposite to the Then, the block speeds up along the groove, when viewed from the disc with a radial acceleration

$$\vec{a_r} = \frac{d^2 r}{dt^2}\hat{r} \tag{6.159}$$

As the block does not lose contact with the groove, its transverse acceleration relative to the disc is zero

$$\vec{a_\theta} = \vec{0} \tag{6.160}$$

As the block does not lose contact with the groove/disc, its vertical acceleration relative to the disc is zero

$$\vec{a_y} = \vec{0} \tag{6.161}$$

Force equations:

The net radial force is

$$\vec{F_r} = m\vec{a_r} \tag{6.162}$$

The net transverse force is

$$\vec{F_\theta} = m\vec{a_\theta} \tag{6.163}$$

The net vertical force is

$$\vec{F_y} = m\vec{a_y} \tag{6.164}$$

Using equations (6.155), (6.160) and (6.162)

$$\Rightarrow \vec{F_{sp}} + \vec{f} + \vec{F_{CF}} = m\vec{a_r}$$

$$\Rightarrow (-kx - f + mr\omega^2)\hat{r} = m\frac{d^2 r}{dt^2}\hat{r}$$

$$\Rightarrow m\frac{d^2 r}{dt^2} + f + kx = mr\omega^2 \text{ Ans.}$$

(b) Using equations (6.156), (6.158), (6.160) and (6.163)

$$\vec{F_\theta} + \vec{F_{Coriolis}} = \vec{0}$$

$$\Rightarrow (F_\theta - 2mv_r\omega)\hat{\theta} = \vec{0}$$

$$\Rightarrow F_\theta = 2mv_r\omega \tag{6.165}$$

Using equations (6.157), (6.161) and (6.164)

$$\vec{F_y} = (N - mg)\hat{j} = \vec{0}$$

$$\Rightarrow N = mg \tag{6.166}$$

From the last two equations, the total reaction force is

$$\overrightarrow{F'} = -f\hat{r} + 2mv_r\omega\hat{\theta} + mg\hat{j} \quad \text{Ans.}$$

Then we have the following force equations of the block relative to the rotating frame (disc).

(c)

(i) When the bead is smooth, friction $f = 0$; so the differential equation is

$$m\frac{d^2r}{dt^2} + kx = mr\omega^2 \quad \text{Ans.}$$

(ii) When the static friction is present, the relative sliding must be absent and so the path of the block is circular. Since the radius $r = \text{constant}$ in circular motion,

$$\frac{d^2r}{dt^2} = 0$$

Then, the differential equation is

$$f_s + kx = mr\omega^2 \quad \text{Ans.}$$

(iii) When the bead is smooth and it does not slide on the groove, we put $f = 0$ and $\frac{d^2r}{dt^2} = 0$ to obtain

$$m\frac{d^2r}{dt^2} + kx = mr\omega^2 \quad \text{Ans.}$$

(iv) When the kinetic friction is present, the relative sliding is present; so, putting

$$\frac{d^2r}{dt^2} = 0,$$

the differential equation is

$$m\frac{d^2r}{dt^2} + f_k + kx = mr\omega^2$$

(v) If the bead is smooth and the spring is absent, putting $f = 0$ and F_{sp} $(=kx) = 0$, we have

$$m\frac{d^2r}{dt^2} = mr\omega^2 \quad \text{Ans.}$$

Method 2 (Ground frame method)

If the block slides, relative to the disc the path of the block is straight (radially outward). But, when we observe the block from the ground, it moves in a spiral path. In the ground frame the inertial or pseudo forces (centrifugal and Coriolis) are absent; then, we have to consider the real forces only. The net radial force acting on the block is

$$-f - kx = ma_r$$

The acceleration of the block along the groove (relative to ground) is given as

$$a_r = \left(= \frac{d^2r}{dt^2} - r\omega^2 \right)$$

So, using the last two equations, we have

$$-f - kx = m\left(\frac{d^2r}{dt^2} - r\omega^2 \right)$$

$$\Rightarrow m\frac{d^2r}{dt^2} + f_k + kx = mr\omega^2$$

So, we can get same differential equations by writing force equations in both reference frames (ground and disc).

N.B.:

1. If the bead (or the groove) is smooth, the net radial force is zero. So, the radial acceleration in the ground frame

$$a_r\left(= \frac{d^2r}{dt^2} - r\omega^2 \right) = 0$$

$$\Rightarrow \frac{d^2r}{dt^2} = r\omega^2$$

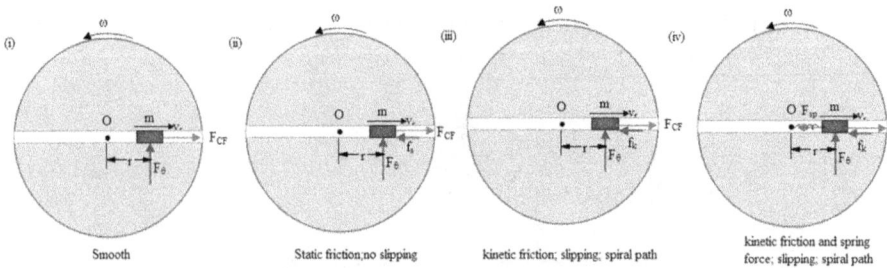

| Smooth | Static friction;no slipping | kinetic friction; slipping; spiral path | kinetic friction and spring force; slipping; spiral path |

2. If the static frictions present or the block does not slip relative to the disc, the path of the block is circular in the ground frame; since in both reference frames, the radial distance remains constant,

$$\frac{d^2r}{dt^2} = 0$$

Then, we have the radial acceleration

$$a_r = \left(\frac{d^2 r}{dt^2} - r\omega^2 \right) = -r\omega^2$$

relative to ground and zero relative to the disc.

The -ve sign of the radial acceleration signifies that it points radially inward; that is what we call centripetal acceleration, which is caused by the radially inward static friction f_s, spring force or resultant of both forces.

3. You should not mis-interpret that the block accelerates radially as its radial velocity increases relative to the ground by the argument

$$a_r = \frac{d^2 r}{dt^2} = \frac{v_r\, dv}{dr}_r \neq 0$$

This is because, in the ground frame, neither $\frac{d^2 r}{dt^2}$ nor $-r\omega^2$ can be called radial acceleration so long as the block slides along the groove. Both the terms $\frac{d^2 r}{dt^2}$ and $-r\omega^2$ contribute to radial acceleration a_r in a polar-coordinate system, which is given as

$$a_r = \left(\frac{d^2 r}{dt^2} - r\omega^2 \right)$$

However, relative to the disc, the radial acceleration can be given as

$$a_r = \frac{d^2 r}{dt^2}$$

4. Relative to the ground, so long as the block does not slip, the path is circular and when it slips, the path will be spiral.
5. So long as the block slips, $v_r \neq 0$; so, the horizontal transverse reaction $F_\theta = 2mv_r\omega$ will appear. This is a pseudo force known as Coriolis force. From the ground frame, you can show that the 'magnitude of v_θ and direction of v_r' change in $+ \theta$- direction. The rate of change in magnitude of v_θ is given as

$$\frac{dv_\theta}{dt} = \frac{d(r\omega)}{dt} = \omega \frac{dr}{dt} = \omega v_r$$

The rate of change in direction of v_r is equal to ωv_r. Since both the accelerations have $+\theta$-direction, adding them, we have the net transverse acceleration of the block which is given as

$$a_\theta = v_r\omega + v_r\omega = 2v_r\omega$$

In order to provide this transverse acceleration, some real force, that is, the transverse horizontal reaction offered by the groove will come into play. So, we can write

$$\Rightarrow F_\theta = ma_\theta = m(2v_r\omega) = 2mv_r\omega$$

If the block moves in a circle, the block does not slip, $v_r = 0$; so, the horizontal transverse reaction $F_\theta = 2mv_r\omega = 0$.

6. So long as the block slips, $v_r \neq 0$; so, the from the ground frame, you can show that the 'direction of v_θ and magnitude of v_r' change in the $+r$-direction. The rate of change in direction of v_θ is equal to ωv_r in the $-r$-direction. The rate of change in magnitude of v_r is

$$\frac{dv_r}{dt} = \frac{d(dr/dt)}{dt} = \frac{d^2r}{dt^2}$$

in the $+r$-direction. So, after adding them, we have the net radial acceleration of the block, which is given as

$$a_r = \left(\frac{d^2r}{dt^2} - r\omega^2\right)$$

So, the net radial force (sum of real forces relative to the ground) is given as

$$F_r = ma_r = m\left(\frac{d^2r}{dt^2} - r\omega^2\right)$$

If the block moves in a circle, the block does not slip, $v_r = 0$; so, the horizontal transverse reaction $F_\theta = 2mv_r\omega = 0$.

Problem 20 A horizontal disc rotating with a constant angular velocity ω has a smooth diametrical groove. (a) When a bead slides along the groove, (i) write the differential equation for the bead relative to the ground. Solve for the differential equation, (ii) equation of motion of the bead relative of ground, (iii) nature of the path followed by the bead relative to ground, and (iv) horizontal transverse reaction. (b) If a kinetic friction is present between the bead and groove so that the bead slides radially away along the groove, write the differential equation of motion of the bead. (c) If a static friction is present between the bead and groove, find the angular velocity of the disc. Assume that the walls of the groove are smooth and the base of the groove is frictional. Assume that μ_k = coefficient of kinetic friction and μ_s = coefficient of static friction and r_0 = initial radial distance of the bead.
Solution

(a)
    (i) *Free-body diagram:*
        The forces acting on the bead are N (up), mg (down), transverse horizontal reaction F_θ

As the groove is smooth, no force acts on the bead along the groove; $F_r = 0$.

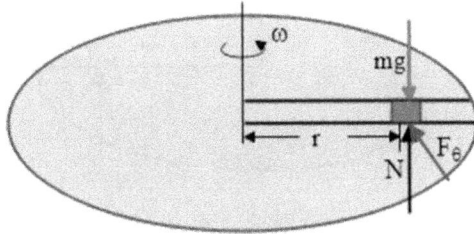

Force equation:

The net radial force is

$$F_r = ma_r$$

Putting $F_r = 0$, the radial acceleration is zero. This means that the bead will not accelerate along the radial line.

Kinematics:

The radial acceleration in the ground frame is

$$a_r = \left(\frac{d^2r}{dt^2} - r\omega^2 \right) = 0,$$

$$\Rightarrow \frac{d^2r}{dt^2} = r\omega^2 \text{ Ans.}$$

(ii) The solution of this differential equation is

$$r = r_0(e^{-\omega t} + e^{+\omega t}) \text{ Ans.}$$

(iii) Putting $\omega t = \theta$, we have

$$r = r_0(e^{-\theta} + e^{\theta})$$

It is a spiral. So, the bead will move in a spiral path.

(iv) The radial velocity is

$$v_r = \frac{dr}{dt} = -r_0\omega(e^{-\omega t} - e^{+\omega t})$$

So, the transverse reaction is

$$F_\theta = 2mv_r\omega = -2mr_0\omega^2(e^{-\omega t} - e^{+\omega t}) \text{ Ans.}$$

(b) Putting the value of kinetic friction which is contributed by the normal reaction N because the walls of the groove are assumed to be smooth. The net radial force is

$$F_r = -f_k = -\mu_k N = -\mu_k mg$$

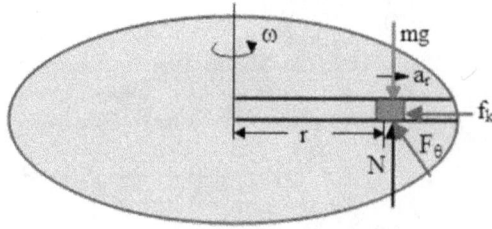

The radial acceleration is

$$a_r = \frac{d^2r}{dt^2} - r\omega^2$$

Force equation:

$$F_r = ma_r$$

Using the last three equations, we have

$$-\mu_k mg = m\left(\frac{d^2r}{dt^2} - r\omega^2\right)$$

$$\Rightarrow \frac{d^2r}{dt^2} - r\omega^2 = -\mu_k g \quad \text{Ans.}$$

(c) As static friction is present, the block does not slip relative to the disc; so, the path of the block is circular in the ground frame; since the radial distance remains constant in circular motion,

$$\frac{d^2r}{dt^2} = 0 \qquad (6.167)$$

So, the acceleration of the bead will be centripetal which is given as

$$a_r = -r\omega^2$$

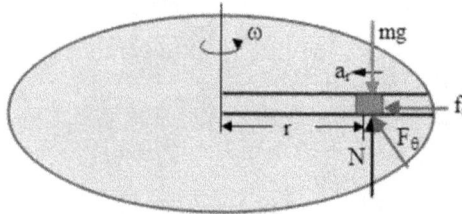

The net radial force is

$$F_r = -f_s$$

The static friction is

$$f_s \leqslant \mu_s N$$

we have the final equation

$$r\omega^2 \leqslant \mu_s g$$

$$\Rightarrow \omega \leqslant \sqrt{\mu_s g / r} \quad \text{Ans.}$$

N.B.: Generally, we think that, since, v_r changes, $\frac{dv_r}{dt} = \frac{d^2r}{dt^2} \neq 0$. So, the acceleration $a_r = \frac{d^2r}{dt^2} \neq 0$, but this is not true, because $a_r = \left(\frac{d^2r}{dt^2} - r\omega^2 \right)$ in polar form, not necessarily equal to either $\frac{d^2r}{dt^2}$ or $r\omega^2$, when $r \neq$ constant. If we introduce static friction, we have $r =$ constant. So, $\frac{d^2r}{dt^2} = 0$ and $a_r = -r\omega^2$. In consequence, the particle moves in a circle, If there is no radial force (smooth surface), the radial acceleration is zero; $a_r = 0$. If f_k is present or any radially inward force is there, i.e., spring force $F_{sp}(=kx)$, kinetic friction f_k, we can write $\sum \vec{F} = m\vec{a}$, where $a = \left(\frac{d^2r}{dt^2} - r\omega^2 \right)$ and $\vec{F_r} = \vec{F_k} + \vec{F_{sp}}$.

Problem 21 At the pole the normal reaction of a body is η times greater than that at the equator. (a) Find the angular velocity of rotation of earth of radius R. (b) Find the minimum velocity relative to Earth with which a projectile must be fired so that it will never return to Earth's surface?
Solution

(a) At the pole P, the centripetal acceleration is zero as the radius of the circle of revolution is zero. So, the net force is $mg - N_1 = 0$, or, $N_1 = mg$. At the equator Q, the centripetal acceleration is zero as the radius of the circle of revolution is zero. So, the net force is

$$mg - N_2 = ma_r = mR\omega^2$$

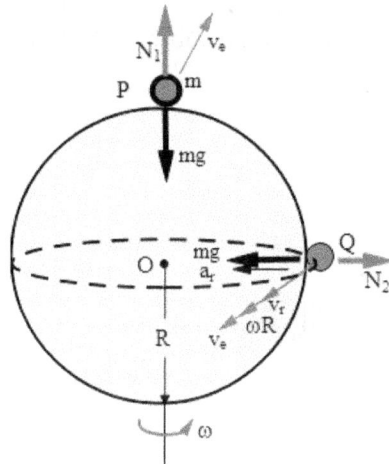

This gives

$$N_2 = mg - mR\omega^2$$

Since $N_1 = \eta N_2$
By putting the values of N_1 and N_2, we have

$$mg = \eta(mg - mR\omega^2)$$

$$\Rightarrow \omega = \sqrt{(\eta - 1)g/R} \quad \text{Ans.}$$

(b) The velocity of projection relative to Earth is minimum only when we fire the projectile in the direction of rotation of Earth. Then the velocity of the projectile is equal to the escape velocity given as

$$v_e = R\omega + v_r.$$

This gives, $v_r = v_e - R\omega$
By putting the obtained value of ω, we have

$$v_r = v_e - \sqrt{(\eta - 1)g/R}, \quad \text{where } v_e = \sqrt{2gR}$$

Then, we have

$$v_r = \sqrt{2gR} - R\sqrt{(\eta - 1)g/R} = \sqrt{2gR}\left\{1 - \sqrt{(\eta - 1)/2}\right\} \quad \text{Ans.}$$

Problem 22 A coin of mass m is resting at the edge of a cylinder of height h and radius R. (a) The cylinder starts spinning at $t = 0$ with a about its vertical axis as shown in the figure. (a) Find the angular velocity of rotation of the cylinder of radius R. (b) If the cylinder is slowly increasing its angular speed, after some time the coin will leave the disc and move in a parabolic path under gravity as a projectile. If the coin strikes at the edge of the smooth disc, find the (i) radius of the disc, (ii) speed of striking of the coin with the disc, (iii) angle made by the velocity at the time of striking of the coin.

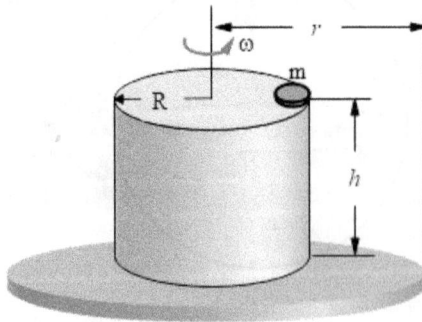

Solution

(a) The coin will slip when friction will be limiting given as fs_{max}. Then, the maximum acceleration of the coin will be $a = fs_{max}$. As the coin moves in a circle, its radial and tangential accelerations are $a_r = R\omega^2$, $a_t = R\alpha$, respectively. Then the total acceleration is $a = \sqrt{a_t^2 + a_r^2}$. Using the last four equations, the maximum angular velocity of the coin till it begins to slip is

$$\omega\sqrt{\frac{\sqrt{\mu_s^2 g^2 - R^2\alpha^2}}{R}}_{max}$$

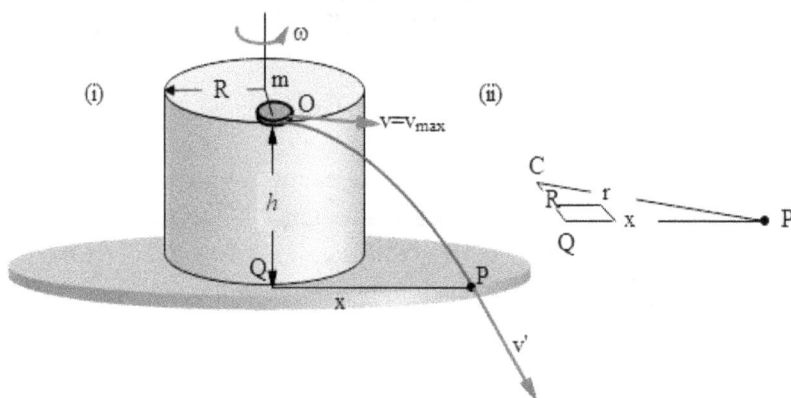

So, the coin will leave the cylinder with a velocity

$$v = v\sqrt{R\sqrt{\mu_s^2 g^2 - R^2\alpha^2}}_{max}$$

The time after which the coin will leave the cylinder is

$$\Rightarrow t = \frac{\omega_{max}}{\alpha\sqrt{(1/R)\sqrt{\mu_s^2 g^2 - R^2\alpha^2}}}$$

The time of free fall is $t' = \sqrt{2h/g}$. So, the time from the beginning of rotation of the disc after which the coin will strike the ground is

$$T = t + t' = \sqrt{2h/g} + \sqrt{(1/R)\sqrt{\mu_s^2 g^2 - R^2\alpha^2}}/\alpha \text{ Ans.}$$

(b)

   (i) If the disc is accelerating slowly, we can neglect its angular acceleration; putting $\alpha = 0$, we have $v = \sqrt{\mu_s gR}$. The horizontal distance covered by the coin is

$$x = QP = vt = \left(\sqrt{\mu_s gR}\right)\left(\sqrt{2h/g}\right) = \sqrt{2\mu_s hR}$$

As the coin strikes at P at the edge of the disc of radius R,

$$CP = r = \sqrt{x^2 + R^2}$$

Putting the value of x, we have

$$r = \sqrt{x^2 + R^2} = \sqrt{2\mu_s hR + R^2} = \sqrt{(2\mu_s h + R)R} \quad \text{Ans.}$$

(ii) As the coin strikes at P at the edge of the disc of radius,

$$v' = \sqrt{\mu_s gR + 2gh}\sqrt{g(\mu_s R + 2h)} \quad \text{Ans.}$$

(iii) The angle made of the velocity with horizontal at the time of striking at P is

$$\theta = \tan^{-1}\frac{v_y}{v_x} = \tan^{-1}\frac{\sqrt{2gh}}{\sqrt{\mu_s gR}} = \tan^{-1}\sqrt{\frac{2h}{\mu_s R}} \quad \text{Ans.}$$

IOP Publishing

Problems and Solutions in Particle Mechanics

Pradeep Kumar Sharma

Chapter 7

Work, energy, power

7.1 Introduction

This chapter deals with three fundamental concepts, namely work, energy and power. They are defined and interpreted by the concept of force, displacement and time. By using Newton's second law and kinematics, we can develop the work–energy theorem that relates *work* with *energy*.

Galileo defined work as the quantity of action. René Descartes and Gottfried Leibniz defined the work as the energy spent in lifting an object through a certain height. This means that work involves a displacement of an object. While work is done, energy is transferred from the *doer* (agent). John Smeaton and Gaspard-Gustave de Coriolis defined power as the rate of doing work.

Energy is the ability of doing work. It can be *potential* and *kinetic*. The term potential energy helps us in defining the conservative forces and distinguishing them from non-conservative forces. Furthermore, the concept of energy forms the basis of explaining the principle of a conservative system.

In science and technology, conservation of energy (energy balance) plays a significant role. While designing a machine (motor, generator, automobiles, etc), the engineer and R&D scientists must account for all sorts of energy transformations happening within the system. The loss of mechanical energy is accountable for efficiency of the system.

The concept of *power* teaches us why it is easier to do any mechanical work taking more time rather than doing the mechanical work rapidly (in less time). The familiar use of this idea is that the hilly roads are made *zig-zag* rather than straight in order to increase the time of climbing the hill; so, the climber (automobile, human being or animals) needs less power to climb the hill. Power supplied is the input power; the power consumed is the output power) and power is lost in the form of heat, light and sound. These are the key factors for the world of science and technology and our daily lives.

doi:10.1088/978-0-7503-6442-3ch7

7.2 Concept of energy and transfer of energy; mechanical work

7.2.1 Energy

You can push or pull an object by the virtue of an invisible agent called *energy* denoted as *E*. So, energy is the ability of doing any work. If the energy comes from the motion of an object we call it kinetic energy, denoted as *K*. If the energy of a system of particles comes solely from the arrangement or configuration of the particles, it is called potential energy, denoted as *U*.

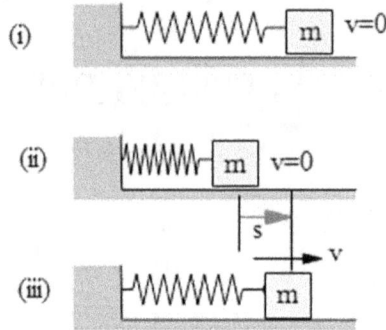

An undeformed spring in fig-(i) cannot push the block;so,it cannot generate kinetic energy; the compressed spring in fig-(ii) can push and displace the block and impart kinetic energy to it.

To understand the concept of kinetic energy, let us take an example of a moving object. If you try to stop a moving object, you will get a push from the object. This pushing ability of the object comes from its motion. Hence, a moving ball possesses a kinetic energy by the virtue of its motion.

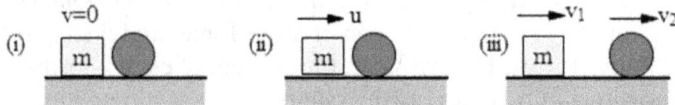

A stationary block of mass *m* in figure (i) cannot push or move the ball because it does not have kinetic energy; but the moving block (block having a speed v) in figure(ii) can displace the ball coming in its way as shown in figure (iii).So, the moving block possesses a kinetic energy.

Let us take another example of a deformed spring. A compressed spring can push you; an elongated spring can pull you; so, a deformed spring gets the energy (pushing or pulling ability) from its deformation (elongation or compression). The deformation of a spring or any object is possible by the deformation of the molecules and atoms of the object. This is known as elasticity of matter. So, the energy stored in the body, that is, deformation energy is called elastic potential energy. We have different types of potential energies, i.e., gravitational, electrostatic, magnetic, etc.

7.2.2 Work

To feed the energy to a system (ball or spring), you need to deliver a speed to the ball or deform the spring. While doing so, you need to apply a force (either push or pull them). The point of an object at which you are pushing or pulling is called *point of application of force F* (*P*, say) that undergoes a displacement *x*, say. This process (or act) is called work done, denoted as '*W*'. During this process, you get exhausted because you lose energy from your body and this energy goes to the system (ball or spring). This energy transfer causes a displacement of the point *P*. In other words, energy transfer takes place from surrounding (or external agent) to the system. Thus, mechanical work is the process of energy transfer.

When you push or pull a system, you supply energy to the system. As a result, the total energy ($U + K = E$) of the system increases. Then, work is done on the system.

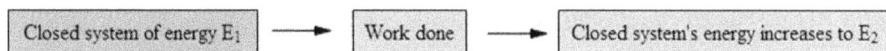

When work is done on a closed system, its energy increases from E_1 to E_2

Work is a process of energy transfer when a *force* is applied to a system. This is accomplished due to the increase in energy (kinetic or potential or both) of the system. Using the idea of work and energy, let us derive their expressions and the relation between work, *KE* and potential energies.

7.3 Relation between work and kinetic energy

Let us pull a smooth box with a constant force *F* on a horizontal surface. The tangential (horizontal) component of the force, that is, $F \cos \theta$ gives an acceleration,

$$a = \frac{F \cos \theta}{m} \tag{7.1}$$

As a result, the box speeds up from v_0 to v with a constant acceleration *a*. Then, by using kinematics of uniforms accelerated motion we can write

$$v^2 - v_0^2 = 2as \tag{7.2}$$

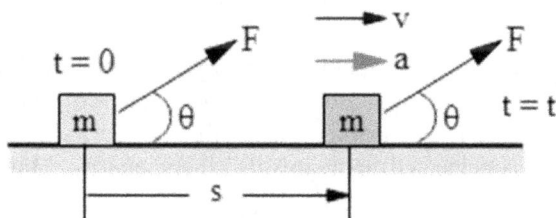

When a force *F* acts on the block, it undergoes
a displacement *s* a work is done is $W = Fs \cos\theta$

Using these equations and rearranging the terms, we get,

$$Fs \cos \theta = \frac{1}{2}mv^2 - \frac{1}{2}mv_0^2 \tag{7.3}$$

7.3.1 Formula of work

Let us look at the two sides carefully. The left side term '$Fs \cos \theta$' tells us about the 'process of pulling the box by a force F through a distance s' which is defined as 'work done by the force F denoted by W_F'; so, we can write

$$W_F = Fs \cos \theta = \vec{F} \cdot \vec{s},$$

where $\vec{s} =$ displacement of the point of application of the force $\vec{F}$ and $\theta =$ angle between force and displacement. This tells us that

The work done by a constant force is equal to the scalar (dot) product of the force and the net displacement. It can be +ve, 0 and −ve.

7.3.2 Formula of kinetic energy

As explained earlier, in the process of pulling the box, energy transfer takes place from your body to the box. So, the box gains energy as given by the right-side terms of equation (7.3). Since this energy comes from the motion of the box, it is called *kinetic energy* (represented by *KE* or *K*). As the *KE* of the box increases from $\frac{1}{2}mv_0^2$ to $\frac{1}{2}mv^2$, the kinetic energy of the box or any particle mass m moving with a speed v is gives as

$$K = \frac{1}{2}mv^2.$$

7.3.3 Work–kinetic energy theorem

As the box speeds up by the application of force F while undergoing a displacement s, the term

$$\frac{1}{2}mv^2 - \frac{1}{2}mv_0^2$$

is known as the change (gain in its kinetic energy) denoted by ΔK.

So, you can say that *KE* of the box increases while work is done by the force (applied by you) on the box. In other words, the net work done on a system (box whose deformation is neglected) increases its kinetic energy. Mathematically,

$$W = \Delta K.$$

This tells us that the net work done by all forces is numerically equal to the change in kinetic energy of the particle. This is known as the work–kinetic energy theorem.

Example 1 A particle undergoes a displacement from a point $P(1, 1, -1)$ m to $Q(-3, -1, 2)$ m by the action of a force $\vec{F} = (2\hat{i} + 3\hat{j} - 4\hat{k})$ N. Find the work done by the force in the corresponding displacement.

Solution

The work done is

$$W = \vec{F} \cdot \vec{s} = \vec{F} \cdot (\vec{r_2} - \vec{r_1})$$
$$= (2\hat{i} + 3\hat{j} - 4\hat{k}) \cdot [(-3\hat{i} - \hat{j} + 2\hat{k}) - (\hat{i} + \hat{j} - \hat{k})]$$
$$= -(2\hat{i} + 3\hat{j} - 4\hat{k}) \cdot (4\hat{i} + 2\hat{j} - 3\hat{k})$$
$$= -8 + 6 + 12 = -26 \text{ J Ans.}$$

Example 2 If you push a smooth box of mass 2 kg by a force of 5 N through a distance of 5 m from rest, so that the maximum work is done. Find the
   (a) work done,
   (b) speed of the box,
   (c) energy delivered to the box,
   (d) increase in KE of the box,
   (e) work done by the normal reaction and gravity.

Solution

   (a) The work done by you on the box is maximum when you push the box horizontally

$$\text{So, } W = Fs \cos 0° = Fs$$
$$= (5 \text{ N})(5 \text{ m}) = 25 \text{ N m} = 25 \text{ J Ans.}$$

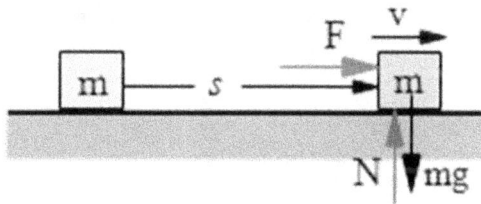

   (b) The increase in KE be $\Delta K = \frac{1}{2}mv^2$ ($\because v_0 = 0$)
      We know that the total or net work done is

$$W = \Delta K = \frac{1}{2}mv^2$$

   By putting $W = 25$ J, $m = 2$ kg, we have $v = 5$ m s^{-1}. Ans.
   (c) The energy delivered to the box = work done on the box = 25 J. Ans.
   (d) The increase in KE of the box = work done on the box = 25 J. Ans.
   (e) Since, $\vec{N} \perp \vec{s}$ and $m\vec{g} \perp \vec{s}$, they do not perform work. It means, work done is zero due to the forces N and mg. Ans.

7.4 Work done by a variable force

You have learnt that the work done by a force depends on the following factors:

  (i) magnitude of force ($|\vec{F}|$),
  (ii) direction of the force (θ),
  (iii) displacement of the point of application of the force.

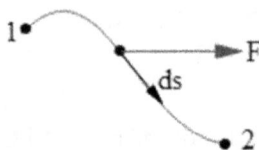

The work done by a variable force F from
the point 1 to 2 is given as $W = \int_{0} \vec{F}.d\vec{s}$

Let us consider a variable force $\vec{F}$ (magnitude or direction or both can change). It can perform work dW for a small displacement ds, given as

$$dW = \vec{F} \cdot d\vec{s}.$$

By integrating both sides, the total work done W by the force $\vec{F}$ from positions 1 to 2 can be given as,

$$W = \int dW$$

$$\Rightarrow W = \int \vec{f} \cdot d\vec{s}$$

In one-dimensional motion, the work done by a variable force can be given as

$$W = \int F ds,$$

where $\int F ds$ = Area under F–s graph.

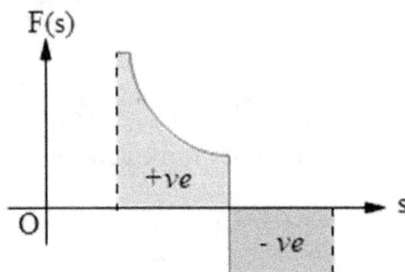

The work done = Area under under F-s
graph = Positive area -Negative area.

This tells us that:

  Area under force–displacement graph gives us work done by that force. If the area lies above the s-axis, W is +ve; if the area lies below the s-axis, W is −ve.

Example 3 Work is done on a system under the action of a variable force F as shown in a F–s graph, where $s = $ displacement of the point of application of force.
(a) Find the work done on the system.
(b) Justify the result.
(c) How much does energy of the system change?

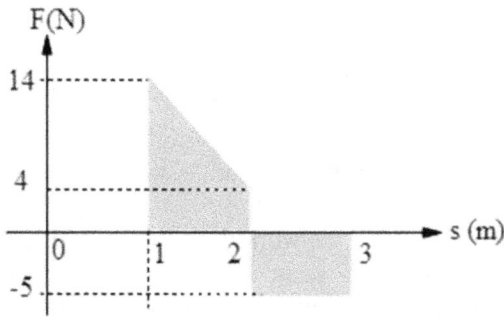

Solution

(a) The area under the F–s graph is an algebraic sum of positive and negative areas given as

$$A = \text{Positive area} + \text{Negative area}$$
$$= \tfrac{1}{2} \times (14 + 4) \times 1 + (-5)(1) = 4 \, \text{N m Ans.}$$

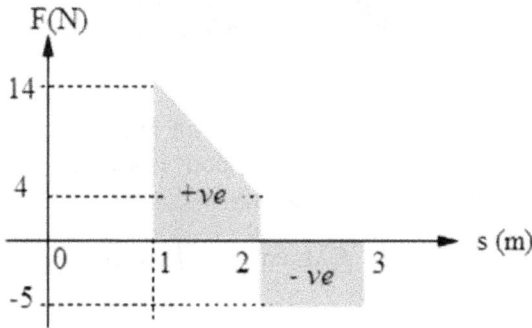

(b) Positive work done on a system means that work is done on the system by the surroundings. Ans.
(c) The energy of the system increases by 4 J because the surroundings do work on the system. Ans.

Example 4 (Work done by a spring)
Derive an expression for work done by a spring.

Solution

When you pull one end of a spring by a distance x, the spring will pull you back by a force kx. If you pull the spring by a small distance dx, the work done by the spring is

$$dW = \vec{F_s} \cdot \vec{dx} = (=k\,x)dx$$

If we pull the spring from its initial deformation x_0 to final deformation x, the total work done by the spring is

$$W = -\int_{x_0}^{x} k\,x\,dx$$

$$\Rightarrow W = -\frac{k}{2}(x^2 - x_0^2)$$

The above expression tells us that the work done by an ideal spring does not depend upon the types of motion of its ends; it depends only on the initial and final deformation of the spring. If the spring is initially undeformed, $x_0 = 0$, the work done by the spring in either compressing or elongating through a distance x is given as

$$W = -\frac{1}{2}kx^2.$$

When one end of the spring is fixed, the work done by the spring does not depend upon the nature of the path followed by the free end of the spring; it only depends upon the initial and final deformation of the spring.

Example 5 (Work done by gravity near Earth's surface)

Find the work done by gravity when we shift a point mass near Earth's surface.

Solution

Method 1

Near Earth's surface, the weight of a body of mass m is vertically downward having a constant magnitude $F_g = mg$. If the body is lowered through an additional vertical distance y, the work done by gravity is

$$W_{gr} = \int_0^s \vec{F_{gr}} \cdot \vec{ds} = \int_0^s F_{gr}ds\,\cos\theta = \int_0^y mgdy = =mg \int_0^y dy$$

$$\Rightarrow W_{gr} = mgy = mg(h_1 - h_2)$$

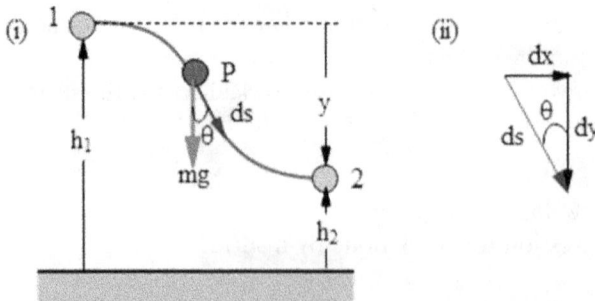

Method 2

The work done due to gravity on the body is

$$W_{gr} = \int_0^s m\vec{g} \cdot \vec{ds} = \int_0^s (-mg\hat{j}) \cdot (dx\hat{i} + dy\hat{j}) = \int_0^y mg\,dy = =mg \int_0^y dy = mgy.$$

Method 2

Since the gravity (weight) of the body is a constant near earth's surface, the work done due to gravity on the body is

$$W_{gr} = m\vec{g} \cdot \vec{s} = (-mg\hat{j}) \cdot (x\hat{i} - y\hat{j}) = mgy.$$

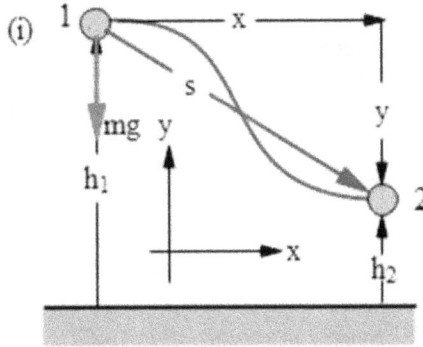

N.B.: If we lower the body, $W_{gr} = +mgy$; so, work done by gravity near Earth's surface is given as

$$W_{gr} = \pm mgy,$$

where y = vertical displacement; work done by gravity W is −ve when the body is lifted up, and the work done by gravity is +ve when the body is lowered.

7.5 Work done in a moving frame

According to Galilean transformations of velocities, an object seems to move with different velocities relative to different reference frames, when the reference frames move relative to each other. This means the same object undergoes different displacements when viewed from two reference frames (observers). Let observers 1 and 2 and the block m be at the same position at $t = 0$ and let them measure the displacements $\vec{s_1}$ and $\vec{s_2}$ of an object after time 't'. Then, the work done by a force $\vec{F}$ relative to observer 1 in time 't' is

$$W_1 = \vec{F} \cdot \vec{s_1}$$

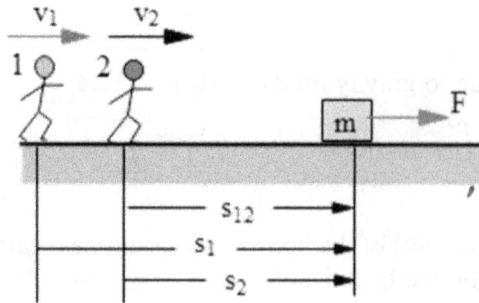

The work done by a force F measured by two different rference frames (observers 1 and 2) are different; $W_1 \neq W_2$

The work done by a force $\vec{F}$ relative to observer 2 in time 't' is

$$W_2 = \vec{F} \cdot \vec{s_2}$$

So, the difference in work performed by the force is

$$W_1 - W_2 = \vec{F} \cdot (\vec{s_1} - \vec{s_2}) = \vec{F} \cdot \vec{s_{21}},$$

where $\vec{s_{21}}$ = relative displacements of the observer (RF) 2 relative to observer (RF) 1 in time t. Since the relative displacement $\vec{s_{21}} \neq \vec{0}$, we can write $W_1 \neq W_2$. This tells us that:

The same force can perform different works in different reference frames.

Example 6 A man is pulling a box with a force of 10 N at an angle $\theta = 37°$ with the horizontal. The box is kept on a trolley car which moves with a velocity of 4 m s^{-1}. The man moves relative to the trolley car with a velocity of 2 m s^{-1}. Find the work done by the man (i) relative to the ground and (ii) relative to the trolley car during $t = 3$ s.

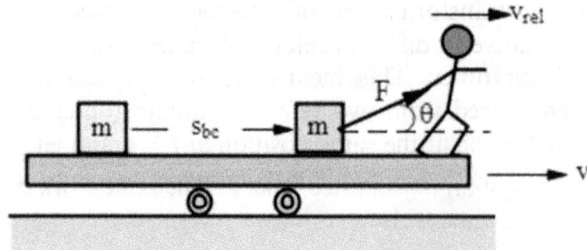

Solution

(i) The displacement of the block relative to the ground is

$$\vec{s}_b = \vec{s}_{bc} + \vec{s}_c$$
$$= \vec{v}_{bc}t + \vec{v}_c t$$
$$= 2 \times 3\hat{i} + 4 \times 3\hat{i} = 18\hat{i}$$

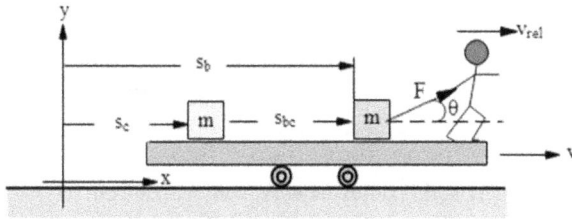

The force applied by the man on the block is

$$\vec{F} = 10\cos 37°\hat{i} + 10\sin 37°\hat{j} = 8\hat{i} + 6\hat{j} \text{ N}$$

Then, the work done by $\vec{F}$ is

$$W = \vec{F} \cdot \vec{s} = (8\hat{i} + 6\hat{j}) \cdot (18\hat{i}) = 144 \text{ J Ans.}$$

(ii) The work done relative to the trolley car is

$$W' = Fs_{bc}\cos\theta$$
$$= Fv_{bc}t\cos\theta = (10)(2)(3)\left(\tfrac{4}{5}\right) = 48 \text{ J Ans.}$$

7.6 Relation between work and potential energy

7.6.1 Work–potential energy theorem

Let us pull the free end of an ideal spring slowly while its other end is fixed. In doing so, positive work is done on the spring and the total energy of the spring increases. Since the mass is negligible for an ideal spring, its kinetic energy is ignored. So, the total energy is almost potential. In other words, how much external work is done on the spring, that much energy will be stored in the spring; so, the work done on the spring is numerically equal to the increase in the elastic potential energy of the spring. If U_0 is the initial potential energy of the spring and you are doing work W_{ext} on the spring, the potential energy of the spring will increase by W; so, the final potential energy of the spring is given as

$$U = U_0 + W_{ext}, \tag{7.4}$$

where U and U_0 are the potential energy of the spring at $x = x_0$ and $x = x$. As you need to balance the spring force while pulling the end of the spring slowly,

$$F_{\text{ext}} = F = -F_{\text{spring}} \qquad (7.5)$$

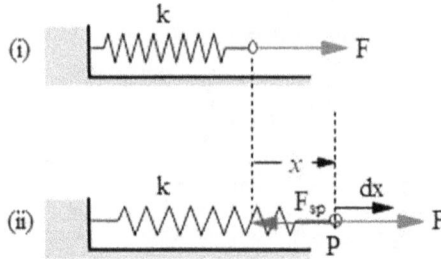

The spring potential energy increases from U_0 to U as an external work is done on it.

So, the spring does equal work on you;

$$W_{\text{ext}} = -W_{\text{spring}} \qquad (7.6)$$

Then using equations (7.4) and (7.6), we can write

$$U_0 - W_{\text{spring}} = U \qquad (7.7)$$

$$\Rightarrow W_{\text{spring}} = -(U - U_0) = -\Delta U$$

This tells us that the spring loses its potential energy while doing positive work (work on the surroundings). The work done by the spring, therefore, can be equated with the loss of its potential energy because work is a process of energy transfer.

In general,

The work done by a conservative force is equal to the loss (negative of change) of the potential energy of the system; mathematically, $W_{cons} = -\Delta U$. This means, potential energy decreases when W_{cons} is positive.

Example 7 Derive potential energy from force.
  Solution
  If a conservative force F acts at a point mass P, it does work

$$W = \int_0^r \vec{F} \cdot \vec{d}\,r$$

when the point moves from positions 1 to 2.

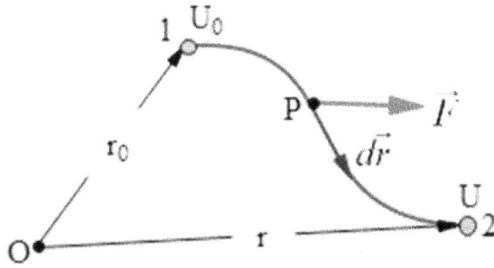

In this process the potential energy of the field changes from U_0 to U.
According to the work–potential energy (W–PE) theorem,

$$W = -(U - U_0)$$

$$\Rightarrow \int_{r_0}^{r} \vec{F} \cdot \vec{dr} = -U + U_0$$

$$\Rightarrow U = U_0 - \int_{r_0}^{r} \vec{F} \cdot \vec{dr}$$

Example 8 Find spring potential energy if its deformation is x.

Solution

Setting the origin at the unstretched position, the work done by the spring for the deformation x is

$$W = -\int_{0}^{x} F dx = -\int_{0}^{x} kx \, dx = \frac{-kx^2}{2}$$

As the potential energy changes from $U_0 = 0$ (at $x = 0$) to U (at $x = x$), the change in potential energy is

$$\Delta U(=-U) = \frac{-kx^2}{2}$$

Then we have $U = \frac{1}{2}kx^2$.

Example 9 If you lift an object of mass m to a height h, prove that its gravitational potential energy is given as $U = mgh$, where $g = 9.8 \text{ m s}^{-2}$. Discuss the necessary assumptions.

Solution

The formula $U = mgh$ is based on three assumptions:

(i) The height h must be negligible compared to the radius of Earth.
(ii) The reference level is Earth's surface.
(iii) The potential energy of the body at Earth's surface is zero; $U_0 = 0$.

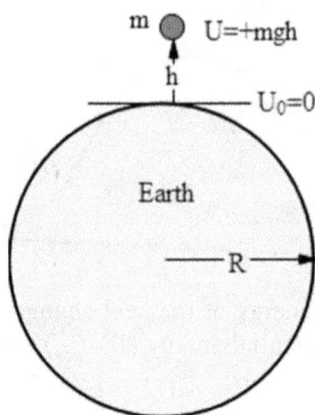

Since $h \ll R$, the gravitational force is taken constant (or uniform) near Earth's surface. Then, the work done by gravity in lifting the body from Earth's surface to a height h is

$$W_{gr} = \vec{F} \cdot \vec{s} = -Fs$$

Putting $F = mg$ and $s = h$, we have

$$W_{gr} = -(mg)(h) = -mgh$$

According to the work–potential energy theorem,

$$W_{gr} = -(U - U_0)$$

Putting $U_0 = 0$, we have $U = -mgh$. Ans.

    N.B.:

      1. $U = mgh$ is an approximated result.

      2. '*mgh*' is the potential energy of the Earth–body combination if we assume $U_0 = 0$.

      3. The accurate formula for *potential energy* of interaction between Earth and a body at a height h can be given by

$$U - U_0 = +\frac{mgh}{1 + \frac{h}{R}}; \quad \text{if } h \ll R.$$

$$U \simeq mgh + U_0 \cong mgh, \quad \text{where } U_0 = 0$$

Example 10 A body of mass $m = 10$ kg is moving from a point $P = (-2\ m, -3\ m, 5\ m)$ to another point $Q = (4\ m, 10\ m, -20\ m)$. Find the work done by gravity.

Solution

The work done by gravity is

$$W_{gr} = - mg(y_2 - y_1)$$
$$= - 10 \times 9.8\{(10) - (-3)\} = 1274 \text{ J Ans.}$$

Example 11 The work done by an external agent in pulling a spring from a deformation of 10 cm to 15 cm is W_1. When pulling the spring from a deformation of 15 cm to 20 cm, the work done is W_2. Find $\frac{W_1}{W_2}$.

Solution

The work done by a spring is

$$W_1 = \frac{1}{2}k(x_2^2 - x_1^2)$$

Putting $x_2 = 15$ and $x_1 = 10$ cm, we have

$$W_1 = \frac{1}{2}k[(15)^2 - (10)^2] = \frac{125}{2}k \tag{7.8}$$

Putting $x_2 = 20$ and $x_1 = 15$ cm, we have

$$W_2 = \frac{1}{2}k[(20)^2 - (15)^2] = \frac{175k}{2} \tag{7.9}$$

From equations (7.8) and (7.9),

$$\frac{W_2}{W_1} = \frac{(175/2)k}{(125/2)k} = \frac{7}{5} \text{ Ans.}$$

Example 12 A spring is elongated from $x = x_1$ to $x = x_2$, the work done is W_1; when elongated from $x = x_2$ to $x = x_3$, the work done is $W_2 = \eta\, W_1$. Find x_2 in terms of x_1 and x_3.

Solution

When elongated from $x = x_1$ to $x = x_2$, the work done by a spring is

$$W_1 = \frac{1}{2}k(x_2^2 - x_1^2) \tag{7.10}$$

When elongated from $x = x_1$ to $x = x_2$, the work done by a spring is

$$W_2 = \frac{1}{2}k(x_3^2 - x_2^2) \tag{7.11}$$

It is given that

$$\frac{W_2}{W_1} = \eta \tag{7.12}$$

From equations (7.10), (7.11) and (7.12),

$$\frac{W_2}{W_1} = \frac{\frac{1}{2}k(x_3^2 - x_2^2)}{\frac{1}{2}k(x_2^2 - x_1^2)} = \frac{(x_3^2 - x_2^2)}{(x_2^2 - x_1^2)}$$

$$\Rightarrow \frac{W_2}{W_1} = \frac{(x_3^2 - x_2^2)}{(x_2^2 - x_1^2)} = \eta$$

$$\Rightarrow (x_3^2 - x_2^2) = \eta(x_2^2 - x_1^2)$$

$$\Rightarrow x_2 = \sqrt{\frac{\eta x_1^2 + x_3^2}{\eta + 1}} \quad \text{Ans.}$$

Example 13 A smooth block of mass m moves up from the bottom to top of a wedge while the wedge is moving with an acceleration a_0. Find the work done by the pseudo-force measured by the person sitting at the edge of the wedge.

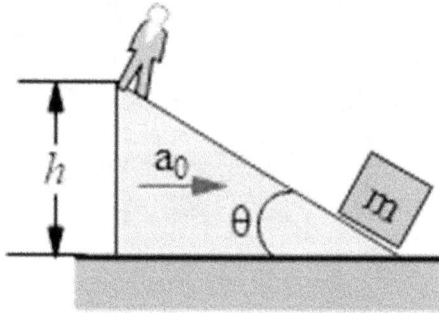

Solution

Since $\vec{F}_{ps} = -m\vec{a}_0$ is a constant force,

$$W_{ps} = \vec{F}_{ps} \cdot \vec{s}$$

$$= -m\vec{a}_0 \cdot \vec{s}$$

$$= -ma(a_0\hat{i}) \cdot (l\hat{i} + b\hat{j}) = ma_0l \quad \text{Ans.}$$

N.B.: If the person moves relative to the wedge with an acceleration a say, will you get the above answer? Ans: No.

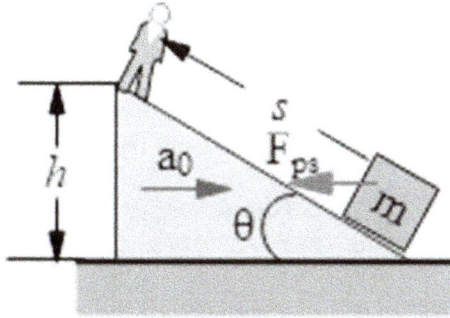

Example 14 Three forces $\vec{F_1} = (-\hat{i} + \hat{j})\,\text{N}$, $\vec{F_2} = (2\hat{j} - \hat{k})\,\text{N}$ and $\vec{F_3} = (3\hat{i} - 2\hat{k})\,\text{N}$ act on a particle which moves from point $P(-2, +1, -3)\,\text{m}$ to $Q(2, -1, 3)\,\text{m}$. Find the net work done.

Solution

The net work done is

$$W_{\text{total}} = \vec{F}_{\text{net}} \cdot \vec{s} \tag{7.13}$$

$$\vec{F}_{\text{net}} = \vec{F_1} + \vec{F_2} + \vec{F_3}$$
$$= (-\hat{i} + \hat{j}) + (2\hat{j} - \hat{k}) + (3\hat{i} - 2\hat{k})$$

$$\Rightarrow \vec{F}_{\text{net}} = 2\hat{i} + 3\hat{j} - 3\hat{k} \tag{7.14}$$

$$\vec{s} = \vec{r_2} - \vec{r_1}$$
$$= (2\hat{i} - \hat{j} + 3\hat{k}) - (-2\hat{i} + \hat{j} - 3\hat{k})$$

$$\Rightarrow \vec{s} = 4\hat{i} - 2\hat{j} + 6\hat{k} \tag{7.15}$$

Substituting $\vec{F}_{\text{net}}$ from equation (7.14) and $\vec{s}$ from equation (7.15) in equation (7.13),

$$W_{\text{total}} = (2\hat{i} + 3\hat{j} - 3\hat{k}) \cdot (4\hat{i} - 2\hat{j} + 6\hat{k})$$
$$= (8 - 6 - 18) = -16\,\text{J}\ \text{Ans.}$$

Example 15 (Projectile motion)

(a) A stone is projected with a speed v_0. Using the W–E theorem, find the speed of the particle as the function of small vertical distance y compared to the radius of Earth.

(b) Apply the formula to find the maximum height attained by the particle if the angle of projection of the stone is θ.

Solution

(a) Let the particle move up through a height y between the points 1 and 2. The work done by gravity in shifting the particle from point 1 to 2 is

$$\Delta K = \frac{1}{2}m(v^2 - v_0^2) \tag{7.16}$$

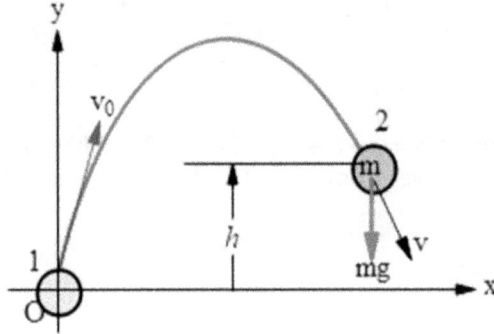

The work done by gravity is

$$W_{gr} = -mgy \tag{7.17}$$

Work–*KE* theorem:

$$W = \Delta K \tag{7.18}$$

Using the last three equations, we have

$$\frac{m}{2}(v^2 - v_0^2) = -mgy$$

$$\Rightarrow v = \sqrt{v_0^2 - 2gy} \quad \text{Ans.}$$

(b) At the maximum height the speed of the stone is

$$v = v_0 \cos \theta \tag{7.19}$$

Putting $v = \sqrt{v_0^2 - 2gy}$ in the last equation, we have

$$v = v_0 \cos \theta = \sqrt{v_0^2 - 2gy_{max}}$$

$$\Rightarrow y_{max} = \frac{v_0^2 \sin^2 \theta}{2g}.$$

N.B.: 1. If point 2 lies below the point 1 of projection, $v = \sqrt{v_0^2 + 2gy}$.
2. Since projectile motion, in general, is a constant accelerated motion, we can apply the formula $v^2 = u^2 + 2\vec{a} \cdot \vec{s}$ to obtain the same answer.

Example 16 A block is released from rest from a height $h = 5$ m. After traveling through the smooth curved surface, it moves on the rough horizontal surface through a length $l = 8$ m and climbs onto the other smooth curved surface through a height h'. If $\mu = 0.5$, find h'.

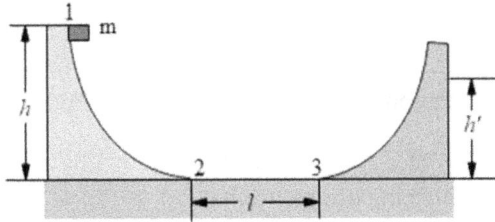

Solution
The forces acting on the block are N, gravity and friction.
$$W_W + W_f + W_{gr} = \Delta K,$$
where, $W_N = 0$, $W_f = -\mu mgl$ and $W_{gr} = mg(h - h')$ and $\Delta K = 0$
$$\Rightarrow -\mu mgl + mg(h - h') = 0$$
$$\Rightarrow h' = h - \mu l = 5 - \frac{1}{2} \times 8 = 1 \text{ m Ans.}$$

N.B: The normal reaction does not perform as N is always perpendicular to elementary displacement ds.

Example 17 (Simple pendulum)
The bob of simple pendulum is projected down initially with a speed v_0, from its (a) horizontal position and (b) lowest position. After swinging through an angle θ, find the speed of the bob.
Solution

(a) There are two forces acting on the bob, i.e., tension T and gravity mg.

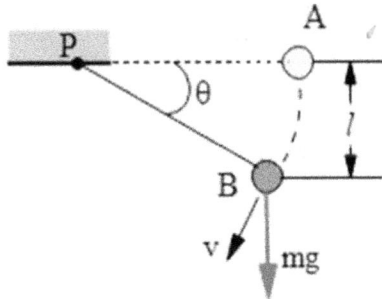

The work done by tension is zero ($W_T = 0$) because the tension is perpendicular to displacement of the bob ($\vec{T} \perp d\vec{s}$). As the bob falls down, the work done by gravity is
$$W_{gr} = mgh$$

Then, the total work is

$$W = W_T + W_{gr} = mgh$$

The change in KE is

$$\Delta K = \frac{1}{2}m(v^2 - v_0^2)$$

Work–energy theorem is

$$W = \Delta K$$

Using the last three equations, we have

$$mgh = \frac{1}{2}m(v^2 - v_0^2)$$

$$\Rightarrow v = \sqrt{v_0^2 + 2gh},$$

where $h = l \sin \theta$

$$\Rightarrow v = \sqrt{v_0^2 + 2gl \sin \theta} \quad \text{Ans.}$$

N.B: If the initial velocity is zero, $v = \sqrt{2gl \sin \theta}$.

(b) The work done by tension is zero $W_T = 0$ because the tension is perpendicular to displacement of the bob ($\vec{T} \perp d\,\vec{s}$). As the bob goes up, the work done by gravity is

$$W_{gr} = -mgh$$

Then, the total work is

$$W = W_T + W_{gr} = mgh$$

The change in KE is

$$\Delta K = \frac{1}{2}m(v^2 - v_0^2)$$

Work–energy theorem is

$$W = \Delta K$$

Using the last three equations, we have

$$-mgh = \frac{1}{2}m(v^2 - v_0^2)$$

$$\Rightarrow v = \sqrt{v_0^2 - 2gh},$$

where $h = l(1 - \cos \theta)$

$$\Rightarrow v = \sqrt{v_0^2 - 2gl(1 - \cos \theta)} \quad \text{Ans.}$$

N.B.: You cannot apply the kinematical equation $v^2 = u^2 + 2as$ to obtain the same answer because the motion of the bob is not uniform accelerated motion.

Example 18 A block of mass m is slowly pulled along a curved surface from position 1 to 2. If the coefficient of kinetic friction between the block and surface is μ, find the work done by the applied force.

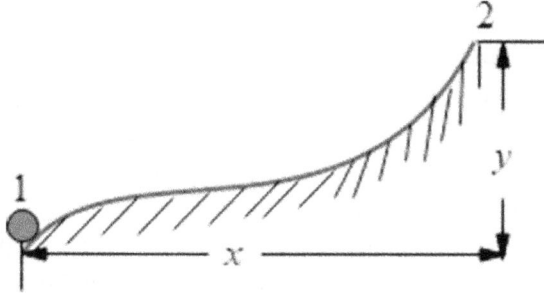

Solution
The forces acting on the block are gravity $mg\downarrow$, $N\nwarrow$, $f_k \nearrow$ and $F\nearrow$.
The W–KE theorem is

$$W = \Delta K$$

$$\Rightarrow W_N + W_{gr} + W_N + W_f = \Delta K \tag{7.20}$$

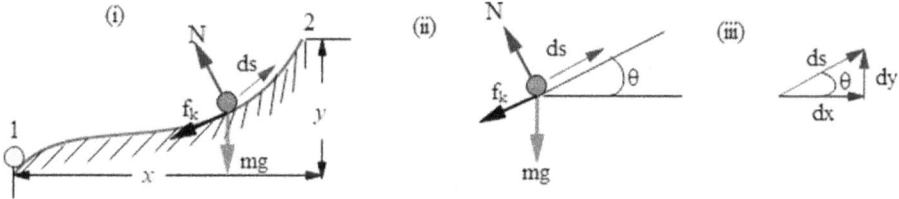

Putting $\Delta K = 0$ (because the body is slowly shifted), $W_N = \vec{N} \cdot d\vec{s} = 0$ $(\because N \perp ds)$ and $W_{gr} = -mgy$ $(\because$ the block moves up) in equation (7.20), we have

$$W_F + W_f = mgy \tag{7.21}$$

The work done by kinetic friction is

$$W_f = \int \vec{F_k} \cdot d\vec{s} = \int f_k \, ds \cos 180°$$
$$= - \int f_k \, ds = - \int \mu N ds$$
$$= - \int \mu(mg \cos \theta) \, ds = - \int_0^x \mu mg \, dx \quad (ds \cos \theta = dx)$$

$$\Rightarrow W_f = -\mu mg \, x \tag{7.22}$$

Using equations (7.21) and (7.22),

$$W_F = mg(\mu x + y) \text{ Ans.}$$

N.B.: The work done by gravity W_{gr} does not depend on x and work done by kinetic friction W_f does not depend on y in this case. If you pull the block with finite constant speed v the same answer will be different because the normal reaction will no longer be equal to $mg \cos \theta$ because the centripetal acceleration is non-zero although the tangential acceleration is zero.

7.7 Conservative and non-conservative forces

7.7.1 Conservative force

If you give a speed v_0 to the block, it moves towards the fixed horizontal spring. As the block compresses the spring to the left, the spring starts pushing the block to the right; so, the block slows down, stops instantaneously after compressing the spring to a maximum amount and then the block speeds up retracing its path. Eventually it leaves the spring with the same initial (speed v_0) kinetic energy. This is because the spring goes on receiving energy of the block and storing this energy in the form of spring elastic potential energy. The spring delivers all its stored energy back to the block during its leftward motion. This tells us that:

The spring conserves (stores) the energy of the block inside it without expending it and gives it back slowly to the block. So, *the spring-force is conservative in nature.*

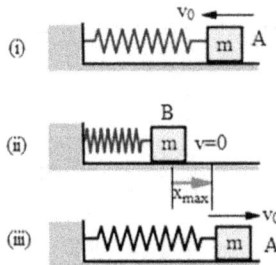

(i)The block starts compressing the undeformed spring with a velocty v_0 at A, (ii) the block delivers all its KE to the spring at B, the spring stores all the energy received from the block in the form of elastic potential energy;the compressed spring realeases all the stored energy delivered by the block from B to A;(iii) eventually the block will regain all its lost kinetic energy from the spring;so, the spring force is conservative.

If you throw a stone vertically up, it gradually loses all its kinetic energy till it reaches the highest point. Then it starts gaining kinetic energy during its downward motion. It regains all its kinetic energy given to the gravitational field as it passes through the point of projection, ignoring other forces such as air drag and air friction etc. Since gravitational force (field) does not dissipate kinetic energy of the stone, it has the capacity of storing (conserving) the energy of the projected body. Hence, *gravitational force (field) is conservative.*

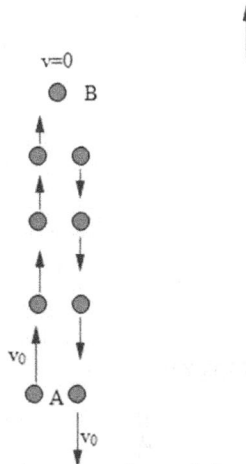

The ball continuously loses all its initial kinetic energy as it ascends from A to B; the KE lost by the ball is stored in the form of gravitational potential energy;During the descent of the ball, gravitational field delivers all the stored energy back to the ball as it descends from B to A; eventually, the ball will regain its initail kinetic energy from the gravity till it comes back to A. As the gravitational field does not absorb the enegy permanently, the gravitational force or field is conservative.

We defined the conservative forces based on simple observations. Let us now talk about the mathematical interpretation of the conservative forces. As proved earlier, the work done by a spring force when its deformation changes from x_1 to x_2, is

$$W_{sp} = -\frac{k}{2}(x_2^2 - x_1^2)$$

It is valid for any arbitrary path followed by the particle P between points 1 and 2.

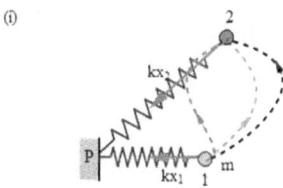

The work done by the spring does not depend upon the path followed by the end points of the spring; it depends only upon the initial and final deformation of the spring.

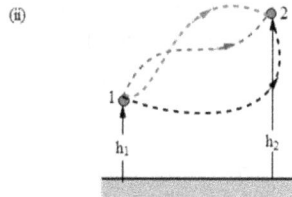

The work done by gravity does not depend upon the path followed by the particle; it depends only upon the initial and final position (heights) of the particle.

Likewise, for any arbitrary path followed by the particle near Earth's surface, between any two points 1 and 2, the work done by gravity is

$$W_{gr} = -mg(y_2 - y_1).$$

In the above two examples, you can notice that the work done by the spring and gravity forces do not depend upon the path followed by the particle between points 1 and 2. This depends only on the initial and final positions 1 and 2 of the moving points. If we put $x_2 = x_1$ for a closed path, we have $W_{sp} = 0$. Similarly, when $y_2 = y_1$, we have $W_{gr} = 0$. This tells us that the work done by the spring and gravity forces are zero if the point of application of force describes a closed loop.

The net work done by a conservative force F in a closed path is zero,

$$W = \oint \vec{F} \cdot d\vec{r} = 0.$$

1. The work done by a conservative force does not depend on the path followed by the point of application of the force. It depends on the initial and final positions only.
2. The work done by a conservative force in a round trip (of the point of application of force) is zero;

$$\oint \vec{F} \cdot d\vec{r} = 0.$$

3. A conservative force must depend on the position only (but not on its velocity and acceleration). Both magnitude and direction remain the same at a fixed point. Mathematically, $F = f(x, y, z)$ for conservative forces and $F = f(x, y, z, t)$ for non-conservative forces.
4. For all conservative forces, a unique scalar potential energy function is defined.

Example 19 What is meant by central force? Give two examples. Prove that a central force field is conservative.

Solution

A force F acting at a point P, is said to be central when the force depends upon the distance r between point P and a fixed point O, say the line joining these points. So, mathematically we can write the force as,

$$\vec{F} = f(r)\hat{r},$$

where $\hat{r}$ = unit vector of the position vector of the point of application P of the force relative to the fixed point (origin O) and $f(r)$ = magnitude of the force as the function or in terms of the distance of separation r between O and P.

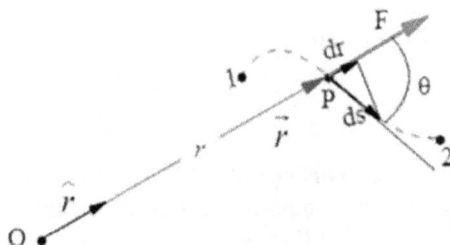

A central force F acting at point P is directed
radially away from the fixed point O along OP = r.

The work done by the force F is given as

$$W_F = \int \vec{F} \cdot d\vec{s} = F\,ds \cos\theta,$$

where $|\vec{F}| = f(r)$ and $|ds|\cos\theta = dr$ (elementary change in radial distance).

So, we have $W_F = \displaystyle\int_1^2 f(r)\,dr = \phi(r_2) - \phi(r_1)$

Since, this expression of work depends only on the initial and final position of the point of application of the force, we conclude that:

A central force is conservative.

Spring force, gravity force between two point masses, electrostatic force between two point charges are central forces.

7.7.2 Non-conservative force

Let us choose two points A and B on a rough horizontal surface and drag an object (a cube, say) slowly and horizontally between these points along paths 1 (from A to B) and 2 (from B to A). It is a matter of daily experience that we feel more tired when we drag a body through greater distance. In other words, work done by friction is more for a longer path. The work done by kinetic friction for paths 1 and 2 can be given as $W_1 = -f\,l_1$, and the work done by kinetic friction from B to A is $W_2 = -f\,l_2$.

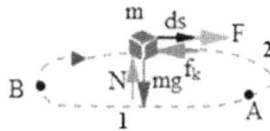

The work done by kinetic friction is $W_f = -\mu N l$ in a round trip ; so, it depends upon the path followed by the cube; from A to B $W_f = -\mu mg l_1$ and from B to A, $W'_f = -\mu mg l_2$.

Since the work done by kinetic friction is path dependent, for a round trip (when we drag the object from position A and B along part 1 and we bring it back slowly to its initial position A along the path 2) the total work done by friction is

$$W = W_1 + W_2 = -f\,l_1 + (-f\,l_2) = -f(l_1 + l_2) = -f\,l,$$

where, $l =$ length of the closed path.

The work done by a non-conservative force in a round trip is non-zero.

Example 20 A force acting on a particle is given as $\vec{F} = \alpha x \hat{j}$. Find the work done by the force when the particle moves from O to B in the path (i) OCB, (ii) OAB, and (iii) $OABCO$.

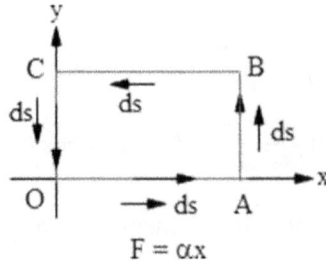

$$F = \alpha x$$

Solution

(i) Since at $x = 0$ and $F = 0, W(O \rightarrow C) = 0$. Since $F \perp dx, W(C \rightarrow B) = 0$. So, no work is done in the path OCB.

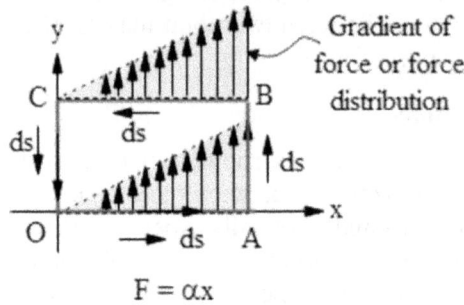

Gradient of force or force distribution

$$F = \alpha x$$

(ii) Since $F \perp ds$ from O to $A \cdot W(O \rightarrow A) = 0$. Since $F = \alpha a$ at $x = a$ and it is parallel to dy, we have

$$W(A \rightarrow B) = \int_0^b F \cdot ds = \int_0^b \alpha a\, dy = \alpha ab$$

So, $W = \alpha ab$ in the path OCB.

(iii) $W(O \rightarrow A \rightarrow B \rightarrow C \rightarrow O) = W_{OAB} + W_{BCO}$, where $W_{OAB} = \alpha ab$, $W_{BCO} = 0$.

So, the work done in the round trip $OABCO$ is αab.

N.B.: In the foregoing example, the force is non-conservative because

$$W = \oint \vec{f} \cdot d\vec{s} \neq 0$$

for the closed path.

7.8 Potential energy

7.8.1 Spring potential energy

We learnt that a deformed spring has. the ability of pushing an object. So it can do work during the displacement of the object. In other words, we can say that the ability of doing work (i.e., energy) comes from the spring . As the spring is massless, its kinetic energy is zero, so, cannot call this ability (energy) kinetic energy. As the ability (energy) comes from the configuration of the spring, we call it 'potential energy'. The spring stores the kinetic energy of the moving block as its potential energy and delivers it back continuously to the block during its return journey.

7.8.2 Gravitational potential energy

When you release or project a body in gravity, it goes on changing its position till it strikes Earth's surface. While the body moves up, gravity does negative work; when the body moves down, gravity does positive work on the body. As you know, the gravitational force depends upon the position (configuration) of the bodies (matter) in space. So, gravity derives its energy (ability of doing work) from the position of the objects. In other words, this energy is known as gravitational potential energy. During the ascent of the object, its kinetic energy is absorbed by the gravitational field and it is released gradually to the body during its descent.

Recapitulating the above facts:

When we use the word potential energy, we have to attribute it to the 'system of interacting bodies' like Earth–body, spring–mass, etc, but not to any object of the system. For instance, when we release a body, it pulls the Earth. So, Earth is also accelerated by the body. Since the mass of Earth is very large compared to that of the body, acceleration, velocity and displacement of Earth is negligible. This means that the change in position (configuration) of an Earth–body system occurs mainly due to the displacement of the body. Following this argument, roughly you can call it 'potential energy of the body,' but in a strict sense that is the 'potential energy of the Earth–body system'.

Similarly, in a block–spring system, as the block is rigid, its deformation is neglected. Hence, the elastic potential energy is mainly contributed by the change in configuration (deformation) of spring.

So, potential energy is defined as the ability of doing work by a conservative force. It arises from the configuration of the system or position of the particles in the system (conservative force fields like spring, gravity, etc).

7.8.3 Work–potential energy theorem

The conservative force always does positive work at the expense of its potential energy stored in its field. When the spring is relaxed, it cannot perform work. When the spring is stretched or compressed, it can pull or push the block and perform positive work. While doing so, the compression or elongation of the spring decreases, the spring force decreases. Consequently, the spring becomes more inactive and the ability of work done by the spring decreases. In this way we can logically interpret that:

Work done by the spring is equal to the loss of its potential energy;

$$W_{sp} = -\Delta U_{sp}$$

A negative sign is used for the loss of potential energy.

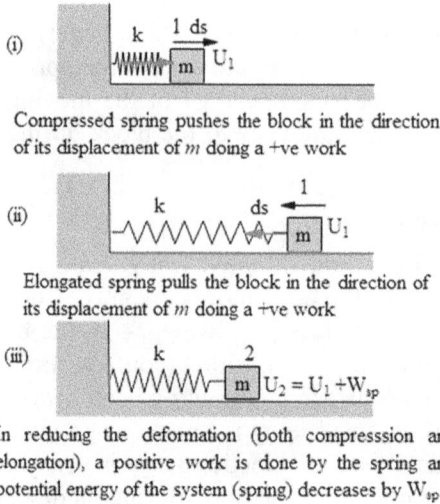

(i)

Compressed spring pushes the block in the direction of its displacement of m doing a +ve work

(ii)

Elongated spring pulls the block in the direction of its displacement of m doing a +ve work

(iii)

In reducing the deformation (both compresssion and elongation), a positive work is done by the spring and potential energy of the system (spring) decreases by W_{sp}

The same logic is also valid for the gravitational field. When the body moves down, its kinetic energy increases and the gravity does a positive work. This means the gravitational potential energy decreases by an amount which is equal to the gravitational work done on the falling body,

$$W_{gr} = -\Delta U_{gr}.$$

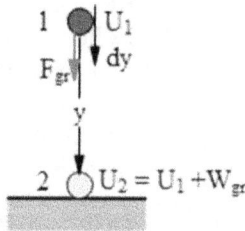

In lowering the object, the positive work is done by gravity and potential energy of the system (earth + object) decreases by W_{gr}

In general, we can state that: any conservative force performs work at the expense of the potential energy of the force field. So, we can write.

$$W_{cons.} = -\Delta U$$

The loss in potential energy is equal to the total work done by the conservative forces.

7.8.4 General definition of potential energy

Let us assume a spring–particle system having zero or no interacting force, at a fixed configuration. If the force is zero, it has zero (no) potential energy at that configuration. Generally, we call it zero potential energy position s_0. Now we shift the particle (in the conservative system) slowly to any configuration given by a position s. Let the configuration have potential energy U. Then the change in potential energy of the configuration is

$$\Delta U = -U_0 = U.$$

Since, $\Delta U = -W_{\text{cons.}}$, where $W_{\text{cons.}} = \int \vec{F} \cdot d\vec{s} =$ work done by all conservative forces substituting $\Delta U = U$, the potential energy of a system can be given as

$$U = -\int_{s_0}^{s} \vec{F} \cdot d\vec{s}$$

This tells us that:

Potential energy of a system is equal to the negative of work done by all conservative forces of the system (to assemble the system to the desired configuration from zero potential energy configuration). Generally, in zero potential energy of two particles system, the particles experience zero or negligible forces.

For the forces obeying inverse square law, zero potential energy refers to infinity because the separation r is very large between the interacting particles. For a spring, zero potential energy refers to the relaxed position of the spring.

7.9 Relation between conservative force and potential energy

Let us consider a conservative force F which does work dW on the surrounding in displacing the point through an elementary displacement dx along the x-axis. The change in potential energy is

$$dU_x = -dW_x = F_x\, dx.$$

Then, we have

$$F_x = -\frac{dU_x}{dx}$$

If the potential energy changes in the x-, y- and z-directions, $U = (x, y, z)$, $\vec{F}_x$ can be written as (a partial derivative of U with respect to s).

$$\vec{F}_x = -\frac{\partial U}{\partial x}\hat{i}$$

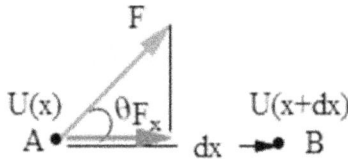

Force is equal to negative of the gradient of potential energy

Similarly, the components of a force F in the y- and z-directions can be given as the negative space derivative (gradient) of the potential measured along the corresponding axes.

$$\vec{F}_y = -\frac{\partial U}{\partial y}\hat{j} \text{ and } \vec{F}_z = -\frac{\partial U}{\partial z}\hat{k}$$

Then, the net force $\vec{F}$ can be given as,

$$\vec{F} = -\left(\frac{\partial U}{\partial x}\hat{i} + \frac{\partial U}{\partial x}\hat{j} + \frac{\partial U}{\partial z}\hat{k}\right) = -\nabla U,$$

where $(\frac{\partial}{\partial x}\hat{i} + \frac{\partial}{\partial x}\hat{j} + \frac{\partial}{\partial z}\hat{k}) = \nabla = $ Del operator.

The above expression tells us that,

The component of a conservative force in any direction is equal to the negative of its potential energy gradient in that direction. In other words, the conservative force points in the direction in which potential energy of that force decreases most rapidly or the potential energy decreases most rapidly with distance in the direction of the given conservative force.

Example 21 The potential energy of a configuration changes in the x- and y-directions as

$$U = kxy,$$

where k is a positive constant. Find the force acting on the particle of the system as the function of x and y.

Solution

The force acting on the particle is

$$F = -\frac{\partial U}{\partial x}\hat{i} - \frac{\partial U}{\partial y}\hat{j} - \frac{\partial U}{\partial z}\hat{k}$$

Putting $U = kxy$ in the last equation

$$F = -\frac{\partial(xy)}{\partial x}\hat{i} - \frac{\partial(xy)}{\partial y}\hat{j} - \frac{\partial(xy)}{\partial z}\hat{k}$$

$$\Rightarrow \vec{F} = -k(y\hat{i} + x\hat{j}) \text{ Ans.}$$

7.10 Stability

If the potential energy U of a particle varies along an axis by analyzing the U–x graph, we can talk about the stability of the particle moving under the influence of the conservative forces. You can see that U is minimum at A, maximum at B and constant near C. So, $\frac{dU}{dt}$, that is, the slope of the U–x graph is zero at these points. Since $F = -\frac{dU}{dx}$, we can conclude that the conservative force F acting on the particle at A, B and C is zero. This means that the particle remains in equilibrium at these points. Now, you can understand more about the nature of equilibrium by just looking at the U–x graph.

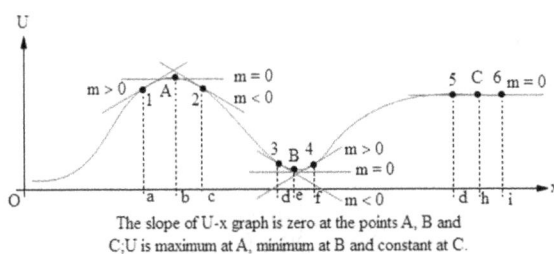

The slope of U-x graph is zero at the points A, B and
C;U is maximum at A, minimum at B and constant at C.

7.10.1 Unstable equilibrium (at *A*)

Let us start from the origin *O* and move to the point *A*; take two points 1 and 2 at both sides of the point *A*. At point 1, the slope $\frac{dU}{dx}$ is positive; so, the force $F = -\frac{dU}{dx}$ is negative at the point 1. This means that the force points to the left (negative *x*-direction). At point 2, the slope is negative; so, the force $F = -\frac{dU}{dx}$ is positive at the point. This means that the force points to the right (positive *x*-direction). This tells us that when we move in the positive *x*-direction, the force changes its direction from −ve to +ve (left to right) when we pass through point *A*. In other words, force *F* points away from point *B* at both sides of point *A*. So, when we displace a particle from *B* either to the left or right, it will never return (accelerate) to point *B*. Eventually, at position *B*, the particle is unstable even though it is in equilibrium at *B*. In other words, the particle is in *unstable equilibrium* at *B*.

7.10.2 Stable equilibrium (at *B*)

Now, let us move to point *B* and take two points 3 and 4 at both sides of the point. At point 3, the slope $\frac{dU}{dx}$ is negative; so, the force $F = -\frac{dU}{dx}$ is positive at the point. This means that the force points to the right (positive *x*-direction). At point 4, the slope is positive; so, the force $F = -\frac{dU}{dx}$ is negative at the point. This means that the force points to the left (negative *x*-direction). This tells us that, when we move in the positive *x*-direction, the force changes its direction from +ve to −ve (left to right) when we pass through point *B*. In other words, force *F* points towards point *B* at both sides of point *A*. So, when a particle is displaced on either side of *A* and released, it will return to point *A*. Since the particle tends to attain a stable position at *A* it is called 'stable equilibrium'. This is the basic condition for a particle to oscillate.

7.10.3 Neutral equilibrium (at *C*)

Finally, let us come to point *C*. At *C*, *F* = 0 as discussed earlier and the particle is in equilibrium at *C*. Let us displace the particle slightly on either side to points 5 and 6. Since, at both the points 5 and 6, $\frac{dU}{dx} = 0$, we can say that the particle experiences zero net force when it is displaced near *C*. In other words, the particle will remain neutral (force-free) at points 5 and 6. So, the particle is said to be neutral (neither accelerating to the left nor right) and at the same time it is in equilibrium. Then, we can say that, at *C*, the particle is said to be in 'neutral equilibrium'.

Recapitulating the above facts, we have the following points.

1. Since the slope m decreases near A,

$$\left(\frac{dm}{dx}\right)_B < 0, \text{ where } m = \frac{dU}{dx} = -F$$

$$\Rightarrow \frac{dF}{dx} > 0 \left(\text{or } \frac{d^2U}{dx^2} < 0\right)$$

$$\Rightarrow \overrightarrow{dF_A} \text{ is parallel to } \overrightarrow{\delta x_A}.$$

As, the unbalanced force $\overrightarrow{\delta F}$ is parallel with the displacement $d\overrightarrow{x}$ near point A. It tends to destabilize the particle at A by pushing it in the direction of displacement $\overrightarrow{\delta x_A}$. Since the work done by the force is positive near point A, the potential energy of the particle is maximum at that point. As the particle cannot return to A when it is displaced from A, it is in unstable equilibrium. The potential energy is maximum at unstable equilibrium.

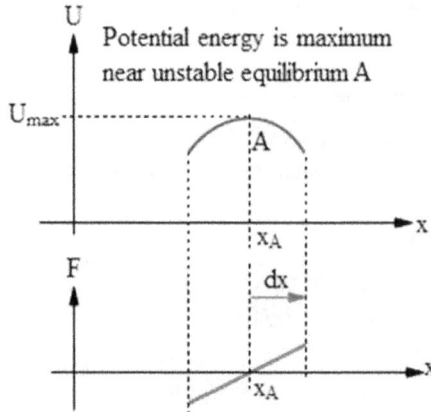

F-x graph is nearly linear about A having a +ve slope; so, the unbalanced or net force dF points away from A when shifted through a distanec dx from the point A in either direction.

2. Since, the slope m increases at B

$$\Rightarrow \left(\frac{dm}{dx}\right)_{x=B} > 0, \text{ where } m = \frac{dU}{dx} = -F$$

$$\Rightarrow \frac{d^2U}{dx^2} < 0$$

So, mathematically, we can say that the second derivative of the U–x graph must be negative at point A

$$\Rightarrow \left(\frac{dF}{dx}\right)_{x=B} < 0$$

$$\Rightarrow \delta \vec{F_B} = -k \delta x_B; \ k = \text{constant.}$$

This tells us that the unbalanced force $\delta \vec{F}$ is antiparallel with the displacement $d \vec{x}$ near point A. In other words, the restoring force $\delta \vec{F}$ tends to stabilize the particle at A. So, the potential energy of the particle is minimum at stable equilibrium.

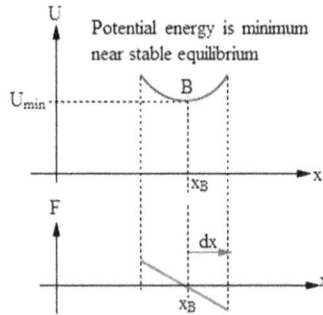

F-x graph is nearly linear about B having a -ve slope; so, the unbalanced or net force dF is restoring when shifted through a distanec dx from the point B in either direction.

3. Since the slope $m = $ constant near C,

$$\left(\frac{dm}{dx} \right)_C = 0, \ \text{where} \ m = \frac{dU}{dm} = -F$$

$$\Rightarrow \delta F = 0 \ \left(\text{or} \ \frac{d^2 U}{dx^2} = 0 \right)$$

This mean that the particle remains neutral experiencing no net force when displaced from C; so, the position of the particle at C is called neutral equilibrium. The potential energy of the particle remains constant at or near neutral equilibrium position.

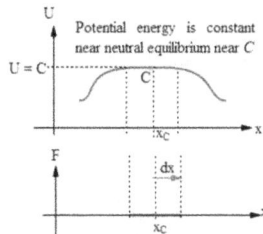

Force is zero about C having a zero slope of U-x curve; so, there is a zero restoring or unbalanced force dF when shifted through a distanec dx from the point C in either direction.

Example 22 The potential energy of a diatomic molecule is given as

$$U = \frac{a}{x^{12}} - \frac{b}{x^6}$$

Find the
  (i) stable equilibrium distance of separation between two atoms,
  (ii) force on each atom as the function of separating distance x,
  (iii) energy required to just dissociate the molecule.

Solution

  (i) Let us assume that the stable equilibrium separation between the atoms is x_0 because the potential energy U is minimum. To find x_0, we take the derivative of U and equate it to zero;

$$\Rightarrow \frac{dU}{dx} = 0 \qquad (7.23)$$

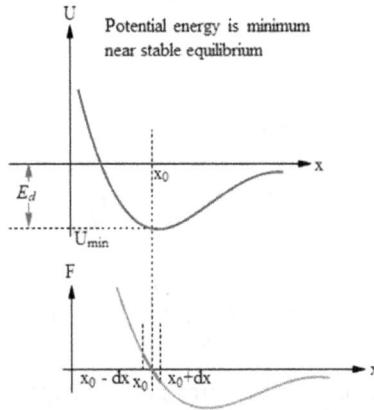

The potential energy is given as

$$U = \frac{a}{x^{12}} - \frac{b}{x^6} \qquad (7.24)$$

Using the last two equations, we have

$$\frac{dU}{dx} = \frac{dU}{dx}\left(U = \frac{a}{x^{12}} - \frac{b}{x^6}\right) = -\frac{12a}{x^{12}} + \frac{6b}{x^6} = 0$$

$$\Rightarrow x = x_0 = \left(\frac{2a}{b}\right)^{1/6} \quad \text{Ans.}$$

(ii) The force between the atoms at any value of x is

$$F = -\frac{dU}{dx} = -\frac{d}{dx}\left(\frac{a}{x^{12}} - \frac{b}{x^6}\right) = \frac{12a}{x^{13}} - \frac{6b}{x^7} \quad \text{Ans.}$$

(iii) By putting $x_0 = (\frac{2a}{b})^{1/6}$ in equation (7.24), the minimum potential energy of the molecule is

$$U_0 = \frac{a}{x_0^{12}} - \frac{b}{x_0^6} = \frac{a}{\left(\frac{2a}{b}\right)^2} - \frac{b}{\frac{2a}{b}} = -\frac{b^2}{4a}$$

Let E_d as dissociation energy, which is defined as the energy required to pull the atoms of the molecules slowly to far apart (infinity) from its stable equilibrium position. By applying energy conservation between stable equilibrium ($r = r_0$) and infinity ($r \to \infty$), we have

$$E = U + E_d$$

At infinity the energy is zero as the separation distance is infinite; putting $E \to 0$ as $r \to \infty$, we have $0 = U + E_d$

$$\Rightarrow E_d = -U = \frac{b^2}{4a} \quad \text{Ans.}$$

7.11 Conservation of energy

In the previous section, you learned that, the work done by the conservative forces in a mechanical system is equal to the change in total mechanical energy of the system.

$$\Rightarrow W_{noncons} = \Delta E$$

If there are no nonconservative forces, or, even if the non-conservative forces are present, if they do not perform work, we can put $W_{noncons} = 0$ in the last equation to obtain

$$\Delta E = 0$$

$$\Rightarrow \Delta U + \Delta K = 0$$

$$\Rightarrow K_f - K_i = U_f - U_i$$

$$\Rightarrow K_i + U_i = K_f + U_f$$

The above equation talks about the principle of conservation of mechanical energy which can be stated as:

If either there are no nonconservative forces or even through the nonconservative forces are present they do not do any work as a whole, the total mechanical energy of the system remains constant.

Example 23 A particle of mass m_1 (a ring, say) is connected with another particle of mass m_2 by an inextensible light string. The ring is free to slide along a smooth horizontal wire. If the system $(m_1 + \text{string} + m_2)$ is released from rest from the given position (i) discuss validity of the principle of conservation of mechanical energy, and (ii) find the maximum speed of the ring. (iii) Describe the motion of m_2 till the string becomes perpendicular to the horizontal wire. Put $\theta_0 = 30°$, $m_1 = m_2$ and $b = 2.5$ m.

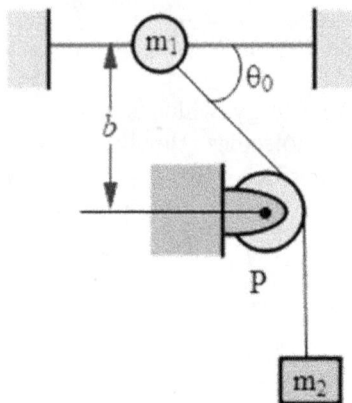

Solution

(i) The total work done by tension during elementary displacement is

$$dW_T = T ds_1 \cos \theta - T ds_2$$
$$= T(ds_1 \cos \theta - ds_2) = 0 \ (\text{because } ds_2 = ds_1 \cos \theta)$$

Since $N \perp ds_1$, the elementary work done by normal reaction is $dW_N = \vec{N} \cdot \vec{ds_1} = 0$.

Then, the net work done is equal to the elementary work done by the gravity which is numerically equal to the elementary change in kinetic energy of the particle, given as

$$dW_g = dK \tag{7.25}$$

According to the work–potential energy theorem,

$$dW_g = -dU \tag{7.26}$$

Using the last two equations,

$$-dU = dk$$

$$\Rightarrow dU + dK = 0 \text{ Ans.}$$

(ii) *Kinematics:* Since $ds_1 \cos \theta = ds_2$ we have

$$v_1 \cos \theta = v_2 \tag{7.27}$$

$$\Delta K = \Delta K_1 + \Delta K_2 = \frac{1}{2}m_1 v_1^2 + \frac{1}{2}m_2 v_2^2$$

When the particle m_1 passes through the position $\theta = 90°$, we have $v_2 = 0$.

$$\Rightarrow \Delta K = \frac{1}{2}m_1 v_1^2 \tag{7.28}$$

The change in gravitational potential of m_2 is

$$\Delta U = -m_2 g y,$$

where $y = b(\operatorname{cosec} \theta_0 - 1)$

$$\Rightarrow \Delta U = -m_2 g b(\operatorname{cosec} \theta_0 - 1) \tag{7.29}$$

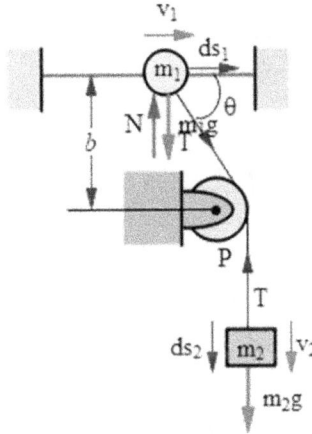

Conservation of energy
    Substituting ΔU and ΔK in the equation

$$\Delta U + \Delta K = 0 \tag{7.30}$$

Using equations (7.27), (7.28), (7.29) and (7.30), we have

$$\frac{1}{2}m_1 v_1^2 - m_2 g b(\operatorname{cosec} \theta_0 - 1) = 0$$

$$\Rightarrow v_1 = \sqrt{\frac{2m_2 g b(\operatorname{cosec} \theta_0 - 1)}{m_1}}$$

$$= \sqrt{\frac{2m(9.8)(2.5)(\operatorname{cosec} 30° - 1)}{m}} = 7 \text{ m s}^{-1} \text{ Ans.}$$

(iii) The velocity of particle 2 is

$$v_2 = v_1 \cos \theta$$

At $\theta = 90°$, $v_2 = v_1 \cos 90° = 0$; so, the body m_2 moves down with zero velocity when the body m_1 acquires a maximum velocity. Ans.

Example 24 A block of mass m strikes a light pan fitted with a vertical spring after falling through a distance h. If the stiffness of the spring is k, find the (a) maximum compression of the spring and (b) maximum compression and equilibrium compression of the spring, if $h = 0$ (c) maximum velocity of the block.
   Solution

(a) Let the maximum compression of the spring be x.
   The change in potential energy of the spring–mass system is:

$$\Delta U = \Delta U_{sp} + \Delta U_{gr}$$
$$= \tfrac{1}{2}kx^2 - mg(h + x)$$

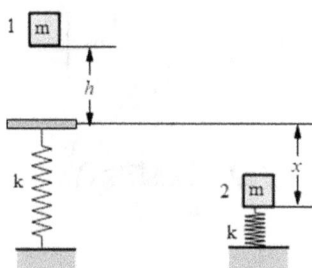

The change in kinetic energy is $\Delta K = 0$ between the highest and lowest position of the block.
   Substituting ΔU and ΔK in the expression

$$\Delta U + \Delta K = 0,$$

we have $\tfrac{1}{2}kx^2 - mg(h + x) = 0$

$$\Rightarrow x^2 - \frac{2mgx}{k} - \frac{2mgh}{k} = 0$$

$$\Rightarrow x = \frac{mg}{k}\left(1 + \sqrt{1 + \frac{2hk}{mg}}\right) \text{ Ans.}$$

(b) Putting $h = 0$, we have

$$\Rightarrow x_{max} = \frac{mg}{k}\left(1 + \sqrt{1 + \frac{2(0)k}{mg}}\right) = \frac{2mg}{k}$$

At equilibrium, the net force is zero; so,

$$F = -\frac{dU}{dx} = -\frac{d}{dx}\left\{\frac{1}{2}kx^2 - mg(h + x)\right\} = mg - kx = 0$$

$$x = x_0 = \frac{mg}{k} \quad \text{Ans.}$$

(c) The velocity is maximum when dv/dt, that is, acceleration is zero that happens at

$$x = x_0 = \frac{mg}{k}.$$

Let the velocity of the block be v at a compression x. Then, the change in kinetic energy is

$$\Delta K = \frac{1}{2}mv^2$$

By conserving energy between position 1 and the position at $x = mg/k$, we have

$$\Delta U + \Delta K = 0$$

$$\Rightarrow \frac{1}{2}kx^2 - mg(h + x) + \frac{1}{2}mv^2 = 0$$

$$\Rightarrow v = \sqrt{2g(h + x) - \frac{k}{m}x^2}$$

Putting $x = mg/k$, we have

$$v = \sqrt{2g(h + mg/k) - \frac{k}{m}(mg/k)^{-2}}$$

$$\Rightarrow v = \sqrt{\frac{mg^2}{2k} + 2gh} \quad \text{Ans.}$$

N.B.:

1. a and x is maximum when $v = 0$.
2. v is maximum when $a = 0$, but not when $x = 0$.
3. If we put $h = 0$ in the expression of velocity, we have our general result, given as

$$v = g\sqrt{\frac{m}{2k}}$$

4. For maximum compression, $v = 0$ but $a \neq 0$.
5. For equilibrium compression, $F = 0 \Rightarrow a = 0$.

7.12 Non-conservative forces and mechanical energy

Recasting the work–energy theorem, we have

$$W_{\text{cons}} + W_{\text{noncons}} = \Delta K$$

If non-conservative forces perform non-zero work, we have

$$W_{\text{noncons}} = \Delta K - W_{\text{cons}}$$

As the conservative forces perform work at the expense (cost) of potential energy of the system, substituting

$$W_{\text{cons}} = -\Delta U$$

in the last equation, we have

$$W_{\text{noncons}} = \Delta U + \Delta K = \Delta E,$$

where E = total energy. Then, we have

$$W_{\text{noncons}} = \Delta E$$

Example 25 A man of mass 50 kg is climbing a ladder with a speed $v = 0.2 \text{ m s}^{-1}$ to a height $h = 5$ m. Find the work done by the man.

Solution

Here the man is a complicated nonconservative system. So, the work done by man is

$$W_{\text{man}} = \Delta U + \Delta K$$

$$= mgh + \tfrac{1}{2}mv^2$$

$$= 50 \times 9.8 \times 5 + \tfrac{1}{2} \times 50 \times \left(\tfrac{2}{10}\right)^2 = 2451 \text{ J} \text{ Ans.}$$

7.13 Motion in a vertical circle

Let us consider a body of mass m tied to one end of a string and made to swing in a vertical circle of radius 'r', which is equal to the length of the string as shown in the figure. Let u be the velocity given to the body in the tangential direction at the lowest position. Let its velocity be 'v' at any angular position θ at point P.

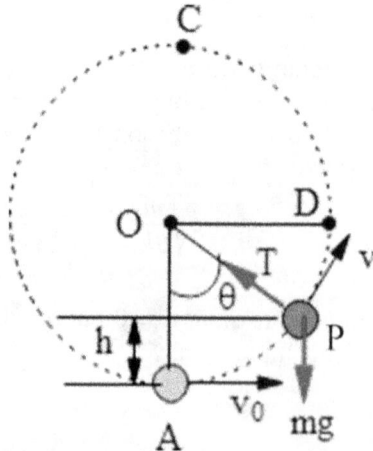

Applying Newton's second law along the string,

$$T - mg \cos \theta = \frac{mv^2}{r}$$

$$\Rightarrow T = mg \cos \theta + \frac{mv^2}{r} \tag{7.31}$$

Since displacement of the body is in a tangential direction and tension is in a radial direction, they are mutually perpendicular and hence the work done by tension is zero. Then, the work is done only by gravitational force. By conserving energy between the lowest and the angular position P,

$$K_i + U_i = K_f + U_f$$

$$\text{Or, } \frac{1}{2}mv_0^2 + 0 = \frac{1}{2}mv^2 + mgh$$

Putting $h = r(1 - \cos \theta)$ in the last expression, we have

$$v^2 = v_0^2 - 2gr(1 - \cos \theta) \tag{7.32}$$

From equation (7.31), you can see that for $0 \leqslant \theta \leqslant 90°$, tension $T > 0$ as $\cos \theta$ is positive. $\cos \theta$ is negative for $\theta > 90°$. So, the tension T can become zero in the upper half of the circle.

Case 1: $(v_0 \geqslant \sqrt{5gr})$

If we want the body to complete the vertical circle then the string must not slacken even at the highest point $C(\theta = \pi)$.

Thus, the tension in the string should be greater than or equal to zero ($T \geqslant 0$) at $\theta = \pi$. Putting $T = 0$ at $\theta = \pi$ substituting in equation (7.31), we have

$$mg = \frac{mv_c^2}{r}$$

$$\Rightarrow v_c = \sqrt{gr} \tag{7.33}$$

Putting $\theta = \pi$ in equation (7.32),

$$v_0^2 = v_c^2 + 2gr(1 - \cos \pi)$$

$$\Rightarrow v_0 = \sqrt{v_c^2 + 4gr} \tag{7.34}$$

Using equations (7.33) and (7.34), we have

$$v_0 = \sqrt{5gr}$$

This is the minimum speed to be given at lowest point A to the body to complete the vertical circle.

So, the condition of successfully completing the vertical circle is $v_0 \geqslant \sqrt{5gr}$.

If it just reaches D, putting $v = 0$ at $\theta = 90°$ in equation (7.32),

$$v_0 = \sqrt{2gr}$$

Case 2: ($\sqrt{2gr} < v_0 < \sqrt{5gr}$); if the velocity of projection lies between $\sqrt{2gr}$ and $\sqrt{5gr}$, the tension in the string will become zero before reaching the highest point. Now the body will leave the circle and move into a parabolic path (projectile trajectory).

Case 3: ($0 < v_0 < \sqrt{2gr}$); if the velocity of projection lies between 0 and $\sqrt{2gr}$, the body will oscillate in the lower half circle ($0° < \theta < 90°$). In this case velocity of the body becomes zero at the two extreme points, but tension in the string is not zero.

Case 4: $v_0 = \sqrt{2gr}$ In this case the body reaches up to point D and the tension and velocity become zero simultaneously at $\theta = 90°$.

Example 26 A small ball is projected with a velocity v_0 from the bottom of a hollow tube such that it moves in a vertical circle. Find the (a) normal reaction at the highest and lowest position of the ball, (b) condition for completing the circular path, and (c) difference in tensions at the top and bottom of the circular path.

Solution

The forces acting on the ball are normal reaction and gravity. The normal reaction does not perform work; so we can conserve mechanical energy of the ball.

(a) Applying Newton's second law along the radial direction at the top

$$N + mg = \frac{mv^2}{r}$$

$$\Rightarrow N = -mg + \frac{mv^2}{r} \tag{7.35}$$

By conservation of mechanical energy between A and B,

$$K_i + U_i = K_f + U_f$$

$$\Rightarrow \frac{1}{2}mv_0^2 + 0 = \frac{1}{2}mv^2 + 2mgr$$

$$\Rightarrow v^2 = v_0^2 - 4gr \tag{7.36}$$

Using equations (7.35) and (7.36), we have

$$\Rightarrow N = -mg + \frac{m(v_0^2 - 4gr)}{r}$$

$$\Rightarrow N = \frac{mv_0^2}{r} - 5mg$$

(b) The circle is completed if $N \geqslant 0$

$$\Rightarrow v_0 \geqslant \sqrt{5gr} \quad \text{Ans.}$$

(c) The normal reaction at the bottom of the ball is given as

$$N_0 - mg = \frac{mv_0^2}{r}$$

$$\Rightarrow N_0 = mg + \frac{mv_0^2}{r}$$

Then, the difference in normal reaction is

$$N_0 - N = mg + \frac{mv_0^2}{r} - \left(\frac{mv_0^2}{r} - 5mg \right) = 6mg \quad \text{Ans.}$$

7.14 Power

7.14.1 Instantaneous power

Power delivered by any force is defined as the work done in a unit time; in other words, it is the rate of work done by the force. If the force $\vec{F}$ performs a work dW in displacing a particle through a distance $d\vec{s}$ during time dt, power delivered by the force $\vec{F}$ is the rate of doing work, given as

$$P = \frac{dW}{dt}$$

Putting the $dW = \vec{F} \cdot d\vec{s}$ in the last expression, we have

$$P = \frac{\vec{F} \cdot d\vec{s}}{dt}$$

Putting $\frac{d\vec{s}}{dt} = \vec{v}$, we have

$$P = \vec{F} \cdot \vec{v} = Fv \cos \theta$$

You can see that the power is expressed as the dot product of force $\vec{F}$ and velocity $\vec{v}$; so, it can be positive, negative and zero with the following conditions:
   (i) P is +ve when θ is acute.
   (ii) P is −ve when θ is obtuse.
   (iii) P is when θ is 90°; $\vec{F} \perp \vec{v}$.

7.14.2 Average power

Power of a force $\vec{F}$ averaged over a time t is defined as the total work done divided by total time.

$$P_{av} = \frac{W}{t}$$

The total work done by a force during a time t is

$$W = \int_0^t P \, dt$$

Using the last two equations, the average power can be given as

$$P_{av} = \frac{\int_0^t P \, dt}{t}$$

Example 27 An insect of mass m moves up along a hanging stationary thread, with acceleration a. (a) Find the power delivered by gravity after a time t. (b) What is the power delivered by tension in the string to the insect?
   Solution

(a) Power delivered by gravity at any instant is

$$P_g = m\vec{g} \cdot \vec{v} = -mgv,$$

where $v =$ instantaneous velocity of the insect.
   Since the insect moves with an upward acceleration a,

$$v = at$$

Using the last two equations, we have

$$P_g = -mgat \quad \text{Ans.}$$

(b) The power delivered by tension in the string to the insect is zero because the velocity of the point of application of static friction on the thread by the insect is zero because the string is stationary. Ans.

N.B: You can show that the net power delivered on the insect $= (ma)(at)$ and its source is the insect itself.

Problem 1 (Work done by f_s on moving surface)
   A block of mass m is kept on a rough plank which moves with a horizontal acceleration a. If the plank was at rest $t = 0$, and the block does not slide relative to the plank. Find the work done by friction on the (a) block, (b) plank, and (c) system (block + plank) during time t.

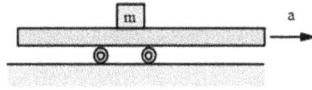

Solution

(a) The block is in contact with the plank. Let us assume that P is the point of contact which has a displacement s during time t.

The work done by static friction on the block is

$$W_{f_1} = +f_s s \qquad (7.37)$$

The net force acting on the block is

$$f_s = ma \qquad (7.38)$$

The displacement of the block is

$$s = \frac{1}{2}at^2 \qquad (7.39)$$

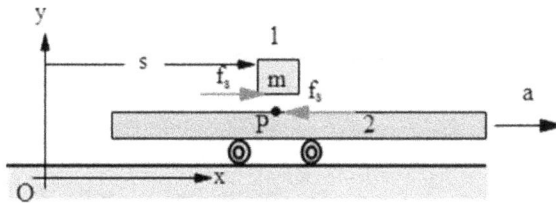

Using the above equations, we have

$$W_{f_1} = \frac{1}{2}ma^2t^2 \ \text{Ans.}$$

(b) The work done by static friction on the plank is

$$W_{f_2} = -f_s s$$
$$= -(ma)\left(\frac{1}{2}at^2\right) = -\frac{1}{2}ma^2t^2 \ \text{Ans.}$$

(c) The net work done by static friction is

$$W_f = W_{f_1} + W_{f_2}$$
$$= (f_s s) + (-f_s s) = 0 \ \text{Ans.}$$

Static friction does positive work on the block and equal negative work on the plank; so, the net work done by static friction (on both surfaces) is zero.

Problem 2 In the previous example, if $a > \mu_s g$, find the work done by (a) friction on the (i) block, (ii) plank, and (iii) net frictional force of the system (plank + block), (b)(i) the external force on the plank, (ii) net force acting on the plank, assuming that M = mass of the plank, as the function of time t. Assume that μ_k = coefficient of kinetic friction between the block and plank and the ground is smooth.

Solution

(a) (i) When the plank moves with an acceleration $a > \mu_s g$, the block will slide relative to the plank. So, kinetic friction comes into play.

The work done by f_k is

$$W_{f_1} = f_k s_1, \tag{7.40}$$

The displacement of the block is

$$s_1 = \frac{1}{2} a_1 t^2 \tag{7.41}$$

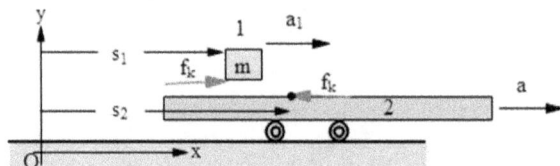

The kinetic friction is

$$f_k = \mu_k mg \tag{7.42}$$

As the force acting on the block is equal to kinetic friction, the acceleration of the block is

$$a_1 = f_k / m = \mu_k g \tag{7.43}$$

Using equations (7.40), (7.41) and (7.43),

$$W_{f_1} = \frac{m}{2} \mu_k^2 g^2 t^2 \quad \text{Ans.}$$

(ii) The work done by kinetic friction on the plank is

$$W_{f_2} = -f_k s_2$$

$$= -(\mu_k mg)\left(\tfrac{1}{2} a t^2\right)$$

$$= -\frac{\mu_k m_1}{2} g a t^2 \quad \text{Ans.}$$

(iii) Summing up the work done by kinetic friction on the block and plank, the net work done by kinetic friction is

$$W_f = \frac{m(a - \mu g)}{2} \mu g t^2 \quad \text{Ans.}$$

(b) (i) Applying Newton's second law,

$$F - m\mu g = Ma$$

$$\Rightarrow F = m\mu g + Ma$$

The work done by the external agent on the system (plank + block) is

$$W_{ext} = Fs_M, \quad \text{where } s_M = \frac{a}{2}t^2$$

$$\Rightarrow W_{ext} = \frac{Fa}{2}t^2$$

Putting the obtained value of F, we have

$$\Rightarrow W_{ext} = \frac{Fa}{2}t^2 = \frac{m\mu g + Ma}{2}at^2 \text{ Ans.}$$

(ii) Total work done on the plank is

$$W_{ext} = F_{net}s_M = Mas_M, \quad \text{where } s_M = \frac{a}{2}t^2$$

$$\Rightarrow W_{ext} = \frac{Ma^2}{2}t^2 \text{ Ans.}$$

N.B:
1. You can also apply the W–KE theorem to solve (c).
2. The net kinetic friction is zero, but the net work done by the kinetic friction is nonzero (negative) and frame independent.

Problem 3 A block of mass m is placed on a block of mass M as shown in the figure. A horizontal constant force $\vec{F}$ acts on M. If the horizontal surface is smooth, assuming no relative sliding between the blocks, find the work done by friction on the blocks.

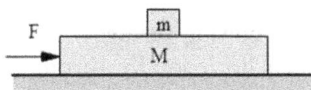

Solution

As there is no relative sliding between the blocks, they move with a common acceleration

$$a = \frac{F}{M + m} \qquad (7.44)$$

Then the static frictional force on the block m is

$$f_1 = ma \qquad (7.45)$$

Using the last two equations,

$$f_1 = m\frac{F}{M + m} \qquad (7.46)$$

The work done by static friction on the upper block is

$$W_{f_1} = \vec{f_1} \cdot \vec{s} = f_1 s \cos 0° = f_1 s \qquad (7.47)$$

The displacement of each block is

$$s = \frac{1}{2}at^2 \qquad (7.48)$$

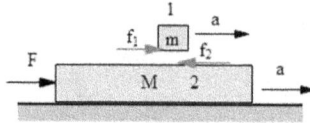

Using the last two equations, we have

$$W_{f_2} = \frac{1}{2}f_1 at^2 \qquad (7.49)$$

Using equations (7.44), (7.46) and (7.49), we have

$$W_{f_1} = \frac{mF^2t^2}{2(M + m)^2} \quad \text{Ans.}$$

Since $W_{f_1} + W_{f_2} = 0$, we have

$$W_{f_2} = -\frac{mF^2t^2}{2(M + m)^2} \quad \text{Ans.}$$

N.B: The net work done by static friction is zero on both surfaces because it is a constraint force.

Problem 6 (Work done by f_k on moving surfaces)

A block of mass m is pulled horizontally through a distance x relative to the moving plank. If the coefficient of kinetic friction between block and plank is μ, assuming a smooth horizontal surface, find the total work done by the kinetic friction between the contacting surfaces.

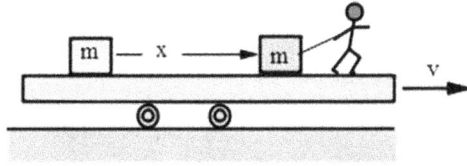

Solution

Referring to the free-body diagram, the frictional forces on m_1 and m_2 are $f_1 \leftarrow$ and $f_2 \rightarrow$, respectively. As discussed in chapter 6, kinetic friction acting on the block and plank are constant, equal in magnitude and oppositely directed as shown in the figure below. Let the block and plank move through distances x_1 and x_2, respectively.

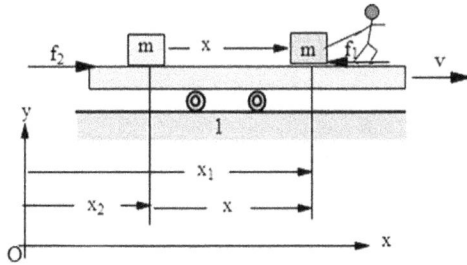

The work done by kinetic friction f_1 on the block is

$$W_{f_1} = -f_1 x_1 \tag{7.50}$$

The work done by kinetic friction f_2 on the plank is

$$W_{f_2} = f_2 x_2 \tag{7.51}$$

Then the total work done by the two frictional forces is

$$W = W_{f_1} + W_{f_2} = -f_1 x_1 + f_2 x_2 \tag{7.52}$$

Law of kinematic friction:

$$f_1 = f_2 = \mu N \tag{7.53}$$

Force equations on the block along the vertical:

$$N = m_1 g \tag{7.54}$$

Using the last five equations, we have

$$W = -\mu m_1 g(x_1 - x_2) \tag{7.55}$$

Constraint equation:

Since the block moves through a distance x relative to the plank, we write

$$x_1 - x_2 = x \tag{7.56}$$

Using equations (7.55) and (7.56) we have,

$$W = -\mu m_1 g x \text{ Ans.}$$

N.B: When two surfaces slide relative to each other, kinetic friction does positive work on one surface and more negative work on the other surface. Hence, the total work done by the kinetic friction on the contacting surfaces is negative.

Problem 7 A block of mass m is pulled slowly by a minimum magnitude of constant unknown force, F (say), on a horizontal surface through a distance x. The coefficient of kinetic friction is μ. Find the work done by (a) F and (b) friction.

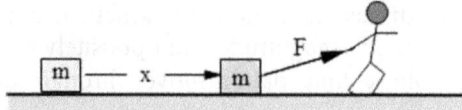

Solution

(a) The work done by the force F is

$$W_F = \int \vec{F}_{\min} \cdot \vec{ds}$$

$$\Rightarrow W_F = F_{\min} x \cos \theta \tag{7.57}$$

Force equation and law of kinetic friction:

$$\sum F_x = ma$$

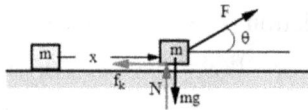

Because for slowly pulling the block its acceleration must be zero.

$$\Rightarrow \sum F_x = F \cos \theta - \mu N = 0 \text{ (zero)}$$

$$\Rightarrow F \cos \theta = \mu N \tag{7.58}$$

$$\sum F_y = mg$$

$$\Rightarrow F \sin \theta + N - mg = 0$$

$$\Rightarrow N = mg - F \sin \theta \tag{7.59}$$

Using equations (7.58) and (7.59),

$$F = \frac{\mu mg}{\sin \theta + \mu \cos \theta}$$

F is minimum when $\tan \theta = \mu$

$$\Rightarrow F_{\min} = \frac{\mu mg}{\sqrt{1 + \mu^2}} \tag{7.60}$$

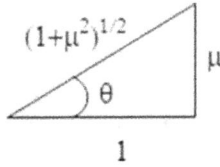

Using equations (7.57) and (7.60),

$$W_F = \frac{\mu mgx}{\sqrt{1 + \mu^2}} \cdot \frac{1}{\sqrt{1 + \mu^2}} = \frac{\mu mgx}{1 + \mu^2} \quad \text{Ans.}$$

The work done by friction is

$$W_f = \vec{f_k} \cdot \vec{s} = -f_k \cdot x$$

Putting $f_k = F_{min} = \frac{\mu mg}{\sqrt{1+\mu^2}}$, we have

$$W_f = -\frac{\mu mgx}{\sqrt{1 + \mu^2}} \quad \text{Ans.}$$

Alternative method:
Since the block is pulled slowly, the change in kinetic energy is zero; so, equating the total work done by all forces such as mg, N, F and f to zero, we have

$$W_f = -W_F = -\frac{\mu mgx}{\sqrt{1 + \mu^2}} \quad \text{Ans.}$$

Problem 8 One end of a fixed spring is pulled by an average force of 18 N through a distance of 5 cm, find the work done by the spring.
Solution
The work done by the spring is

$$W = -\frac{k}{2}(x_2^2 - x_1^2)$$
$$= -\frac{k}{2}(x_2 + x_1)(x_2 - x_1)$$
$$= -\left(\frac{kx_2 + kx_1}{2}\right)(x)$$
$$= -F_{av}x$$
$$= -18 \times \frac{5}{100} = -0.9 \, \text{N m} \quad \text{Ans.}$$

N.B.: Since the external force $F = kx$, the average force can be given as,

$$F_{av} = k\left(\frac{x_2 + x_1}{2}\right)$$

Problem 9 A block of mas m_1 moves with an acceleration a_{12} relative to a plank 2, as shown in the figure below. (a) What is the work done by the inertial force, as observed by a man, on the block during time t? Assume zero initial velocities of the bodies and observer. (b) Find the work done by the inertial force on the block (relative to the man or observer) during time t if the observer stands on the (i) plank and (ii) the ground.

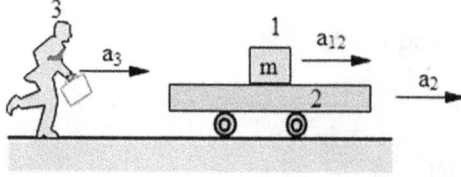

Solution

(a) As the man moves with an acceleration $\vec{a} = a_3\hat{i}$, the pseudo-force acting on the block is

$$\vec{F}_{\text{pseudo}} = -m\vec{a} = -ma_3\hat{i} \tag{7.61}$$

The work done by the pseudo-force is

$$W_{\text{pseudo}} = \vec{F}_{\text{pseudo}} \cdot \vec{s} = -m\vec{a_3} \cdot \vec{s}_{13} \tag{7.62}$$

The displacement of the block relative to the man is

$$\vec{s}_{13} = \frac{1}{2}\vec{a}_{13}t^2 \tag{7.63}$$

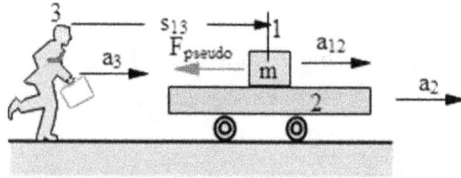

The acceleration of the block relative to the man is

$$\vec{a}_{13} = \vec{a}_1 - \vec{a}_3 = \vec{a}_{12} + \vec{a}_2 - \vec{a}_3 = (a_{12} + a_2 - a_3)\hat{i} \tag{7.64}$$

Using the last three equations, we have

$$W_{\text{pseudo}} = -\frac{ma_3}{2}(a_{12} + a_2 - a_3)t^2 \text{ Ans.}$$

(b)

(i) If the observer (man) stands on the plank, the acceleration of the relative to the ground is $\vec{a}_{\text{man}} = \vec{a}_2$; so, the pseudo-force acting on the block is

$$\vec{F}_{\text{pseudo}} = -m\vec{a_2} = -ma_2\hat{i} \tag{7.65}$$

The work done by the pseudo-force is

$$W_{\text{pseudo}} = \vec{F}_{\text{pseudo}} \cdot \vec{s} = -m\vec{a_2} \cdot \vec{s}_{13} \tag{7.66}$$

The displacement of the block relative to the man is

$$\vec{s}_{13} = \frac{1}{2}\vec{a}_{13}t^2 \tag{7.67}$$

The acceleration of the block relative to the man is

$$\vec{a}_{13} = \vec{a}_1 - \vec{a}_3 = \vec{a}_{12} + \vec{a}_2 - \vec{a}_3 = (a_{12} + a_2 - a_2)\hat{i} = a_{12}\hat{i} \tag{7.68}$$

Using the last three equations, we have

$$W_{\text{pseudo}} = -\frac{ma_2}{2}a_{12}t^2 \text{ Ans.}$$

(ii) If the observer (man) stands on the ground, the acceleration of the man relative to the ground is $\vec{a}_{\text{man}} = \vec{0}$; so, the pseudo-force acting on the block is zero. So, the work done by the pseudo-force is

$$W_{\text{pseudo}} = \vec{F}_{\text{pseudo}} \cdot \vec{s} = 0 \text{ Ans.}$$

Problem 10 A block of mass m, welded with a light spring of stiffness k is placed on an inclined plane when the spring is undeformed. The angle of inclination of the plane is θ and the coefficient of kinetic friction between the block and inclined plane is μ. Find the (a) speed of the block as the function of elongation x of the spring if F remains constant. (b) The force applied by the external agent if the block is pulled slowly through a distance x. Assume that initially the spring is undeformed.

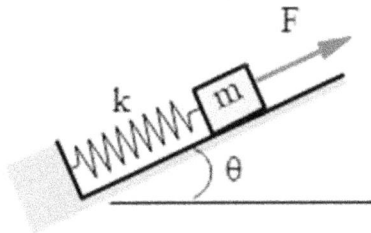

Solution

(a) Referring to the free-body diagram, the forces acting on the block are $N \nwarrow$, $mg \downarrow$, $f_k \swarrow$ and $F_s \nearrow$. Let us find the work done by each force for an elementary displacement $d\vec{s}$.

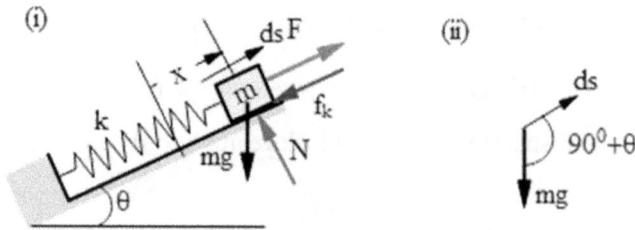

The work done by the normal reaction is

$$W_N = \vec{N} \cdot \vec{s} = 0 \text{ (because } \vec{N} \perp \vec{s}) \tag{7.69}$$

The work done by the kinetic friction is

$$W_{f_k} = \vec{f_k} \cdot \vec{s} = -f_k x \text{ (because } \vec{f} \text{ opposes } d\vec{s}) \tag{7.70}$$

Putting $f_k = \mu mg \cos \theta$, we have

$$W_{f_k} = (-\mu mg \cos \theta) x = -\mu\, mgx \cos \theta \tag{7.71}$$

The work done by gravity is

$$W_{gr} = m\,\vec{g} \cdot \vec{s} = mg\, s \cos(90° + \theta) = -mg\, x \sin \theta \tag{7.72}$$

The work done by the spring force is

$$W_s = -\frac{1}{2}kx^2 \tag{7.73}$$

The work done by the applied force F is

$$W_F = Fx \tag{7.74}$$

Then adding work done by all forces, we have

$$W = Fx - \mu mg\, x \cos \theta - mgx \sin \theta - \frac{kx^2}{2} \tag{7.75}$$

The change in kinetic energy of the block is

$$\Delta K = \frac{1}{2}mv^2 \tag{7.76}$$

According to the work–energy theorem

$$W = \Delta K \tag{7.77}$$

Using the last three equations, we have

$$\frac{1}{2}mv^2 = Fx - \mu mg\, x \cos \theta - mgx \sin \theta - \frac{kx^2}{2} \tag{7.78}$$

$$\Rightarrow v = \sqrt{2Fx/m - \mu g\, x \cos \theta - gx \sin \theta - \frac{kx^2}{2m}}.$$

(b) If block is pulled slowly through a distance x, putting $a = 0$ in equation (7.78), we have

$$\vec{F}_{net} = F - (mg \sin \theta + \mu mg \cos \theta + kx)\hat{i} = m\vec{a} = \vec{0}$$

$$\Rightarrow F = \mu mg \cos \theta + mg \sin \theta + kx.$$

N.B.: You can verify that the net work done is equal to the work done by the net force.

Problem 11 (a) Let us pull the free end of a spring of mass m with a velocity v_0. Find the kinetic energy of the spring. (b) If we attach a particle of mass M to the (i) end and (ii) mid-point of the spring, find the kinetic energy of the spring–particle system.
   Solution

(a) Since the element of mass dm of the spring at a distance x moves with a velocity

$$v = (x/l)v_0 \tag{7.79}$$

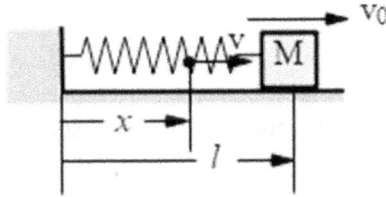

the kinetic energy of the element is

$$dK = \frac{1}{2}dmv^2 \tag{7.80}$$

Using the last two equations,

$$dK = \frac{v_0^2 x^2}{l^2}dm \tag{7.81}$$

Then, the kinetic energy of the spring is

$$K = \int dK = \frac{v_0^2}{l^2} \int_0^l x^2 \, dm,$$

where $dm = \frac{m}{l}dx$

$$\Rightarrow K = \frac{1}{6}mv_0^2 \text{ Ans.}$$

(b)

(i) The spring of mass m is equivalent to a massless spring connected with a mass $m/3$ at its free end. In other words, the effective mass of the spring is $m_{\text{eff}} = m/3$. If we connect a block of mass M at the end of the spring, its velocity will be equal to

$$v = (x/l)v_0 = \{(l)/l\}v_0 = v_0$$

Then, its kinetic energy is

$$K = \frac{1}{2}Mv_0^2$$

Then, the total kinetic energy of the spring–block system is

$$\Rightarrow K = \frac{1}{2}(M + m/3)v_0^2$$

(ii) If we connect a block of mass m at the mid-point of the spring, its velocity will be equal to $v = (x/l)v_0 = \{(l/2)/l\}v_0 = v_0/2$
Then, its kinetic energy is

$$K = \frac{1}{2}M(v_0/2)^2 = \frac{1}{8}Mv_0^2$$

Then, the total kinetic energy of the spring–block system is

$$\Rightarrow K = \frac{1}{2}(M/4 + m/3)v_0^2 \text{ Ans.}$$

Problem 12 A force F acts on a block of mass m at an angle θ with the horizontal. If the coefficient of friction between block and ground is μ, find the speed of the block as the function of horizontal distance x moved by it. Assume that the block does not loose contact with the surface.

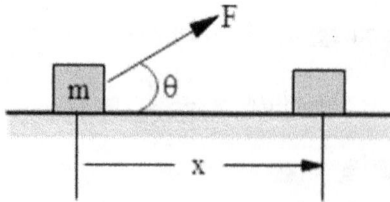

Solution

Referring to the free-body diagram, the forces acting on the block are $F \nearrow$, $mg\downarrow$, $N\uparrow$ and $f = f_k \leftarrow$.

The work done by the force F is

$$W_F = Fx \cos \theta \qquad (7.82)$$

The work done by gravity is

$$W_{gr} = 0 \qquad (7.83)$$

The work done by the normal reaction is

$$W_N = 0 \qquad (7.84)$$

The work done by friction is

$$W_f = -\mu N x = -\mu(mg - F \sin \theta)x \text{ (beacuse } N = mg - F \sin \theta) \qquad (7.85)$$

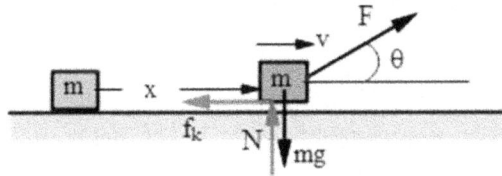

Using the last four equations, the total work done is

$$W = W_F + W_{gr} + W_N + W_f = Fx \cos \theta - \mu(mg - F \sin \theta)x \qquad (7.86)$$

The change in kinetic energy is

$$\Delta K = \frac{1}{2}mv^2 \qquad (7.87)$$

Applying the W–E theorem,

$$W = \Delta K \qquad (7.88)$$

Using the last three equations,

$$Fx \cos \theta - \mu(mg - F \sin \theta)x = \frac{1}{2}mv^2$$

$$\Rightarrow v = \sqrt{\frac{2[F(\cos \theta + \mu \sin \theta) - \mu mg]x}{m}} \quad \text{Ans.}$$

Problem 13 A force horizontal F acts on a block of mass M. The block is connected with a spring of stiffness k. The left end of the spring is attached to the wall. Another block of mass m is connected with the block of mass M by an inextensible string that passes over a smooth pulley P. The coefficient of friction between block and the ground is μ. If block m is released from rest when the spring is relaxed and the string is just taut, find the speed of the block as the function of horizontal distance x moved by it.

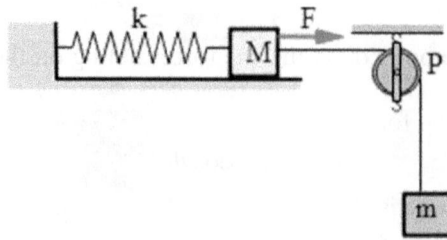

Solution

The forces acting on block M are $F\rightarrow$, $Mg\downarrow$, $N\uparrow$ and $f = f_k \leftarrow$. The forces acting on block m are, $mg\downarrow$ and tension $T\uparrow$.

The work done by gravity and normal reaction N on the block M is zero. The tension is a constraint force; so, its net work done is zero.

The work done by the force F on block M is

$$W_F = Fx \tag{7.89}$$

The work done by the kinetic friction f_k on block M is

$$W_f = -\mu N x = -\mu m g x \tag{7.90}$$

The work done by gravity on block m is

$$W_{gr} = mgx \tag{7.91}$$

The work done by the spring on block M is

$$W_{gr} = -\frac{1}{2}kx^2 \tag{7.92}$$

Using the last four equations, the total work done is

$$W = W_F + W_{gr} + W_{sp} + W_f = Fx + mgx - \mu Mgx - \frac{1}{2}kx^2 \tag{7.93}$$

The change in kinetic energy is

$$\Delta K = \frac{1}{2}mv^2 \tag{7.94}$$

Applying W–E theorem,

$$W = \Delta K \tag{7.95}$$

Using the last three equations,

$$Fx + mgx - \mu Mgx - \frac{1}{2}kx^2 = \frac{1}{2}mv^2$$

$$\Rightarrow v = \sqrt{\frac{2F + 2mg - 2\mu Mg - kx]x}{m}} \quad \text{Ans.}$$

Problem 14 A block of mass m is connected rigidly with a smooth wedge (plank) by a light spring of stiffness k. If the wedge is moved with constant velocity v_0, find the work done by the external agent till the maximum compression of the spring.

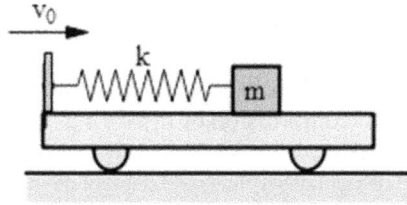

Solution

Let us take wedge + spring + block, as a system. Referring to the free-body diagram, the forces responsible for performing work are spring force $kx(\leftarrow)$ and the external force $F(\rightarrow)$.

Applying the W–E theorem for the system (block + spring + plank) relative to the ground, the sum of work done by the external and spring force is

$$W_{ext} + W_{sp} = \Delta K,$$

where W_{sp} = the total work done by the spring on wedge and block $= -\frac{1}{2}kx^2$ and $\Delta K=$ change in KE of the block (because the plank does not change its kinetic energy).

$$\Rightarrow W_{ext} = \frac{1}{2}kx^2 + \Delta K \tag{7.96}$$

As the block was initially stationary and it will acquire a velocity v_0 equal to that of plank at the time of maximum compression of the spring, the change in kinetic energy of the block relative to ground is

$$\Delta K = \frac{1}{2}mv_0^2 \tag{7.97}$$

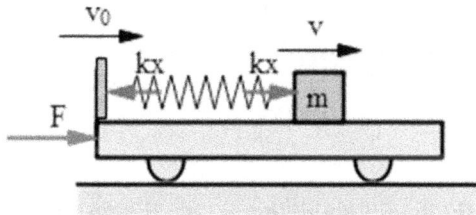

Using the last two equations, we have

$$W_{ext} = \frac{1}{2}kx^2 + \frac{1}{2}mv_0^2 \tag{7.98}$$

Let us now apply the W–E theorem for the system (block + spring + plank) relative to the plank. Since the plank moves with constant velocity, there is no pseudo-force acting on the block. Then the net work done on the system (block + plank) due to the spring is,

$$W_{sp} = -\frac{1}{2}kx^2$$

As the relative velocity between the observer (plank) and block decreases from v_0 to zero at the time of maximum compression of the spring, the change in kinetic energy of the block is

$$\Delta K = -\frac{1}{2}mv_0^2$$

Applying the work–energy theorem we have,

$$W = \Delta K,$$

where $W = -\frac{1}{2}kx^2$ and $\Delta K = -\frac{1}{2}mv_0^2$

$$\Rightarrow \frac{1}{2}kx^2 = \frac{1}{2}mv_0^2 \tag{7.99}$$

Using equation (7.98) in equation (7.99), we have

$$W_{ext} = mv_0^2 \quad \text{Ans.}$$

Problem 15 A small ball is placed at the top of a smooth hemispherical wedge of radius R. If the wedge is accelerated with an acceleration a, find the (a) velocity of the ball relative to the wedge as the function of θ, and (b) angle θ at which the ball loses contact with the wedge.

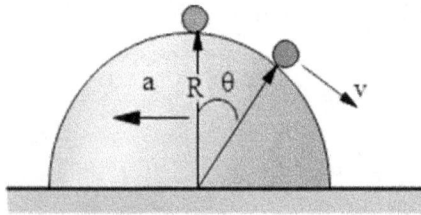

Solution

(a) Referring to the free-body diagram, relative to the accelerating wedge the forces acting on the ball are pseudo-force $ma\rightarrow$, $mg\downarrow$ and $N \nearrow$.
The work done by the pseudo-force is

$$W_{sp} = max \tag{7.100}$$

The work done by gravity is

$$W_{gr} = mgy \tag{7.101}$$

The work done by the reaction force is

$$W_N = \vec{N} \cdot d\,\vec{s}_{bw} = 0, \text{ (beacuse } \vec{N} \perp d\,\vec{s}_{bw}), \tag{7.102}$$

where $d\,\vec{s}_{bw}$ is the elementary displacement of the ball relative to the wedge. Using the last three equations, total work done is

$$W = max + mgy \tag{7.103}$$

If v = velocity of the ball relative to the wedge, the change in kinetic energy of the ball relative to the wedge is

$$K = \frac{1}{2}mv^2 \tag{7.104}$$

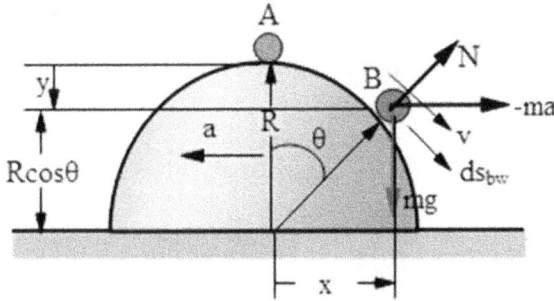

Applying the work–energy theorem relative to the wedge,

$$W = \Delta K \tag{7.105}$$

Using the last three equations, we have

$$max + mgy = \frac{1}{2}mv^2$$

Putting $x = R\sin\theta$ and $y = R(1 - \cos\theta)$, we have

$$v = \sqrt{2\{gR(1 - \cos\theta) + aR\sin\theta\}} \quad \text{Ans.}$$

(b) The net force acting on the ball in the radial direction is

$$mg\cos\theta - N - ma\sin\theta = mv^2/R \tag{7.106}$$

Putting the obtained value $f\,v$ in equation (7.106), we have

$$mg\cos\theta - N - ma\sin\theta = 2m\{g(1 - \cos\theta) + a\sin\theta\}$$

$$\Rightarrow N = m\{g(3\cos\theta - 2) - 3a\sin\theta\} \quad \text{Ans.}$$

(c) When the ball leaves the wedge, $N = 0$; so, we have

$$N = m\{g(3\cos\theta - 2) - 3a\sin\theta\} = 0$$

$$\Rightarrow (3\cos\theta - 2) = 3(a/g)\sin\theta$$

Put $a/g = x$; then, we have

$$\Rightarrow 3\cos\theta - 2 = 3x\sin\theta$$

Squaring both sides,

$$9\cos^2\theta - 12\cos\theta + 4 = 9x^2\sin^2\theta$$

$$\Rightarrow 9\cos^2\theta - 12\cos\theta + 4 = 9x^2\sin^2\theta$$

$$\Rightarrow 9\cos^2\theta - 12\cos\theta + 4 = 9x^2(1 - \cos^2\theta)$$

$$\Rightarrow 9(1 + x^2)\cos^2\theta - 12\cos\theta + (4 - 9x^2) = 0$$

$$\Rightarrow \theta = \cos^{-1}\left\{\frac{2 + x\sqrt{5 + 9x^2}}{3(1 + x^2)}\right\}, \quad \text{where } x = \frac{a}{g} \quad \text{Ans.}$$

Problem 16 Two blocks of masses m_1 and m_2 interconnected by a spring of stiffness k are placed on a horizontal surface. If a constant horizontal force F acts on the block m_1 it slides through a distance x while m_2 remains stationary. If the coefficient of friction between all contacting surfaces is μ, find the (a) speed of the block m_1 as the function x and (b) minimum value of F when block m_2 starts sliding.

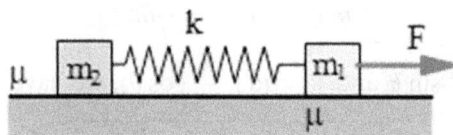

Solution

(a) Since block m_2 does not move, the force acting on m_2 does not perform work. As $N_1 \perp ds$ and $m_1 g \perp ds$, N_1 and $m_1 g$ do not work on m_1. Then, the forces F_{sp}, F and f_k do work on block m_2.

Let $v =$ speed of m_1. Applying the work–energy theorem,

$$W_{sp} + W_{N_1} + W_{gr_1} + W_{f_k} + W_{ext} = \Delta K$$

$$\Rightarrow \left(-\frac{1}{2}kx^2\right) + (0) + (0) - \mu Nx + Fx = \frac{m_1 v^2}{2}, \qquad (7.107)$$

where $N = m_1 g$

$$\Rightarrow v = \sqrt{2\left(\frac{F}{m_1} - \mu g\right) - \frac{k}{m_1}x^2} \quad \text{Ans.}$$

(b) Let block m_2 start sliding at the elongation x of the spring; then, putting $v = 0$ in equation (7.107), we have

$$F = \frac{1}{2}kx + \mu N_1 + \frac{1}{2}mv^2/x \qquad (7.108)$$

Putting $N_1 = m_1 g$, we have

$$F = \frac{1}{2}kx + \mu m_1 g + \frac{1}{2}m_1 v^2/x \qquad (7.109)$$

Since block m_2 is at the verge of sliding,

$$kx = \mu m_2 g \qquad (7.110)$$

Using the last two equations,

$$F = \frac{1}{2}\mu m_2 g + \mu m_1 g + \frac{1}{2}mv^2/x = \mu g(m_2/2 + m_1) + \frac{1}{2}m_1 v^2/x$$

For minimum value of F, put $v = 0$; then we have

$$F = \mu g(m_2/2 + m_1) \quad \text{Ans.}$$

N.B.: If you apply Newton's second law, and blindly put $a = 0$ and $kx = \mu m_2 g$ you will get $F = \mu(m_1 + m_2)g$ which is a wrong answer.

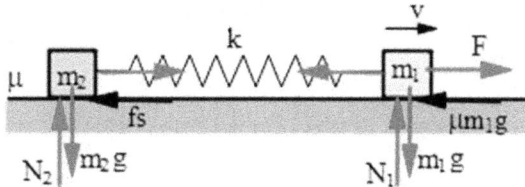

Problem 17 (a) The bob of a simple pendulum swings with a horizontal displacement x. The length of the ideal string is l. (a) Find the gravitational potential energy of a pendulum bob of mass m as the function of x. (b) If $x \ll l$, find (i) $U = f(x)$ (ii) $U = f(y)$. (c) Describe the stability of the pendulum bob.

Solution

(a) Taking the lowest position of the bob as a reference point, the potential energy of the bob at any angular position θ is

$$U = mgy = mgl(1 - \cos\theta), \tag{7.111}$$

where

$$\cos\theta = \frac{\sqrt{l^2 - x^2}}{l} \tag{7.112}$$

Using equations (7.111) and (7.112),

$$U = mgl\left(1 - \sqrt{1 - \frac{x^2}{l^2}}\right) \text{ Ans.}$$

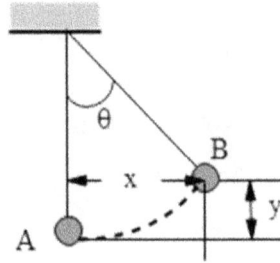

(b) (i) For $x \ll l$, x/l tends to zero or negligible; for this condition, the binomial expression,

$$(1 + x)^n \cong 1 + nx$$

The last expression can be written as

$$U = mgl\left\{1 - \sqrt{1 - \frac{x^2}{l^2}} \right\} = mgl\left\{1 - \left(1 - \frac{x^2}{l^2}\right)^{1/2}\right\}$$

$$\Rightarrow U \cong mgl\left\{1 - \left(1 - \frac{x^2}{2l^2}\right)\right\}$$

$$\Rightarrow U \cong \frac{mgx^2}{2l} \text{ Ans.}$$

(ii) The gravitational potential energy verses y is

$$U = mgy \text{ Ans.}$$

(c) The gravitational potential energy verses y is

$$U \simeq mg\frac{x^2}{2l} \ (\text{for } x \ll l)$$

(i)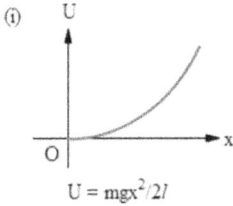

$U = mgx^2/2l$

(ii)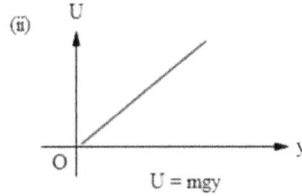

$U = mgy$

The restoring force acting on the bob near the mean position is

$$dF \cong -\frac{mg}{l}dx$$

The stiffness of the effective spring for the bob to oscillate simple harmonically is

$$k_{eff} = -\frac{dF}{dx} = \frac{mg}{l}$$

The stiffness of the equivalent spring is

$$-\frac{dF}{dx} = -\frac{d}{dx}\left(-\frac{dU}{dx}\right) = \frac{d^2U}{dx^2} = \frac{mg}{l} = k_{eff}$$

which is a positive quantity. So, the potential energy is minimum at $x = 0$. In other words, the bob possesses stable equilibrium at $x = 0$. So, the bob will oscillate about its lowest position ($x = 0$) along the x-axis but not along the y-axis. For this purpose, we have to find the restoring force by differentiating the potential energy with respect to x but not y. Recapitulating, we have to write potential energy U as the function of x and then differentiate with x. Ans.

Problem 18 A block of mass m is oscillating on a vertical massless spring of stiffness k. Find the (a) potential energy of spring–block system as the function of x and draw the U–x graph, (b) net force acting on the particle as the function of x, where $x =$ deformation of the spring, (c) speed of the block as the function of x, (d) maximum velocity of the block, (e) displacement–time equation, and (f) the nature of motion of the block with the relevant graphs.

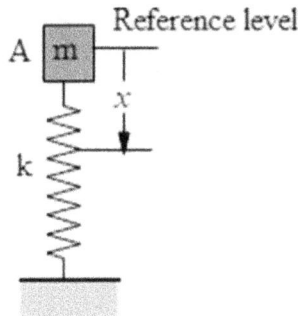

Reference level

A m

x

k

Solution

(a) The potential energy of spring + mass + Earth is
$$U = U_{sp} + U_g,$$

where $U_{sp} = \frac{1}{2}kx^2$ and $U_g = -mgx$

$$\Rightarrow U = -mgx + \frac{1}{2}kx^2 \quad \text{Ans.}$$

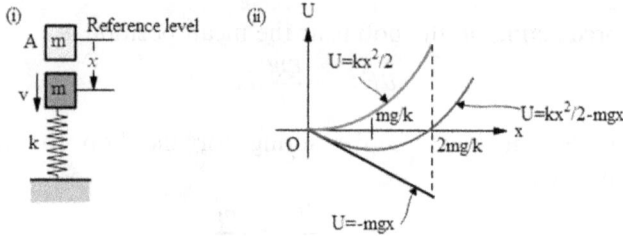

(b) The net force acting on the block is
$$F = -\frac{dU}{dx}$$

$$\Rightarrow F = -\frac{d}{dx}\left(-mgx + \frac{kx^2}{2}\right)$$

$$\Rightarrow F = mg - kx \quad \text{Ans.}$$

(c) As the block falls from rest, the change in the kinetic energy is
$$\Delta K = \frac{1}{2}mv^2$$

The change in the potential energy of the spring–block system is
$$\Delta U = U - U_A = U - 0 = -mgx + \frac{1}{2}kx^2$$

Conserving the energy of the system,
$$\Delta U + \Delta K = 0$$

Using the last three equations, we have
$$-mgx + \frac{1}{2}kx^2 + \frac{1}{2}mv^2 = 0$$

$$\Rightarrow v = \sqrt{2gx - \frac{k}{m}x^2} \quad \text{Ans.}$$

(d) The speed is maximum, when $dv/dx = 0$ or $v\,dv/dx = 0$ or $a = 0$ or $F_{net} = 0$. Putting $F = mg - kx = 0$, we have

$$x = mg/k$$

which is called stable equilibrium. Putting this value in the obtained expression of speed, we have

$$\Rightarrow v_{max} = \sqrt{2g\left(\frac{mg}{k}\right) - \frac{k}{m}\left(\frac{mg}{k}\right)^2} = g\sqrt{\left(\frac{m}{k}\right)} \quad \text{Ans.}$$

(e) The speed of the block is

$$v = \frac{dx}{dt} = \sqrt{2gx - \frac{k}{m}x^2}$$

Separating the variables,

$$\frac{dx}{\sqrt{2gx - \frac{k}{m}x^2}} = dt$$

Integrating both sides,

$$\int_o^x \frac{dx}{\sqrt{2gx - \frac{k}{m}x^2}} = \int_0^t dt$$

$$\Rightarrow \int_o^x \frac{dx}{\sqrt{\left(2g - \frac{k}{m}x\right)x}} = \int_0^t dt$$

Let $x = u^2$; then, $2u\,du = dx$. Putting the values of x and dx in the integration, we have

$$I = \int_o^x \frac{dx}{\sqrt{\left(2g - \frac{k}{m}x\right)x}} = \int_o^u \frac{2u\,du}{\sqrt{\left(2g - \frac{k}{m}u^2\right)u^2}} = t$$

$$\Rightarrow 2\int_o^u \frac{du}{\sqrt{\left(2g - \frac{k}{m}u^2\right)}} = t$$

$$\Rightarrow 2\sqrt{\frac{m}{k}}\,\sin^{-1}\sqrt{\frac{k}{2mg}}\,u = t$$

$$\Rightarrow x = \frac{2mg}{k}\sin^2\sqrt{\frac{k}{m}}\frac{t}{2}$$

$$x = \frac{mg}{k}\left(1 - \cos\sqrt{\frac{k}{m}}\,t\right) \text{ Ans.}$$

(f) The nature of motion of the block is simple harmonic having the displacement from the mean position, given as

$$y = \left|x - \frac{mg}{k}\right| = \frac{mg}{k}\left(1 - \cos\sqrt{\frac{k}{m}}\right)$$

The amplitude of oscillation is given as

$$A = \pm\frac{mg}{k} \text{ Ans.}$$

The angular frequency of oscillation is

$$\omega = \sqrt{\frac{k}{m}} \text{ Ans.}$$

The frequency of oscillation is

$$f = \frac{\omega}{2\pi} = \frac{1}{2\pi}\sqrt{\frac{k}{m}} \text{ Ans.}$$

The time of oscillation is

$$T = \frac{1}{f} = 2\pi\sqrt{\frac{m}{k}} \text{ Ans.}$$

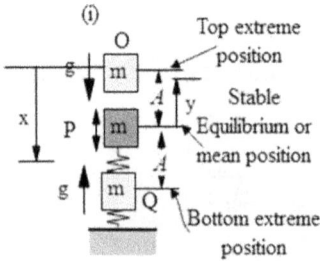

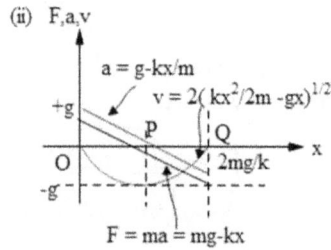

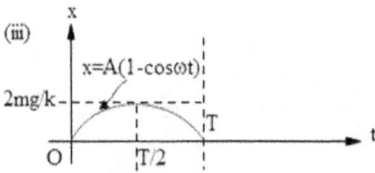

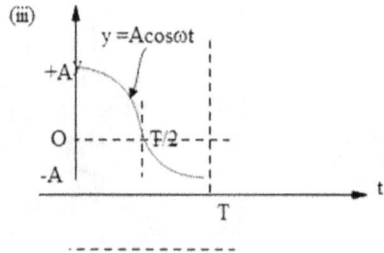

Magnitude of displacement x versus time

The displacement (magnitude and direction) y versus time

N.B.: The maximum compression of the spring is $x = 2mg/k$, when the speed of the block is $v = 0$ and acceleration is $a = g$ (up) at the lowest position. The minimum compression of the spring is $x = 0$, when the speed of the block is $v = 0$ and acceleration is $a = g$ (down) at the highest position. The compression the spring is $x = mg/k$, when the velocity of the block is $v = v_{max}$ and acceleration is $a = 0$ at the equilibrium position.

Problem 19 A pendulum of length l having a bob of mass m is hanging from a rigid support. If the bob is projected horizontally with a velocity v_0, such that the bob will leave the circular path and pass through the point of suspension, find the speed of the pendulum.

Solution

Let the bob pass through θ with a speed v. The change in potential energy of the bob is

$$\Delta U = mgl(1 + \cos \theta) \tag{7.113}$$

The change in kinetic energy of the bob is

$$\Delta K = \frac{1}{2}m(v^2 - v_0^2) \tag{7.114}$$

According to energy conservation

$$\Delta U + \Delta K = 0 \tag{7.115}$$

Using the last three equations, we have

$$mgl(1 + \cos \theta) + \frac{1}{2}m(v^2 - v_0^2) = 0$$

$$\Rightarrow v = \sqrt{v_0^2 - 2gl(1 + \cos \theta)} \tag{7.116}$$

Referring to the free-body diagram the forces on the particle are $mg\downarrow$ and $T \nearrow$

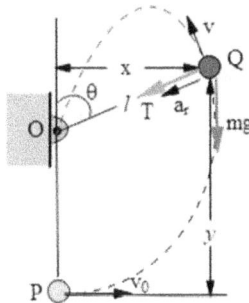

Applying Newton's second law radially we have

$$\sum F_r = mg \cos \theta - T = ma_r \tag{7.117}$$

The radial acceleration is

$$a_r = \frac{v^2}{l} \tag{7.118}$$

Using the last two equations, we have

$$mg \cos \theta - T = m\frac{v^2}{l} \tag{7.119}$$

For the string to be slack, putting $T = 0$ in the last equation, we have

$$\Rightarrow v^2 = gl \cos \theta$$

$$\Rightarrow v = \sqrt{gl \cos \theta} \tag{7.120}$$

Using equations (7.117) and (7.119),

$$v_0^2 - 2gl(1 + \cos \theta) = gl \cos \theta$$

$$v_0 = \sqrt{gl(2 + 3 \cos \theta)} \tag{7.121}$$

If the bob leaves the circular path and passes through the point of suspension following a parabolic (projectile) path.

Assuming $t =$ time of flight from Q to P, the range along the inclined plane is

$$l = \frac{1}{2}(g \cos \theta)t^2, \tag{7.122}$$

The time of flight from Q to O is given as

$$0 = vt - \frac{1}{2}(g \sin \theta)t^2$$

$$t = \frac{2v}{g \sin \theta} \tag{7.123}$$

Using the last two equations,

$$\Rightarrow v = \sqrt{\frac{gl \sin \theta}{2 \cos \theta}} \tag{7.124}$$

By using equations (7.120) and (7.124),

$$\sqrt{gl \cos \theta} = \sqrt{\frac{gl \sin \theta}{2 \cos \theta}}$$

$$\Rightarrow \tan \theta = \sqrt{2}$$

$$\Rightarrow \cos \theta = 1/\sqrt{3} \tag{7.125}$$

By using equations (7.121) and (7.125),

$$v_0 = \sqrt{gl(2 + 3(1/\sqrt{3}))} = \sqrt{(2 + \sqrt{3})gl} \quad \text{Ans.}$$

Problem 20 A pendulum of length l having a bob of mass m is hanging from a rigid support. If the bob is projected horizontally with a velocity v_0, such that the bob will leave the circular path and pass through the point of suspension, find the speed of the pendulum.

Solution

The locus equation of a projectile is

$$y = (\tan \theta)x - \frac{gx^2}{2v^2 \cos^2 \theta} \qquad (7.126)$$

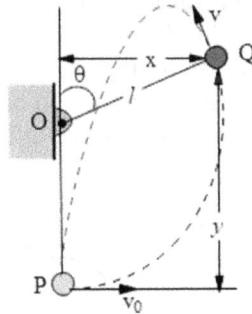

If the bob will leave the circular path and pass through the point of projection following a parabolic (projectile) path from Q to P, put $x = x = l \sin \theta$, $y = -l(1 + \cos \theta)$ and $v = \sqrt{gl \cos \theta}$ (equation (7.120) of the last problem), we have

$$-l(1 + \cos \theta) = (\tan \theta)(l \sin \theta) - \frac{g(l \sin \theta)^2}{2(gl \cos \theta)\cos^2 \theta}$$

$$\Rightarrow -l(1 + \cos \theta) = \frac{\sin^2 \theta}{\cos \theta} - \frac{\sin^2 \theta}{2 \cos^3 \theta}$$

$$\Rightarrow -l(1 + \cos \theta) = \frac{\sin^2 \theta}{\cos \theta}\left(1 - \frac{1}{2 \cos^2 \theta}\right)$$

$$\Rightarrow -l(1 + \cos \theta) = \frac{l(1 + \cos \theta)(1 - \cos \theta)}{\cos \theta}\left(1 - \frac{1}{2 \cos^2 \theta}\right)$$

$$\Rightarrow -1 = \frac{(1 - \cos \theta)}{\cos \theta}\left(1 - \frac{1}{2 \cos^2 \theta}\right)$$

$$\Rightarrow -2 \cos^3 \theta = (2 \cos^2 \theta - 1)(1 - \cos \theta)$$

$$\Rightarrow -2\cos^3\theta = 2\cos^2\theta - 2\cos^3\theta - 1 + \cos\theta$$

$$\Rightarrow 2\cos^2\theta + \cos\theta - 1 = 0$$

$$\Rightarrow \theta = 60° \text{ Ans.}$$

Equation (7.121) of the last problem is given as

$$v_0 = \sqrt{gl(2 + 3\cos\theta)}$$

Putting $\theta = 60°$ in the above equation, we have

$$v_0 = \sqrt{7gl/2} \text{ Ans.}$$

Problem 21 A smooth small ball is released from a height H along a tube. The tube is bent in the form of a semi-circle of radius R. Find the (a) velocity of the ball, (b) reaction (pressing) force offered by the tube on the ball when it reaches the highest point of the circular path, and (c) variation of reaction N offered by the tube to the ball with the change in H by taking three different values.

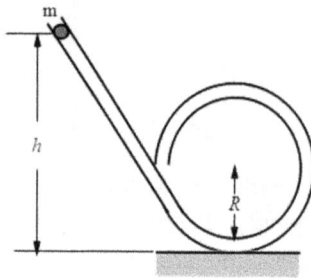

Solution

(a) When the ball falls from A to B, the change in gravitational potential energy is

$$\Delta U = -mgh = -mg(H - 2R) \tag{7.127}$$

The change in gravitational potential energy is

$$\Delta K = \frac{1}{2}mv^2 \tag{7.128}$$

Since the reaction force N does not perform work because $\vec{N} \perp d\vec{s}$. Hence, we can conserve mechanical energy of the ball. According to energy conservation,

$$\Delta U + \Delta K = 0 \tag{7.129}$$

Using the last three equations,

$$\frac{1}{2}mv^2 - mg(H - 2R) = 0$$

$$\Rightarrow v = \sqrt{2g(H - 2R)} \tag{7.130}$$

(b) Referring to the free-body diagram, let the pressing force $N\downarrow$ along with gravity $mg\downarrow$ accelerate the ball with $a_r\downarrow$.

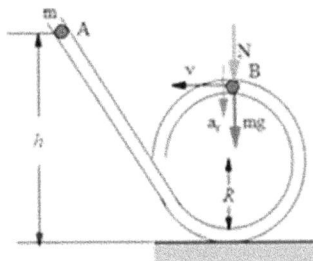

The net force equation on m at the top is

$$\sum F_r = N + mg = ma_r \tag{7.131}$$

The radial acceleration of the ball is

$$a_r = \frac{v^2}{R} \tag{7.132}$$

Using equations (7.131) and (7.132), we have

$$N + mg = \frac{mv^2}{R} \tag{7.133}$$

Using equations (7.130) and (7.133), we have

$$N = mg\left(\frac{2H}{R} - 5\right) \text{ Ans.}$$

(c) If $H = \frac{5R}{2}$, $N = 0$; if $H > \frac{5R}{2}$, N is vertically downward; if $2R < H < \frac{5R}{2}$, N is vertically upward

Problem 22 (a) In the ideal pulley–particle system, the mass m_2 is connected with a vertical spring of stiffness k. If the mas m_2 is released from rest, when the spring is undeformed, find the maximum compression of the spring. (b) If we lower body 2 slowly by an external agent till it remains at rest after withdrawing the external force, find the work done by the external agent.

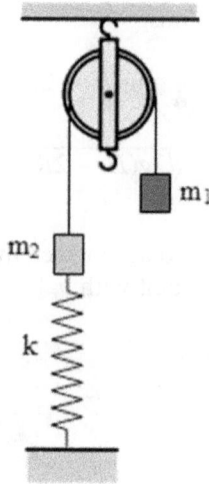

Solution

(a) Since work done by the constraint forces, tension and normal reaction is zero, $\Delta E = 0$.

Let x = maximum compression if the spring.

Then, the change in potential energy of the spring is

$$\Delta U_{sp} = \frac{k}{2}x^2 \tag{7.134}$$

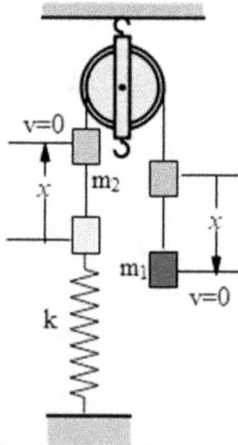

As the mass m_1 descends by a distance x, the change in its gravitational potential energy is

$$\Delta U_{gr_1} = -m_1 g x \tag{7.135}$$

As the mass m_2 ascends by a distance x, the change in its gravitational potential energy is

$$\Delta U_{gr_2} = m_2 g x \qquad (7.136)$$

Since, both m_1 and m_2 come to rest instantaneously,

$$\Delta K = 0 \qquad (7.137)$$

Conserving of energy of the pulley–particle system,

$$\Delta E = \Delta U_{sp} + \Delta U_{gr_1} + \Delta U_{gr_2} + \Delta K = 0 \qquad (7.138)$$

Using all the equation,

$$\Rightarrow \frac{1}{2} k x^2 - m_1 g x + m_2 g x + 0 = 0$$

$$\Rightarrow x = \frac{2(m_2 - m_1)g}{k} \quad \text{Ans.}$$

(b) As the lower body 2 is slowly lowered till it remains at rest, we must apply a vertical external force; so, we cannot conserve the mechanical energy of the system. However, we can apply the work–energy principle. The work done by the spring, gravity. As the mass m_1 descends by a distance x and the mass m_2 ascends by a distance x, the net work done by gravity is

$$W_{gr} = W_{gr_1} + W_{gr_1} = +m_1 g x - m_2 g x = (m_1 - m_2)g x \qquad (7.139)$$

The work done by the spring $= -$ change in potential energy of the spring is

$$W_{sp} = -\Delta U_{sp} = -\frac{k}{2} x^2 \qquad (7.140)$$

Let the work done by the external agent be W. Adding all works, the total work done is

$$W_{total} = W_{gr} + W_{sp} + W_{ext} = (m_1 - m_2)g x - \frac{k}{2} x^2 + W_{ext} \qquad (7.141)$$

As both m_1 and m_2 are moved slowly, change in kinetic energy is zero,

$$\Delta K = 0 \qquad (7.142)$$

According to the work–energy theorem,

$$W_{total} = \Delta K \qquad (7.143)$$

Using the last two equations,

$$W_{total} = 0 \qquad (7.144)$$

Using last equations (7.140) and (7.144),

$$W_{total} = (m_1 - m_2)gx - \frac{k}{2}x^2 + W_{ext} = 0$$

$$W_{ext} = -(m_1 - m_2)gx + \frac{k}{2}x^2 \qquad (7.146)$$

If we lower body 2 slowly till it remains at rest after removal of the external force, the net force acting on m_2 is zero; writing force equation on m_2, we have

$$kx + m_2g = m_1g$$

$$\Rightarrow x = \frac{(m_2 - m_1)g}{k} \qquad (7.147)$$

Using the last two equations, we have

$$W_{ext} = -\frac{(m_1 - m_2)gx}{2} = -\frac{(m_1 - m_2)g}{2}\left\{\frac{(m_1 - m_2)gx}{k}\right\}$$

$$\Rightarrow W_{ext} = -\frac{(m_1 - m_2)^2g}{2k} \quad \text{Ans.}$$

When we slowly release, $x = \frac{(m_2 - m_1)g}{k}$; when we suddenly release $x_{max} = \frac{(m_2 - m_1)g}{k}$.

Problem 23 A smooth block of mass m is released from rest from a height h. It slides and compresses a spring of stiffness k. Find the (a) velocity as the function of compression x of the spring, (b) maximum compression of the spring and (c) time of contact with the spring.

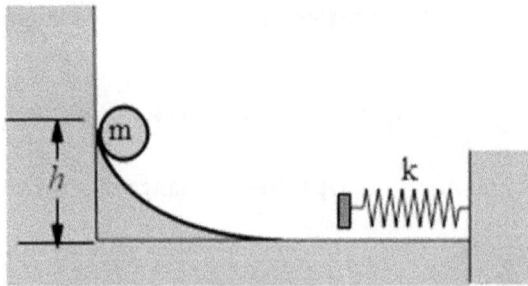

Solution

(a) Let $x =$ maximum compression of the spring. Then, the change in potential energy of the mass–spring system is

$$\Delta U = \Delta U_{sp} + \Delta U_{gr} \qquad (7.148)$$

As the spring is deformed from $x = 0$ to $x = x$, the change in potential energy of the spring is

$$\Delta U_{sp} = \frac{k}{2}x^2 \qquad (7.149)$$

As the block falls down through a vertical distance h, the change in gravitational potential energy is

$$\Delta U_{gr} = -mgh \qquad (7.150)$$

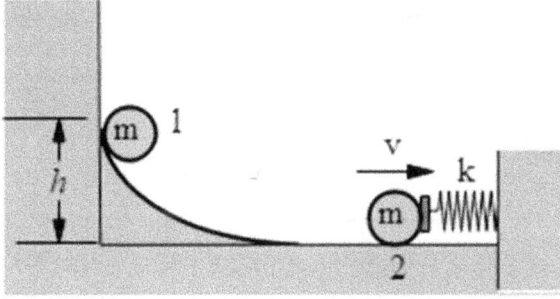

Using the last three equations, we have

$$\Delta U = \frac{k}{2}x^2 - mgh \qquad (7.151)$$

Since, the block has velocity v at the time of compression x, the change in kinetic energy is

$$\Delta K = \frac{1}{2}mv^2 \qquad (7.152)$$

Using the last two equations

$$\Delta U + \Delta K = 0,$$

we have

$$\frac{k}{2}x^2 - mgh = \frac{1}{2}mv^2$$

$$\Rightarrow v = \sqrt{2gh - \frac{k}{m}x^2} \quad \text{Ans.}$$

(b) At the time of maximum compression of the spring, $v = 0$,

$$\Rightarrow x = \sqrt{\frac{2mgh}{k}} \quad \text{Ans.}$$

(c) The time of contact of the block with the spring is equal to half of the time period of oscillation of mass–spring system, given as

$$t_{contact} = T/2 = \pi\sqrt{\frac{m}{k}} \quad \text{Ans.}$$

Problem 24 An external unknown horizontal force F pulls a smooth block of mass m attached with an ideal spring of stiffness k. Find the work done by the force F if the block moves with a velocity v when the deformation of the spring is x. Put $k = 25$ N m^{-1}, $v = 1$ m s^{-1}, $m = 1$ kg, $\mu = 0.5$ and $x = 20$ cm.

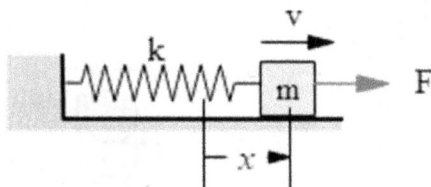

Solution

The forces involved in the spring–block system are normal reaction N, friction f, spring force and the weight of the block. Gravity and N do not perform work. The work–energy theorem states that the net work done by all nonconservative and conservative forces is equal to the change in kinetic energy of the block, which is given as

$$W_{total} = W_{ext} + W_f + W_N + W_{sp} + W_{gr} = \Delta K$$

$$\Rightarrow W_{ext} - \mu mgx + 0 - \frac{1}{2}kx^2 + 0 = \frac{1}{2}mv^2$$

$$\Rightarrow W_{ext} = \mu mgx + \frac{1}{2}kx^2 + \frac{1}{2}mv^2$$

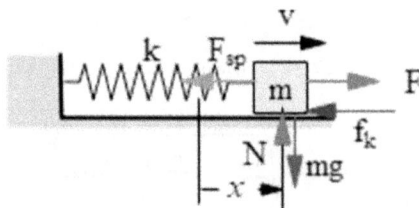

Putting $k = 25$ N m^{-1}, $v = 1$ m s^{-1}, $m = 1$ kg and $x = 20$ cm., we have

$$W_{ext} = (0.5)(1)(10)\left(\frac{1}{5}\right) + \frac{1}{2} \times 25 \times \left(\frac{1}{5}\right)^2 + \frac{1}{2} \times 1 \times 1 = 2 \text{ J} \quad \text{Ans.}$$

Problem 25 A smooth ring of mass m is fitted with a spring of stiffness k, which passes through a light rod, as shown in the figure below. The natural length of the spring is l_0. If the rod is rotated with a constant angular velocity ω, for an elongation x of the spring, find the work done by an external agent.

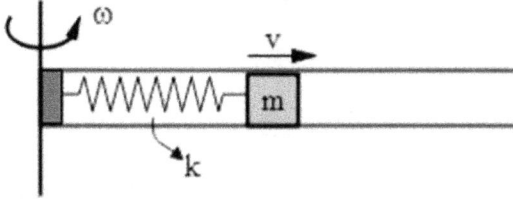

Solution

The sleeve experiences gravity, normal reactions and $F_{sp}(=kx)\leftarrow$; the rod experiences $\tau_{ext} \nwarrow F_{sp}\rightarrow$. Since, the weight $mg\downarrow$ of the ring and the normal contact force N perform no work, only the spring force and the external agent who is rotating the rod perform work.

Applying the work–kinetic energy theorem for the system (external agent + rod + spring + ring) relative to ground, the total work done by the external agent is

$$W_{ext} = \Delta U + \Delta K \tag{7.153}$$

The change in potential energy of the spring is

$$\Delta U = \frac{1}{2}kx^2 \tag{7.154}$$

The change in kinetic energy of the system is

$$\Delta K = \frac{1}{2}m(v^2 - v_0^2) + \frac{1}{2}mv_r^2 \tag{7.155}$$

Using the last three equations,

$$W_{ext} = \frac{1}{2}kx^2 + \frac{1}{2}m(v_0^2 - v_0^2) + \frac{1}{2}mv_r^2 \tag{7.156}$$

Putting $v_0 = l_0\omega$, $v = l\omega$ in equation (7.158), we have

$$W_{ext} = \frac{k}{2}x^2 + \frac{m}{2}[(l\omega)^2 - (l_0\omega)^2] + \frac{1}{2}mv_r^2 \tag{7.157}$$

Applying the work–energy theorem relative to the rod (rotating frame), we have

$$W' = W_{cf} + W_{sp} \tag{7.158}$$

The work done by the pseudo (centrifugal force) is

$$W_{cf} = \int_{l_0}^{l} F_{cf}\,dr = \int_{l_0}^{l} mr\omega^2 dr = \frac{m}{2}\{(l\omega)^2 - (l_0\omega)^2\}$$

$$\Rightarrow W_{cf} = \frac{m\omega^2}{2}(l^2 - l_0^2) \tag{7.159}$$

The work done by the spring force is

$$W_{sp} = -\frac{1}{2}kx^2 \tag{7.160}$$

Using the last two equations, the total work done by the spring force and centrifugal force is

$$W' = W_{sp} + W_{cf} = \Delta K'$$

$$\Rightarrow W' = \frac{m\omega^2}{2}(l^2 - l_0^2) - \frac{1}{2}kx^2 \tag{7.161}$$

As the ring is initially at rest and finally will stop sliding relative to the rod, the change in kinetic energy of the ring relative to the rod is

$$\Delta K' = \frac{1}{2}mv_r^2 = 0 \tag{7.162}$$

According to the work–energy theorem relative to the rotating frame, we have

$$W' = \Delta K' \tag{7.162}$$

Using the last three equations, we have

$$W' = \frac{m\omega^2}{2}(l^2 - l_0^2) - \frac{1}{2}kx^2 = 0$$

$$\frac{m\omega^2}{2}(l^2 - l_0^2) = \frac{1}{2}kx^2 \tag{7.164}$$

Using equations (7.157) and (7.164),

$$W_{ext} = \frac{k}{2}x^2 + \frac{k}{2}x^2 + \frac{1}{2}mv_r^2 = kx^2 + \frac{1}{2}mv_r^2$$

Since the ring will stop sliding, putting $v_r = 0$, we have

$$W_{ext} = kx^2 = m\omega^2(l^2 - l_0^2) = m\omega^2\{(l_0 + x)^2 - l_0^2\} \tag{7.165}$$

$$\Rightarrow W_{ext} = m\omega^2(l_0 + x)^2 - m\omega^2 l_0^2 = m\omega^2 x(2l_0 + x) = kx^2 \quad \text{Ans.}$$

When the ring remains at rest relative to the rod, the net force acting on the ring in the rotating frame is zero; by equating the forces on the ring, we have

$$F'_{net} = m\omega^2(l_0 + x) - kx = ma = 0$$

$$\Rightarrow m\omega^2(l_0 + x) = kx$$

$$\Rightarrow x = \frac{m\omega^2 l_0}{k - m\omega^2}$$

Putting this value in equation (7.165), we have

$$W_{ext} = kx^2 = k\left(\frac{m\omega^2 l_0}{k - m\omega^2}\right)^2 \quad \text{Ans.}$$

Problem 26 A smooth ring of mass m is fitted with a spring of stiffness k, which passes through a light rod. Find the work done by an external agent when the rod is spun slowly to an angular speed ω. Assume that the natural length of the spring is l_0.

Solution

The sleeve experiences gravity, normal reactions and $F_{sp}(=kx)\leftarrow$; the rod experiences $\tau_{ext} \searrow F_{sp}\rightarrow$. Since, the weight $mg\downarrow$ of the ring and the normal contact force N perform no work, only the spring force and the external agent who is rotating the rod perform work.

Applying work–kinetic energy theorem for the system (external agent + rod + spring + ring) relative to ground, the total work done by the external agent is

$$W_{ext} = \Delta U + \Delta K \tag{7.166}$$

The change in potential energy of the spring is

$$\Delta U = \frac{1}{2}kx^2 \tag{7.167}$$

The change in kinetic energy of the system is

$$\Delta K = \frac{1}{2}m(v^2 - v_0^2) + \frac{1}{2}mv_r^2 \tag{7.168}$$

Using the last three equations,

$$W_{ext} = \frac{1}{2}kx^2 + \frac{1}{2}m(v_0^2 - v_0^2) + \frac{1}{2}mv_r^2 \tag{7.169}$$

Putting $v_0 = 0$, $v = l\omega$ and $v_r = 0$ (as the ring is slowly moving because of slow speeding up of the rod), in equation (7.169), we have

$$W_{ext} = \frac{k}{2}x^2 + \frac{m}{2}(l\omega)^2 = \frac{k}{2}x^2 + \frac{m}{2}l^2\omega^2$$

$$\Rightarrow W_{\text{ext}} = \frac{k}{2}x^2 + \frac{m}{2}(l_0 + x)^2\omega^2 \tag{7.170}$$

When the ring remains nearly at rest relative to the rod, the net force acting on the ring in the rotating frame is zero; by equating the forces on the ring, we have

$$F'_{\text{net}} = m\omega^2(l_0 + x) - kx = ma = 0$$

$$\Rightarrow m\omega^2(l_0 + x) = kx$$

$$\Rightarrow x = \frac{m\omega^2 l_0}{k - m\omega^2} \tag{7.171}$$

Using equations (7.170) and (7.171), we have

$$\Rightarrow W_{\text{ext}} = \frac{k}{2}x^2 + \frac{m}{2}(l_0 + x)^2\omega^2$$

$$W_{\text{ext}} = \frac{kml_0^2\omega^2(k + ml_0^2\omega^2)}{2(k - m\omega^2)^2} \quad \text{Ans.}$$

Problem 27 A block of mass m is connected with a rigid wall by a light spring of stiffness k. If the block is pushed with a velocity v_0, it oscillates back and forth and stops. (a) Find the work done friction till the block stops. Find the maximum compression of the spring. Assume that $\mu =$ coefficient of kinetic friction between block and ground,

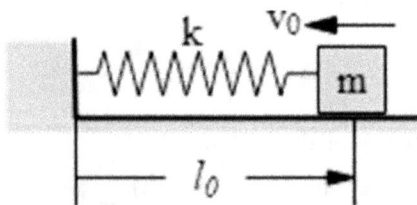

Solution
(a) Referring to the free-body diagram, as the block moves, the spring force $F_s(=kx)$ increases till it nullifies $f_k(=\mu mg)$. In consequence, the block will stop. So, the change in kinetic energy of the block is

$$\Delta K = -\frac{1}{2}mv_0^2 \tag{7.172}$$

When the block stops finally, let us assume that the spring is deformed by an amount x.

Then, the change in potential energy of the spring is

$$\Delta U = \frac{kx^2}{2} \tag{7.173}$$

Applying the work–energy theorem, the work done by friction W_f is given as

$$W_f = \Delta K + \Delta U_{sp} \tag{7.174}$$

Using the last three equations,

$$W_f = \frac{kx^2}{2} - \frac{1}{2}mv_0^2 \tag{7.175}$$

We can calculate x as follows.

Since the block is at the verge of sliding, the net force acting on the block is

$$F_{net} = kx - \mu mg = 0$$

$$\Rightarrow x = \frac{\mu mg}{k} \tag{7.176}$$

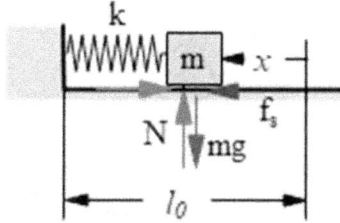

Using the last two equations, we have

$$W_f = \frac{\mu^2 m^2 g^2}{2k} - \frac{1}{2}mv_0^2 \quad \text{Ans.}$$

(b) At the time of maximum compression of the spring, the velocity of the block is zero. So, the change in kinetic energy is zero. The work done by friction is

$$W_f = -\mu mgx$$

Equating this with the obtained expression, we have

$$W_f = \frac{kx^2}{2} - \frac{1}{2}mv_0^2 = -\mu mgx$$

$$\Rightarrow kx^2 + 2\mu mgx - \frac{1}{2}mv_0^2 = 0$$

Solving this quadratic equation, we have

$$x = \frac{\mu mg}{k}\left(-1 + \sqrt{1 + \frac{kv_0^2}{m\mu^2 g^2}}\right) \quad \text{Ans.}$$

Problem 28 A smooth block of mass m is placed at the bottom of a massless smooth wedge which is placed on a horizontal surface. When we push the wedge with a constant unknown force (F, say), the block moves up the wedge. Find the work done by the external agent when the block has a speed v and is situated at a height y on the wedge.

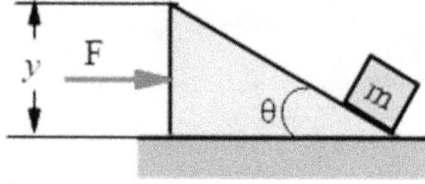

Solution

As the block moves up, work done by gravity is

$$W_{gr} = -mgy.$$

The reaction forces do not perform work as a whole. Since, the wedge moves horizontally, gravity does not perform work on it.

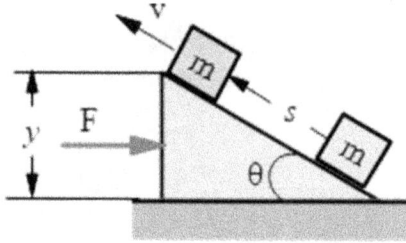

The external agent performs work W_{ext}, say. Then the total work done on the wedge–block system is

$$W = W_{ext} + W_{gr} = W_{ext} - mgy$$

This is equal to change in kinetic energy, that is, ΔK of the block because the wedge is massless.

Applying the work–energy theorem for the wedge–block system, we have

$$W_{ext} - mgy = \Delta K$$

$$\Rightarrow W_{ext} = mgy + \Delta K \qquad (7.177)$$

The kinetic energy of the block changes by

$$\Delta K = \frac{1}{2}mv^2 \qquad (7.178)$$

Using equations (7.177) and (7.178), we have

$$W_{ext} = \frac{1}{2}mv^2 + mgy \quad \text{Ans.}$$

Problem 29 In the above example, if a constant horizontal force F acts on a wedge, using the work–energy theorem on a block, find the (a) velocity of the block relative to the wedge at its top in terms of F, m, x and y. (b) If the block will just reach at the top of the wedge, find the applied force F.

Solution

(a) If a constant horizontal force F acts on the wedge, the wedge moves with a constant acceleration, A, say. Let us now find the work done by each force acting on the block relative to the wedge.

As the block moves up, work done by gravity is

$$W_{gr} = -mgy \tag{7.179}$$

The reaction forces do not perform work as a whole. Since, the wedge moves horizontally, gravity does not perform work on it. Let us use the work–energy theorem sitting on the wedge. As the wedge moves with an acceleration A, the pseudo-force mA acts on the block to the left. The work done by the pseudo-force on the block is

$$W_{pseudo} = mAx \tag{7.180}$$

The work done by normal reaction on the block is zero, because normal reaction N is perpendicular to the displacement of the block relative to the wedge as the block moves along the slant relative to the wedge.

The external agent (force F) performs work W_{ext}, say. Using the last two equations, the total work done on the block system is

$$W' = W_{pseudo} + W_{gr} + W_N = mAx - mgy + 0$$

$$\Rightarrow W' = mAx - mgy \tag{7.181}$$

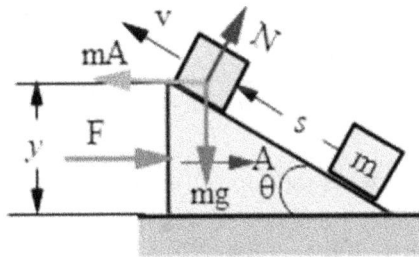

This is equal to change in KE, that is, ΔK of the bead because the wedge is massless. If v_r = velocity of the block relative to the wedge, the kinetic of the block changes by

$$\Delta K' = \frac{1}{2}mv_r^2 \tag{7.182}$$

Using the work–energy theorem relative to the wedge, we have

$$W' = \Delta K' \tag{7.183}$$

Using the last three equations, we have

$$\Rightarrow mAx - mgy = \frac{1}{2}mv_r^2$$

$$\Rightarrow v_r = \sqrt{2(Ax - gy)} \qquad (7.184)$$

The net force acting on the system $(M + m)$ is

$$F = MA + ma_x \qquad (7.185)$$

As the horizontal acceleration of the block is

$$a_x = (A \sin \theta)\sin \theta + (g \sin \theta)\cos \theta$$

$$\Rightarrow a_x = \sin \theta (A \sin \theta + g \cos \theta) \qquad (7.186)$$

Using the last two equations, we have

$$F = MA + m \sin \theta (A \sin \theta + g \cos \theta)$$

$$A = \frac{F - mg \sin \theta \cos \theta}{M + m \sin^2 \theta} \qquad (7.187)$$

Using equations (7.184) and (7.187), the velocity of the block relative to the wedge at the time of its escaping is

$$v_r = \sqrt{2\left(\frac{F - mg \sin \theta \cos \theta}{M + m \sin^2 \theta} x - gy\right)}$$

Putting $x = y \cot \theta$, we have

$$v_r = \sqrt{2\left(\frac{F - mg \sin \theta \cos \theta}{M + m \sin^2 \theta} y \cot \theta - gy\right)}$$

$$v_r = \sqrt{2\left(\frac{F - mg \cos^2 \theta}{M + m \sin^2 \theta} - g\right)y} \quad \text{Ans.}$$

(b) If the block just reaches the top of the wedge,

$$v_r = \sqrt{2\left(\frac{F - mg \cos^2 \theta}{M + m \sin^2 \theta} - g\right)y} = 0$$

$$\Rightarrow \frac{F - mg \cos^2 \theta}{M + m \sin^2 \theta} - g = 0$$

$$\Rightarrow F - mg \cos^2 \theta = (M + m \sin^2 \theta)g$$

$$\Rightarrow F = (M + m)g \quad \text{Ans.}$$

Problem 30 A body is thrown with a velocity v_0 at an angle θ_0 with the horizontal. Find the (a) instantaneous power delivered by gravity after a time t measured from the instant of projection, (b) average power delivered by gravity during time t, and (c) ratio of average power delivered by gravity over time t measured from the instant of projection and instantaneous power delivered by gravity at time t, for $t = \frac{v_0 \sin \theta_0}{2g}$.

Solution

(a) Power delivered by gravity after a time t when the particle has velocity v is:

$$P_g = m\,\vec{g}\cdot\vec{v},$$

where $\vec{g} = -g\,\hat{j}$ and $\vec{v} = v_0 \cos \theta_0 \hat{i} + (v_0 \sin \theta_0 - gt)\hat{j}$

$$\Rightarrow P_g = (-g\,\hat{j}) \cdot \{\vec{v} = v_0 \cos \theta_0 \hat{i} + (v_0 \sin \theta_0 - gt)\hat{j}\,\}$$

$$\Rightarrow P_g = -mg(v_0 \sin \theta_0 - gt) \text{ Ans.}$$

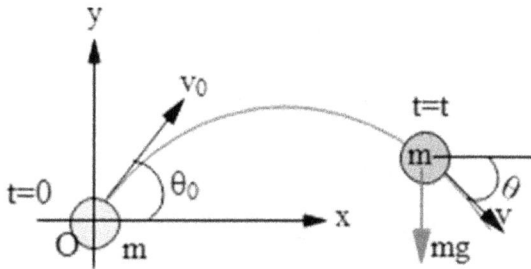

(b) The average power is

$$P_{av} = \frac{\int_0^t P\,dt}{t}$$

Putting $P = P_g = -mg(v_0 \sin \theta_0 - gt)$, we have

$$P_{av} = -\frac{mg}{t} \int_0^t (v_0 \sin \theta_0 - gt)\,dt$$

$$\Rightarrow P_{av} = -mg\left(v_0 \sin \theta_0 - \frac{gt}{2}\right) \text{ Ans.}$$

(c) Putting $t = \frac{v_0 \sin \theta_0}{2g}$, the gravitational power delivered at time t is

$$P = -mg(v_0 \sin \theta_0 - gt) = -mg(v_0 \sin \theta_0 - v_0 \sin \theta_0/2)$$

$$\Rightarrow P = -mg(v_0 \sin \theta_0 - v_0 \sin \theta_0/2)$$

Putting $t = \frac{v_0 \sin \theta_0}{2g}$, the gravitational power delivered is

$$P' = -mg(v_0 \sin \theta_0 - gt/2) = -mg(v_0 \sin \theta_0 - v_0 \sin \theta_0/4) = -3mgv_0 \sin \theta_0/4$$

Then, $P'/P = (-3mgv_0 \sin \theta_0/4)/(-mgv_0 \sin \theta_0/2) = 3/2$ Ans.

Problem 31 A horizontal constant force F pulls a block of mass m placed on a horizontal surface. If the coefficient of kinetic friction between the block and ground is μ, find the power delivered by the (a) external agent, (b) friction, and (c) net force as the function of time t measured from the beginning of action of the force.
  Solution

(a) Referring to the free-body diagram, the forces acting on the block are $F\rightarrow$, kinetic friction $f\leftarrow$, $mg\downarrow$ and $N\uparrow$. The power delivered by the external agent (force F) is

$$P_{\text{ext}} = \vec{F} \cdot \vec{v} = Fv \cos 0° = Fv, \tag{7.188}$$

The velocity v of the block after a time t is

$$v = \int_0^t a \, dt \tag{7.189}$$

The acceleration a is given as

$$a = \frac{F - f}{m} \tag{7.190}$$

Using equations (7.188), (7.189) and (7.190), we have

$$P_{\text{ext}} = F\left(\int \frac{F - f}{m} dt\right) = F\left(\frac{F - f}{m}\right)t \tag{7.191}$$

Law of kinetic friction:

$$f = \mu mg \tag{7.192}$$

Using equations (7.191) and (7.192), we have

$$P_{\text{ext}} = \frac{F(F - \mu mg)t}{m} \quad \text{Ans.}$$

(b) The power delivered by friction is

$$P_f = \vec{f} \cdot \vec{v} = fv \cos 180° = -fv \tag{7.193}$$

Putting $f = -\mu mg$ and $v = at$ in equation (7.193), we have

$$P_f = -\mu mg(at) \tag{7.194}$$

Using equations (7.191) and (7.193),

$$P_f = -\mu mg\left(\frac{F-f}{m}\right)t \tag{7.195}$$

Putting $f = -\mu mg$ in equation (7.191), we have

$$P_f = -\mu mg\left(\frac{F-\mu mg}{m}\right)t = \mu g(F - \mu mg)t \quad \text{Ans.}$$

(c) The power delivered by the net force is

$$P_{\text{net}} = \vec{F}_{\text{net}} \cdot \vec{v} = m\vec{a} \cdot \vec{v} = mav\cos 0° = mav \tag{7.196}$$

Putting $v = at$ in equation (7.193), we have

$$P_{net} = ma(at) = ma^2t \tag{7.197}$$

Using equations (7.191) and (7.197),

$$P_{net} = ma^2t = m\left(\frac{F-f}{m}\right)^2 t \tag{7.198}$$

Putting $f = -\mu mg$ in equation (7.198), we have

$$P_{net} = ma^2t = m(F/m - \mu g)^2t \quad \text{Ans.}$$

N.B.: Power delivered by a force F is given as $P_F = \vec{F} \cdot \vec{v}$, where $\vec{v} = \vec{v}_0 + \int a\, dt$, where $a = \frac{F_{\text{net}}}{m}$ (but not F/m)

Problem 32 The potential energy of a 2 kg particle free to move along the x-axis is given by

$$U(x) = \left(\frac{x}{b}\right)^4 - 5\left(\frac{x}{b}\right)^2 \text{J},$$

where $b = 1$ m. Find the (a) minimum potential energy, (b) maximum speed by assuming the total mechanical energy (i) 36 J and (ii) –4 J. (c) Plot the potential energy, identifying the extremum points and the values of x in both sides of the origin where the particle can be found.

Solution

(a) Putting $b = 1$ m, the potential energy is

$$U(x) = x^4 - 5x^2 \tag{7.199}$$

$$\Rightarrow \frac{dU}{dx} = 4x^3 - 10x \qquad (7.200)$$

$$\Rightarrow \frac{d^2U}{dx^2} = 12x^2 - 10 \qquad (7.201)$$

For extremum points, $\frac{dU}{dx} = 0$

$$\Rightarrow \frac{dU}{dx} = 4x^3 - 10x = 0$$

$$\Rightarrow (2x^2 - 5)x = 0$$

$$\Rightarrow x = 0, \sqrt{\frac{5}{2}} \text{ are the extremum points.}$$

Putting $x = 0$ in equation (7.201), we have

$$\Rightarrow \frac{d^2U}{dx^2} = 12x^2 - 10 = -10, \text{ at } x = 0$$

So, at $x = 0$ is a local maximum.

Putting $x = \pm\sqrt{\frac{5}{2}}$ in equation (7.201), we have

$$\Rightarrow \frac{d^2U}{dx^2} = 12x^2 - 10 = 12\left(\frac{5}{2}\right) - 10 = 20$$

So, at $x = x'(\text{say}) = \pm\sqrt{\frac{5}{2}}$ is a minimum possessing minimum potential energy which is given by putting the value of $x = x'\pm\sqrt{\frac{5}{2}}$ in equation (7.199). So, we have

$$U_{\min} = \left(\pm\sqrt{\frac{5}{2}}\right)^4 - 5\left(\pm\sqrt{\frac{5}{2}}\right)^2 = -\frac{25}{4} \text{ J} = -6.25 \text{ J Ans.}$$

(b) (i) For maximum kinetic energy, the potential energy must be minimum; then the maximum kinetic energy is

$$K_{\max} + U_{\min} = K_{\max} - \frac{25}{4} \text{ J} = E(=36 \text{ J})$$

$$\Rightarrow K_{\max} = \frac{25}{4} + 36 = \frac{169}{4}$$

$$\Rightarrow \frac{1}{2}mv_{\max}^2 = \frac{169}{4}$$

$$\Rightarrow \frac{1}{2}(2)v_{\max}^2 = \frac{169}{4}$$

$$\Rightarrow v_{\max} = 6.5 \text{ m s}^{-1} \text{ Ans.}$$

(ii) Now, $E = K + U = -4$ J

The particle is in the region $-2 < x < -1$ and $1 < x < 2$,

$$x = \pm\sqrt{\frac{5}{2}}, \quad U_{min} = -6.25 \text{ J}$$

$$\therefore K + (-6.25) = -4$$

$$K = 2.25 \text{ J}$$

$$\frac{1}{2}mv_{max}^2 = 2.25$$

Or, $v_{max} = 1.5$ m s^{-1}. Ans.

(c) Thus, in a region $-3 < x < 3$ we have $U_{min} = -6.25$ J and $U_{max} = 36$ J. Using the data, the following graph is drawn.

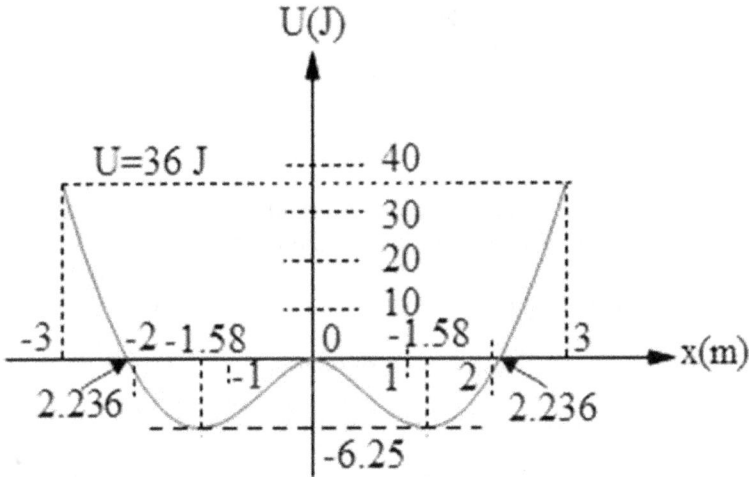

We can see that $U = 0$ at $x = 0$ and $x = +2.236$ and -2.236; U is minimum at $x = 1.58$ and -1.58; U is locally maximum at $x = 0$ and then its positive values increase to infinite as x increases in both sides. The graph is drawn between $x = +3$ and -3 for which $U = 36$ J. Ans.

Problem 33 A block of mass m is sliding relative to a wedge through a distance l downwards (along the slope). If the wedge is moved with constant velocity, assuming the coefficient of kinetic friction between wedge and block as μ, find the total work performed by the (a) normal reaction, and (b) friction forces between the block and wedge.

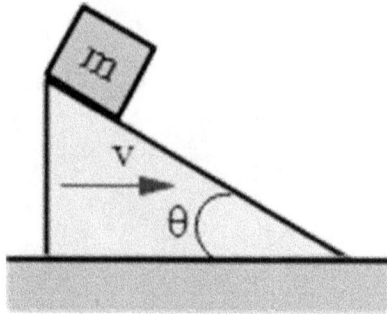

Solution

(a) Referring to the free-body diagram, the forces acting on the block and wedge are $N_1 \nearrow$ and $N_2 \swarrow$ (normal reactions), $f_1 \nwarrow$ and $f_2 \searrow$ (frictional forces). The elementary work done by the normal reactions on the block and wedge are given as follows:

$$dW_{N_1} = \vec{N_1} \cdot d\, \vec{s_1}$$

$$dW_{N_2} = \vec{N_2} \cdot d\, \vec{s_2},$$

where $d\, \vec{s_1}$ and $d\, \vec{s_2}$ are the elementary displacements of the wedge and block, respectively. Then, the sum of these elementary works is

$$dW = dW_{N_1} + dW_{N_2} = \vec{N_1} \cdot d\, \vec{s_1} + \vec{N_2} \cdot d\, \vec{s_2}.$$

Substituting $d\, \vec{s_2} = d\, \vec{s_{21}} + d\, \vec{s_1}$, we have

$$dW_N = \vec{N_1} \cdot d\, \vec{s_1} + \vec{N_1} \cdot (d\, \vec{s_{21}} + d\, \vec{s_1})$$

Rearranging the terms, we have

$$dW_N = (\vec{N_1} + \vec{N_2}) \cdot d\, \vec{s_1} + \vec{N_2} \cdot d\, \vec{s_2}$$

Since, $\vec{N_1} + \vec{N_2} = 0$ (Newton's third law) and $d\, \vec{s_{21}} \perp \vec{N_2}$ ($d\, \vec{s_{21}} =$ elementary displacement of the block relative to wedge), we have $dW_N = 0$. So, the total work done by the constraint forces (N_1 and N_2) is zero. Ans.

(b) Since the block slides on the wedge, f_1 and f_2 (kinetic friction) are not the constraint forces. However, the static friction is a constraint force as it prevents sliding.

Since the block moves down relative to the wedge by a distance l, as described earlier, the total work done by kinetic friction is,

$$W_f = W_{f_1} + W_{f_2} = -f\, l$$

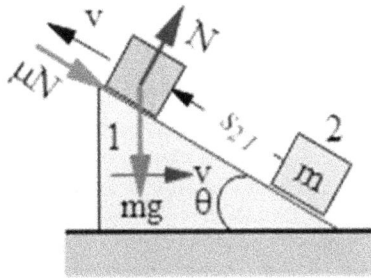

Law of kinetic friction: $f = \mu N$, where $N = mg \cos \theta$, because accleration of the wedge is zero. So, we have

$$f = \mu \, mg \cos \theta$$

Using the last two equations, the total work done by the interacting forces between the block and wedge is

$$W = -\mu mg \, l \cos \theta \quad \text{Ans.}$$

N.B.: The total work done by kinetic friction is frame-independent; it does not matter if the wedge is moving or fixed.

Problem 34 A person 1 stands on an elevator moving with an initial velocity v and upward acceleration a. A person 2 is standing on the elevator. What is the work done by normal contact force and gravity on person 1 as observed by person 2 standing on the elevator and person 3 standing on the ground?

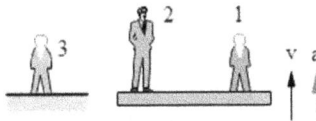

Solution

The forces on the people are $mg\downarrow$ and $N\uparrow$ pseudo-force $ma\downarrow$. Since the displacement of person 1 relative to person 2 is zero, the forces do not perform work relative to person 2. However, the displacement of person 1 relative to person 3 (ground) is not zero. So, the forces will perform non-zero work relative to the ground (person 3).

The work done by the normal reaction is

$$W_N = \int \vec{N} \cdot d\, \vec{s}_{13} = \vec{N} \cdot \int d\, \vec{s} = \vec{N} \cdot \vec{s} \tag{7.202}$$

The force equation on m:

$$\vec{N} = m\,(g + a)\hat{j} \tag{7.203}$$

Kinematics:

$$\vec{s} = \vec{v}_{13}t + \frac{1}{2}\vec{a}_{13}t^2 = \left(vt + \frac{1}{2}at^2\right)\hat{j} \qquad (7.204)$$

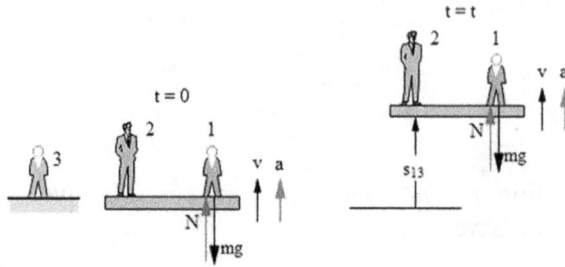

Using equations (7.202), (7.203) and (7.204) we have

$$W_N = m(g + a)\left(vt + \frac{1}{2}at^2\right) \text{ Ans.}$$

Similarly, the work done by gravity is

$$W_{gr} = m\,\vec{g}\cdot\vec{s},$$

where $\vec{g} = -g\,\hat{j}$ and $\vec{s} = (vt + \frac{1}{2}at^2)\hat{j}$

$$\Rightarrow W_{gr} = -mg\cdot\left(vt + \frac{1}{2}at^2\right) \text{ Ans.}$$

Problem 35 A smooth ring of mass m is moved along a rigid wire made in the shape of a curve, given as $y(x) = A \sin kx$. If the curve lies in the vertical plane, find the gravitational potential energy U as the function of x and test for the stability of the ring.

$$U(x) = A \sin kx,$$

where A, b and k are positive constants. Plot the potential energy and test for the stability of the ring.

Solution

(a) The relation between U and x is given as

$$U(x) = mgy = mgA \sin kx$$

The first derivative with respect to x is

$$U'(x) = \frac{dU}{dx} = mgAk \cos kx \qquad (7.205)$$

7-94

The second derivative relative to x is

$$U''(x) = \frac{dU'(x)}{dx} = -mgAk^2 \sin kx \qquad (7.206)$$

For equilibrium of the ring

$$\frac{dU}{dx} = \frac{d}{dx}(mgA \sin kx) = 0$$

$$\Rightarrow \cos kx = 0$$

$$\Rightarrow kx = \frac{\pi}{2}, 3\frac{\pi}{2}, 5\frac{\pi}{2}, 7\frac{\pi}{2}, \ldots, (2n-1)\frac{\pi}{2}; \quad n = 1, 2, 3, \ldots$$

$$\Rightarrow x = \frac{\pi}{2k}, 3\frac{\pi}{2k}, 5\frac{\pi}{2k}, 7\frac{\pi}{2k}, \ldots, (2n-1)\frac{\pi}{2}; \quad n = 1, 2, 3, \ldots \qquad (7.207)$$

Putting $x = \frac{\pi}{2k}, 5\frac{\pi}{2k}, 9\frac{\pi}{2k}, \ldots, (2n-1)\frac{\pi}{2}; \quad n = 1, 3, 5, \ldots$, in equation (7.206), we have

$$U''(x) = -mgAk^2(+1) = -mgAk^2$$

Since the above value is less than zero, U is maximum at all the above locations (values of x) and the ring has unstable equilibrium.

Putting $x = 3\frac{\pi}{2k}, 7\frac{\pi}{2k}, 11\frac{\pi}{2k}, \ldots, (2n-1)\frac{\pi}{2}; \quad n = 2, 4, 6, \ldots$, in equation (7.206), we have

$$U''(x) = -mgAk^2(-1) = +mgAk^2$$

Since the above value is greater than zero, U is minimum at all the above locations (values of x) and the ring has stable equilibrium. So, the ring has stable equilibrium at the lowest points such as 2, 4, 6, ... and unstable equilibrium at the top points such as 1, 3, 5, ... of the curved surface, as shown in the following figure.

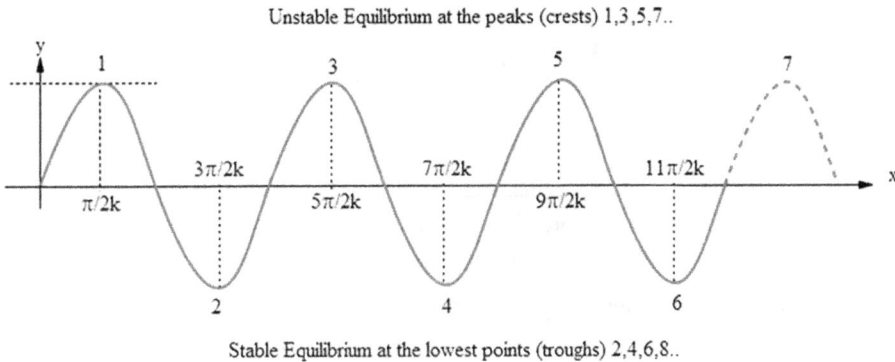

Unstable Equilibrium at the peaks (crests) 1,3,5,7..

Stable Equilibrium at the lowest points (troughs) 2,4,6,8..

Problem 36 Five bodies are placed at the points A, B and C of the smooth surfaces as shown in the figure below. Plot the gravitational potential energy and test for the stability of the bodies.

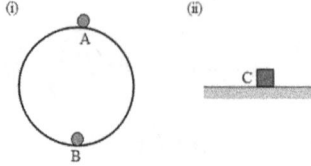

Solution

(a) The gravitational potential energy is given as
$$U(x) = mgy$$

The shape of the surface (curve) is given as
$$U(x) = mgy(x)$$

The maximum and minimum potential energy can be found by locating the highest and lowest points of the given curves. At point A height is maximum; so, gravitational potential energy U is maximum. Then, at A the body has unstable equilibrium. If we shift the body to either to the left or right, the unbalanced force due to gravity is nearly pointing in the direction of displacements; so, the body will never return to point A.

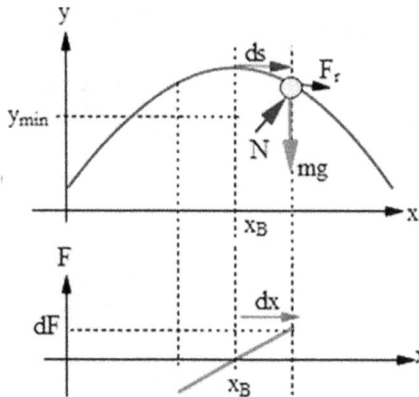

(i) Potential energy is maximum at the highest point A of the bowl. (ii) The net force is $+dF$ that acts to right for a displacement dx (to right). The ball does not come back to A.

At point B, the potential energy is minimum; so, the body has stable equilibrium at B.

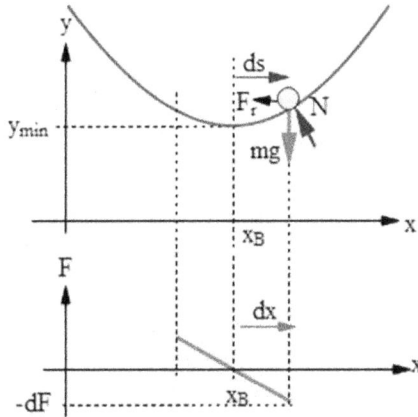

(i) Potential energy is minimum at the lowest point B of the bowl.(ii) the net force is -dF that acts to left(-ve) for a displacement dx (to right); dF and dx are opposite;so, the ball returns to B.

If you shift the particle to the left or right of C, height remains constant; so, the gravitational potential energy does not change near C. Then, the particle is said to be in neutral equilibrium at C having a zero net force.

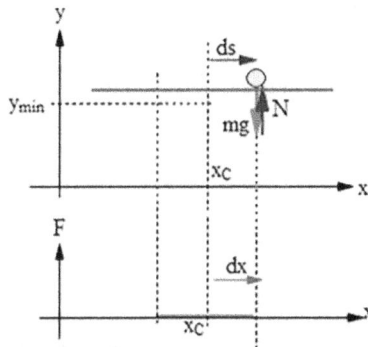

(i) Potential energy remains constant about the point C of the surface.
(ii) the zero net force acts an the ball after a displacement dx (to right or left);so, the ball stays there without returning to C.

Problem 37 A particle moves in a force field where the potential energy of the particle due to that field changes as $U = \frac{k}{2}(x^2 + y^2)$. Prove that the force acting on the particle is $\overrightarrow{F} = -k\,\overrightarrow{r}$, where $\overrightarrow{r}$ = position vector of the particle. Can you call this force central?

Solution

The potential energy of the particle is

$$U = \frac{k}{2}(x^2 + y^2)$$

The force acting on the particle is

$$\vec{F} = -\left(\frac{\partial U}{\partial x}\hat{i} + \frac{\partial U}{\partial x}\hat{j} + \frac{\partial U}{\partial z}\hat{k}\right)$$

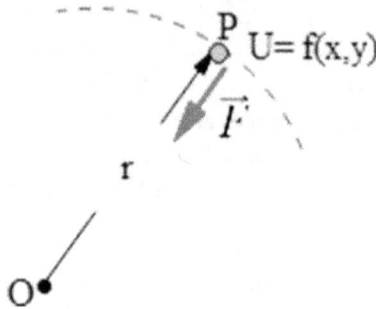

Using the last two equations,

$$\vec{F} = -\left(\frac{\partial\left\{\frac{k}{2}(x^2 + y^2)\right\}}{\partial x}\hat{i} + \frac{\partial\left\{\frac{k}{2}(x^2 + y^2)\right\}}{\partial x}\hat{j} + \frac{\partial\left\{\frac{k}{2}(x^2 + y^2)\right\}}{\partial z}\hat{k}\right)$$

$$\Rightarrow \vec{F} = -k(x\hat{i} + y\hat{j})$$

$$\Rightarrow \vec{F} = k(x\hat{i} + y\hat{j}) = -k\vec{r},$$

where $\vec{r} = (x\hat{i} + y\hat{j})$. So, the force is central as it points towards the fixed point (origin) O.

Problem 38 A small smooth ball of mass m is released from a height h. After sliding down the track AE, it moves in a vertical circular arc EBC of radius R. If the ball can move from C to D without bouncing find the (a) range of h (b) range of H (c) critical value of θ (d) possibility of the given condition of climbing the track CD without bouncing if $\theta = 60°$ and $\theta = 37°$ (e) range of H for $\theta = 37°$ (f) change in normal reaction just before and after crossing point C for $\theta = 37°$ (g) change in radial acceleration just before and after crossing point C for $\theta = 37°$ (h) change in tangential just before and after crossing point C for $\theta = 37°$.

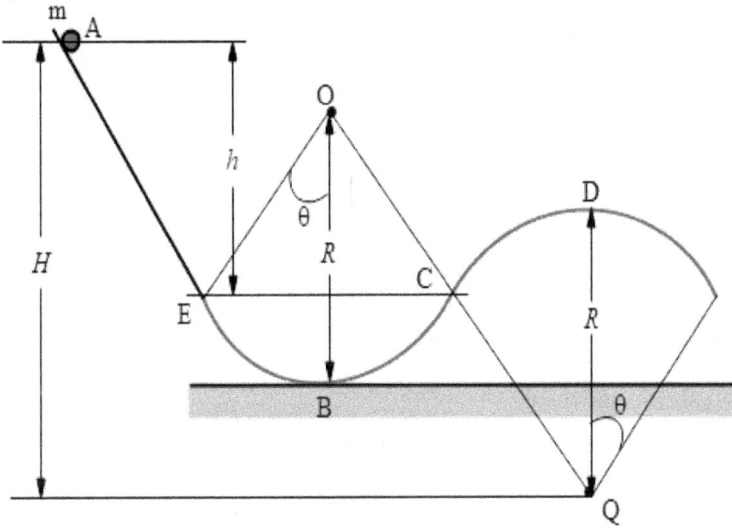

Solution

(a) In order not to lose contact, the normal reaction

$$N = mg \cos\theta - \frac{mv^2}{R} \geqslant 0$$

$$\Rightarrow v \leqslant \sqrt{gR \cos\theta} \qquad (7.208)$$

To just go to the top D of the vertical circular track, conserving the energy between C and D, we have

$$\frac{mv'^2}{2} + mgR(1 - \cos\theta) = \frac{mv^2}{2}$$

$$\Rightarrow v \geqslant \sqrt{2gR(1 - \cos\theta)} \qquad (7.209)$$

From last two equations, the value of velocity v at C must lie in the interval,

$$\sqrt{gR \cos\theta} \geqslant v \geqslant \sqrt{2gR(1 - \cos\theta)} \quad \text{Ans.}$$

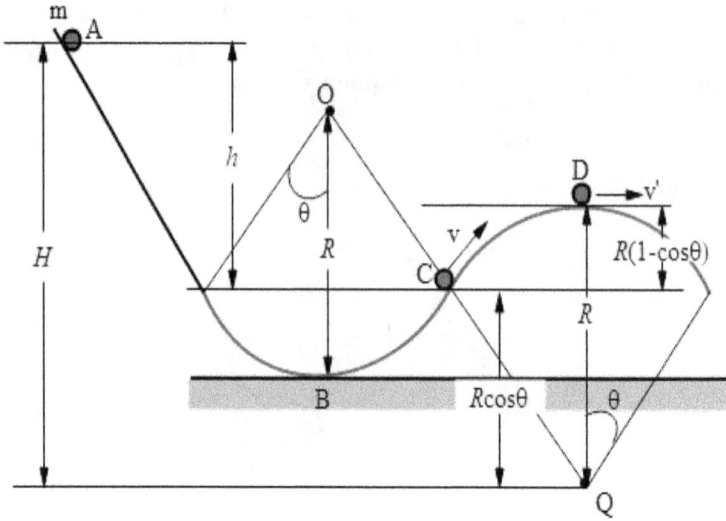

(b) **By conserving energy between** A **and** C, **we have**

$$v = \sqrt{2gh}$$

Using last two expressions, the values of h can be given as

$$h \leqslant \frac{R\cos\theta}{2} \quad \text{and} \quad h \geqslant R(1 - \cos\theta)$$

So, the values of h must lie between

$$R(1 - \cos\theta) \leqslant h \leqslant \frac{R}{2}\cos\theta \quad \text{Ans.}$$

(c) If we want to find the values of H, putting $h = H - R\cos\theta$, we have

$$R(1 - \cos\theta) \leqslant (H - R\cos\theta) \leqslant \frac{R}{2}\cos\theta$$

$$\Rightarrow R \leqslant H \leqslant \frac{3R}{2}\cos\theta \quad \text{Ans.}$$

(d) If we put $R\cos\theta = h = \frac{3R}{2}\cos\theta$, we get $\cos\theta = 2/3$; or, $\theta = 48.19°$ (nearly), which is a critical angle at which both velocities will be equal. For the inequality to hold good, $\cos\theta > 2/3$; or, θ must be less than $48.19°$ Ans.

(e) If $\theta = 60°$ $\cos\theta = 1/2 < 2/3$; so the above inequality does not hold good. If $\theta = 37°$, $\cos\theta = 4/5 < 2/3$; so the above inequality holds good and the value of H can be given by the inequality

$$R \leqslant H \leqslant \frac{3R}{2}\cos 37°$$

$$\Rightarrow R \leqslant H \leqslant 6R/5$$

So, the value of H must lie between R and $6R/5$ so that the ball will climb the hump (circular track) without bouncing. Ans

(f) If $H = R$, then, the velocity at C is

$$v = \sqrt{2gh} = \sqrt{2gR(1 - \cos\theta)}$$

$$\Rightarrow v^2/R = 2g(1 - \cos\theta)$$

As the normal reaction just before crossing point C is

$$N = m(g\cos\theta + v^2/R)$$

As the normal reaction just before crossing point C is

$$N' = m(g\cos\theta - v^2/R)$$

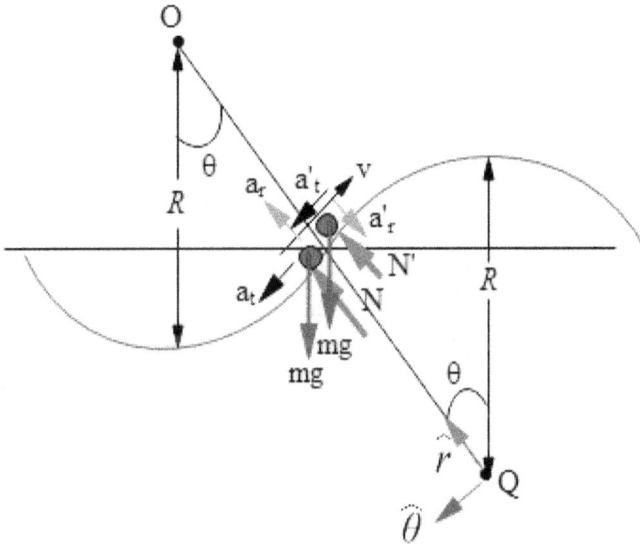

Let us assume that the centre of curvature Q is the reference point (origin); then, the normal reaction and radial or centripetal acceleration of the ball is +ve if it is radially away and −ve if it is radially inward. Using this coordinate system, the magnitude of normal reaction changes (decreases) during crossing of point C, whereas its direction remains unchanged if we assume that it does not lose contact at C. So, the change in normal reaction is

$$\vec{N'} - \vec{N} = m(g\cos\theta - v^2/R)\hat{r} - m(g\cos\theta + v^2/R)\hat{r} = -\frac{2mv^2}{R}\hat{r}$$

Putting $v^2/R = 2g(1 - \cos\theta)$, we have

$$\vec{N}' - \vec{N} = -\frac{2mv^2}{R}\hat{r} = -2mg(1 - \cos\theta)\hat{r}$$

Putting $\cos\theta = 4/5$, we have $\vec{N}' - \vec{N} = -2mg(1 - 4/5)\hat{r} = -\frac{2mg}{5}\hat{r}$ Ans.
The negative sign shows that normal reaction changes towards the centre of curvature Q. Ans.
(g) As centripetal acceleration reverses its direction during the crossing of point C, the change in normal reaction is

$$\vec{a}'_r - \vec{a}_r = -(v^2/R)\hat{r} - (+v^2/R)\hat{r} = -\frac{2v^2}{R}\hat{r}$$

Putting $v^2/R = 2g(1 - \cos\theta)$, we have

$$\vec{a}'_r - \vec{a}_r = -\frac{4g(1 - \cos\theta)}{R}\hat{r}$$

Putting $\cos\theta = 4/5$, we have $\vec{a}'_r - \vec{a}_r = -\frac{4g(1 - 4/5)}{R}\hat{r} = -\frac{4g}{5R}\hat{r}$ Ans.
(h) As tangential acceleration remains unchanged during the crossing of point C, the change in tangential acceleration is $\vec{a}'_t - \vec{a}_t = mg\sin\theta\hat{\theta}$ $-mg\sin\theta\hat{\theta} = \vec{0}$ Ans.

Problem 39 A small ball of mass m is released from a height h. After sliding down a smooth track AM, it moves in a vertical circle of radius R which is cut at an angular position θ from A to B. If the ball leaves the circular track at B and moves as a projectile and again launches on to the circular track at B, find the angle θ. (b) If the ball moves as a projectile acquiring a maximum possible height H, find the angle θ.

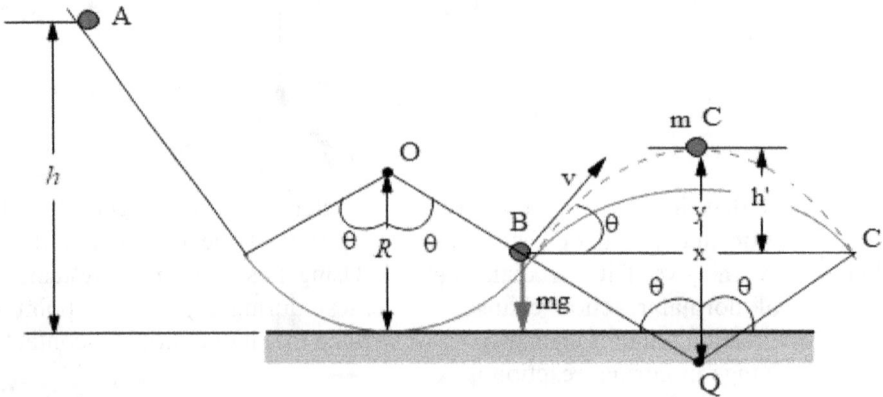

Solution

(a) The horizontal range is

$$R_H = x = \frac{2v^2 \sin\theta \, \cos\theta}{g}$$

Also, from the geometry of the figure,

$$x = 2R \sin\theta$$

From last two equations,

$$\frac{2v^2 \sin\theta \, \cos\theta}{g} = 2R \sin\theta$$

$$\Rightarrow v = \sqrt{\frac{gR}{\cos\theta}} \quad \text{Ans.}$$

Let the velocity be $u > v$, so that the ball will lose contact at P. Its maximum height is

$$h' = \frac{v^2 \sin^2\theta}{2g}$$

The maximum height from the centre of curvature Q is

$$y = h' + R \cos\theta$$

$$\Rightarrow y = \frac{v^2 \sin^2\theta}{2g} + R \cos\theta$$

For the maximum value of the maximum height

$$\frac{dy}{d\theta} = \frac{2v^2 \sin\theta \, \cos\theta}{2g} - R \sin\theta = 0$$

$$\Rightarrow \cos\theta = \frac{gR}{v^2} \tag{7.210}$$

By conserving the energy between A and B, we have

$$mg\{h - R(1 - \cos\theta)\} = \frac{mv^2}{2}$$

$$\Rightarrow v = \sqrt{2g\{h - R(1 - \cos\theta)\}} \tag{7.211}$$

Using last two equations,

$$\cos\theta = \frac{gR}{2g\{h - R(1 - \cos\theta)\}} = \frac{R}{2\{h - R(1 - \cos\theta)\}}$$

$$\Rightarrow \cos\theta[2\{h - R(1 - \cos\theta)\}] - R = 0$$

$$\Rightarrow 2R\cos^2\theta + 2(h - R)\cos\theta - R = 0$$

$$\Rightarrow \cos\theta = \frac{-2(h - R) \pm \sqrt{4(h - R)^2 + 8R^2}}{4R}$$

$$\cos\theta = \frac{-(h - R) + \sqrt{(h - R)^2 + 2R^2}}{2R}$$

Putting $h = 3R$, we have

$$\cos\theta = \frac{\sqrt{(3R - R)^2 + 2R^2} - (3R - R)}{2R} = \frac{\sqrt{6} - 2}{2} = 0.225$$

$$\Rightarrow \theta = 77° \text{ Ans.}$$

Problem 40 A small ball of mass m is released from a height h. After sliding down a smooth track AM, it moves in a vertical circle of radius R which is cut at an angular position θ from A to B. (a) If the ball loses contact at A, moves as a projectile and again launches on to the circular track at B, find the angle θ. (b) If the ball moves as a projectile to acquire a maximum possible height H, find the angle θ. (c) Find the maximum possible height attained by the ball from the ground.

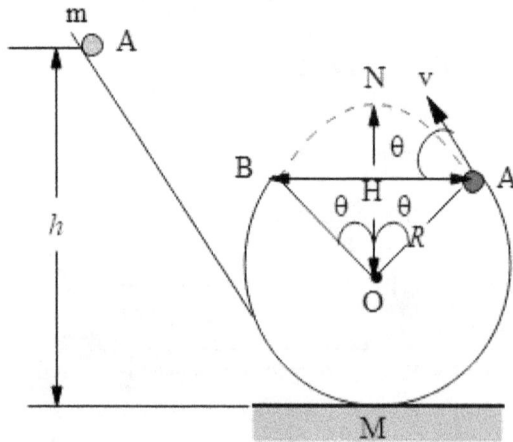

Solution

(a) The horizontal range is

$$R_H = x = \frac{2v^2 \sin\theta \cos\theta}{g}$$

Also, from the geometry of the figure,

$$x = 2R \sin\theta$$

From the last two equations,

$$\frac{2v^2 \sin\theta \cos\theta}{g} = 2R \sin\theta$$

$$\Rightarrow v = \sqrt{\frac{gR}{\cos\theta}} \text{ Ans.}$$

(b) Let the velocity be v, so that the ball leaves the track at P. Its maximum height is

$$h' = \frac{v^2 \sin^2\theta}{2g}$$

The maximum height from the ground is

$$y = h' + R(1 + \cos\theta)$$

$$\Rightarrow y = \frac{v^2 \sin^2\theta}{2g} + R(1 + \cos\theta)$$

For the maximum value of the maximum height

$$\frac{dy}{d\theta} = \frac{2v^2 \sin\theta \cos\theta}{2g} - R \sin\theta = 0$$

$$\Rightarrow \cos\theta = \frac{gR}{v^2} \tag{7.212}$$

By conserving the energy between A and B, we have

$$mg\{h - R(1 + \cos\theta)\} = \frac{mv^2}{2}$$

$$\Rightarrow v = \sqrt{2g\{h - R(1 + \cos\theta)\}} \tag{7.213}$$

Using the last two equations,

$$\cos\theta = \frac{gR}{2g\{h - R(1 + \cos\theta)\}} = \frac{R}{2\{h - R(1 + \cos\theta)\}}$$

$$\Rightarrow \cos\theta[2\{h - R(1 + \cos\theta)\}] - R = 0$$

$$\Rightarrow 2R\cos^2\theta - 2(h - R)\cos\theta + R = 0$$

$$\Rightarrow \cos\theta = \frac{(h - R) \pm \sqrt{(h - R)^2 - 2R^2}}{2R}$$

Putting $h = 3R$, we have

$$\cos\theta = \frac{\sqrt{(3R - R)^2 - 2R^2} \pm (3R - R)}{2R} = \frac{\sqrt{2} \pm 2}{2}$$

$$\Rightarrow \cos\theta = \frac{2 - \sqrt{2}}{2} = 0.292$$

$$\Rightarrow \theta = 72.97° \text{ Ans.}$$

(c) The horizontal range is

$$R_H = x = \frac{2v^2 \sin\theta \cos\theta}{g}$$

Also, from the geometry of the figure,

$$x = 2R \sin\theta$$

From the last two equations,

$$\frac{2v^2 \sin\theta \cos\theta}{g} = 2R \sin\theta$$

$$\Rightarrow v = \sqrt{\frac{gR}{\cos\theta}}$$

Following the previous procedure, we can find $\theta = 72.97°$.
(d) So, in this case the ball will launch onto the circular track again and at the same time acquire a maximum possible height H.

Putting $v = \sqrt{gR/\cos\theta}$ in $y = \frac{v^2\sin^2\theta}{2g} + R(1 + \cos\theta)$, we have

$$y = \frac{gR\sin^2\theta}{2g\cos\theta} + R(1 + \cos\theta)$$

$$\Rightarrow y = \left\{ 1 + \frac{(1 - \cos\theta)}{2\cos\theta} \right\} R(1 + \cos\theta)$$

Putting the value of $\theta = 72.97°$ in the last equation

$$y = \left\{ 1 + \frac{(1 - 0.292)}{2(0.292)} \right\}$$

$$y = 2.21 \text{ R } \text{ Ans.}$$

N.B:

1. For losing contact with the track at A, the normal reaction

$$N = mg\cos\theta - \frac{mv^2}{R} = 0$$

So, the minimum velocity is

$$v = \sqrt{gR\cos\theta}$$

2. After leaving the track at A, for the ball to attain maximum possible height y_{max} from the ground, the velocity of the ball at A must be $v = \sqrt{gR/\cos\theta}$ the normal reaction.
3. After leaving the track at A, for the ball to launch at B, the velocity of the ball at A must be $v = \sqrt{gR/\cos\theta}$.
4. So, if the ball leaves the track at A with a velocity $v = \sqrt{gR/\cos\theta}$, the ball will attain a maximum possible height so also it will launch at B.

www.ingramcontent.com/pod-product-compliance
Lightning Source LLC
Chambersburg PA
CBHW082121210326

41599CB00031B/5827